NASA/TP–2009–213146–REVISION A

International Space Station Science Research Accomplishments During the Assembly Years: An Analysis of Results from 2000-2008

Cynthia A. Evans and Julie A. Robinson
Office of the International Space Station Program Scientist
NASA Johnson Space Center, Houston, Texas

Judy Tate-Brown, Tracy Thumm, and Jessica Crespo-Richey
Engineering & Science Contract Group, Houston, Texas

David Baumann and Jennifer Rhatigan
NASA Johnson Space Center, Houston, Texas

June 2009

THE NASA STI PROGRAM OFFICE . . . IN PROFILE

Since its founding, NASA has been dedicated to the advancement of aeronautics and space science. The NASA Scientific and Technical Information (STI) Program Office plays a key part in helping NASA maintain this important role.

The NASA STI Program Office is operated by Langley Research Center, the lead center for NASA's scientific and technical information. The NASA STI Program Office provides access to the NASA STI Database, the largest collection of aeronautical and space science STI in the world. The Program Office is also NASA's institutional mechanism for disseminating the results of its research and development activities. These results are published by NASA in the NASA STI Report Series, which includes the following report types:

- TECHNICAL PUBLICATION. Reports of completed research or a major significant phase of research that present the results of NASA programs and include extensive data or theoretical analysis. Includes compilations of significant scientific and technical data and information deemed to be of continuing reference value. NASA's counterpart of peer-reviewed formal professional papers but has less stringent limitations on manuscript length and extent of graphic presentations.

- TECHNICAL MEMORANDUM. Scientific and technical findings that are preliminary or of specialized interest, e.g., quick release reports, working papers, and bibliographies that contain minimal annotation. Does not contain extensive analysis.

- CONTRACTOR REPORT. Scientific and technical findings by NASA-sponsored contractors and grantees.

- CONFERENCE PUBLICATION. Collected papers from scientific and technical conferences, symposia, seminars, or other meetings sponsored or cosponsored by NASA.

- SPECIAL PUBLICATION. Scientific, technical, or historical information from NASA programs, projects, and mission, often concerned with subjects having substantial public interest.

- TECHNICAL TRANSLATION. English-language translations of foreign scientific and technical material pertinent to NASA's mission.

Specialized services that complement the STI Program Office's diverse offerings include creating custom thesauri, building customized databases, organizing and publishing research results . . . even providing videos.

For more information about the NASA STI Program Office, see the following:

- Access the NASA STI Program Home Page at http://www.sti.nasa.gov

- E-mail your question via the internet to help@sti.nasa.gov

- Fax your question to the NASA Access Help Desk at (301) 621-0134

- Telephone the NASA Access Help Desk at (301) 621-0390

- Write to:
  NASA Access Help Desk
  NASA Center for AeroSpace Information
  7121 Standard
  Hanover, MD 21076-1320

NASA/TP–2009–213146–REVISION A

International Space Station Science Research Accomplishments During the Assembly Years: An Analysis of Results from 2000-2008

Cynthia A. Evans and Julie A. Robinson
Office of the International Space Station Program Scientist
NASA Johnson Space Center, Houston, Texas

Judy Tate-Brown, Tracy Thumm, and Jessica Crespo-Richey
Engineering & Science Contract Group, Houston, Texas

David Baumann and Jennifer Rhatigan
NASA Johnson Space Center, Houston, Texas

June 2009

Available from:

NASA Center for AeroSpace Information
7115 Standard Drive
Hanover, MD 21076-1320
Phone: 301-621-0390 or
Fax: 301-621-0134

National Technical Information Service
5285 Port Royal Road
Springfield, VA 22161
703-605-6000

This report is also available in electronic form at http://ston.jsc.nasa.gov/collections/TRS/

CONTENTS

Introduction 1

Executive Summary 6
International Space Station Research Accomplishments During the Assembly Years: 2000–2008
International Space Station Research Results 6
    Technology Development for Exploration 6
    Physical Sciences in Microgravity 7
    Biological Sciences in Microgravity 8
    Human Research Program 8
    Observing the Earth and Educational Activities 9
    Science from International Space Station Observations 10
International Space Station Research: Benefits to Life on Earth 10
    Spin-offs and Patents 10
Supporting Future Exploration: Science After the First 10 Years 11
References 12

Summary of NASA Experiments and Their Results
Technology Development
Analyzing Interferometer for Ambient Air (ANITA) 18
Active Rack Isolation System-ISS Characterization Experiment (ARIS-ICE) 19
Dust and Aerosol Measurement Feasibility Test (DAFT) 21
Elastic Memory Composite Hinge (EMCH) 22
In-space Soldering Experiment (ISSI) 23
Lab-on-a-Chip Application Development-Portable Test System (LOCAD-PTS) 24
Middeck Active Control Experiment-II (MACE-II) 26
Microgravity Acceleration Measurement System (MAMS) and Space Acceleration
    Measurement System II (SAMS-II), Two Investigations 27
Maui Analysis of Upper Atmospheric Injections (Maui) 30
Materials International Space Station Experiment-1 and -2 (MISSE-1 and -2) 31
Materials on the International Space Station Experiment-3 and -4 (MISSE-3 and -4) 33
Materials International Space Station Experiment-5 (MISSE-5) 34
Ram Burn Observations (RAMBO) 36
Smoke and Aerosol Measurement Experiment (SAME) 38
Serial Network Flow Monitor (SNFM) 39
Synchronized Position Hold, Engage, Reorient, Experimental Satellites (SPHERES) 40
Space Test Program-H2 (STP-H2): Three Investigations 42
    Atmospheric Neutral Density Experiment (ANDE) 42
    Microelectromechanical System-(MEMS) Based Picosat Inspector (MEPSI) 42
    Radar Fence Transponder (RAFT) 42

Physical Sciences in Microgravity
Advanced Protein Crystallization Facility (APCF), Eight Investigations 47
Binary Colloidal Alloy Test-3 (BCAT-3), Three Investigations 49
Capillary Flow Experiment (CFE) 52
Commercial Generic Protein Crystal Growth-High Density (CPCG-H) 55
Coarsening in Solid Liquid Mixtures-2 (CSLM-2) 57
Dynamically Controlled Protein Crystal Growth (DCPCG) 59
ExPRESS Physics of Colloids in Space (EXPPCS) 60
Fluid Merging Viscosity Measurement (FMVM) 62
Viscous Liquid Foam-Bulk Metallic Glass (Foam) 64
Investigating the Structure of Paramagnetic Aggregates from Colloidal Emulsions (InSPACE) 66
Miscible Fluids in Microgravity (MFMG) 67
Protein Crystal Growth-Enhanced Gaseous Nitrogen (PCG-EGN) 69
Protein Crystal Growth-Single Locker Thermal Enclosure System (PCG-STES), Nine Investigations ... 70

Toward Understanding Pore Formation and Mobility During Controlled Directional Solidification
    in a Microgravity Environment (PFMI) 74
Solidification Using Baffle in Sealed Ampoules (SUBSA) 76
Zeolite Crystal Growth (ZCG) 77

Biological Sciences in Microgravity
Avian Development Facility (ADF), Two Investigations 81
Advanced Astroculture (AdvAsc) 83
Biomass Production System (BPS) Technology Validation Test (TVT) 85
Cellular Biotechnology Operations Support System (CBOSS), Seven Investigations 87
Commercial Generic Bioprocessing Apparatus (CGBA), Three Investigations 90
Fungal Pathogenesis, Tumorigenesis, and Effects of Host Immunity in Space (FIT) 93
Threshold Acceleration for Gravisensing (Gravi) 94
Microencapsulation of Anti-tumor Drugs (MEPS) 96
Effect of Space Flight on Microbial Gene Expression and Virulence (Microbe) 97
Molecular and Plant Physiological Analyses of the Microgravity Effects on Multigeneration Studies
    of *Arabidopsis thaliana* (Multigen) 99
The Optimization of Root Zone Substrates (ORZS) for Reduced-Gravity Experiments Program 100
Photosynthesis Experiment and System Testing and Operation (PESTO) 101
Plant Generic Bioprocessing Apparatus (PGBA) 103
Passive Observatories for Experimental Microbial Systems (POEMS) 104
Streptococcus pneumoniae Expression of Genes in Space (SPEGIS) 106
StelSys Liver Cell Function Research (StelSys) 107
Analysis of a Novel Sensory Mechanism in Root Phototropism (Tropi) 108
Yeast-Group Activation Packs (Yeast-GAP) 109

Human Research and Countermeasure Development for Exploration
Advanced Diagnostic Ultrasound in Microgravity (ADUM) 113
Anomalous Long-Term Effects in Astronauts' Central Nervous System (ALTEA) 115
Bonner Ball Neutron Detector (BBND) 116
Effect of Prolonged Space Flight on Human Skeletal Muscle (Biopsy) 118
Commercial Biomedical Testing Module (CBTM): Effects of Osteoprotegerin (OPG)
    on Bone Maintenance in Microgravity 120
Commercial Biomedical Test Module-2 (CBTM-2) 122
Cardiovascular and Cerebrovascular Control on Return from ISS (CCISS) 124
Cell Culture Module-Immune Response of Human Monocytes in Microgravity
    (CCM-Immune Response) 125
Cell Culture Module-Effect of Microgravity on Wound Repair: In Vitro Model of New
    Blood Vessel Development (CCM-Wound Repair) 126
Chromosomal Aberrations in Blood Lymphocytes of Astronauts (Chromosome) 127
Dosimetric Mapping (DOSMAP) 128
Space Flight-induced Reactivation of Latent Epstein-Barr Virus (Epstein-Barr) 130
A Study of Radiation Doses Experienced by Astronauts in EVA (EVARM) 131
Foot/Ground Reaction Forces During Space Flight (Foot) 133
Hand Posture Analyzer (HPA) 134
Effects of Altered Gravity on Spinal Cord Excitability (H-Reflex) 135
Crew Member and Crew-ground Interactions during International Space Station
    Missions (Interactions) 137
Behavioral Issues Associated with Isolation and Confinement: Review and Analysis of
    ISS Crew Journals (Journals) 140
Incidence of Latent Virus Shedding During Space Flight (Latent Virus) 141
Test of Midodrine as a Countermeasure Against Postflight Orthostatic Hypotension-
    Short-duration Biological Investigation (Midodrine-SDBI) 143
Promoting Sensorimotor Response Generalizability: A Countermeasure to Mitigate
    Locomotor Dysfunction After Long-duration Space Flight (Mobility) 144
Nutritional Status Assessment (Nutrition) 145

Bioavailability and Performance of Promethazine During Space Flight (PMZ)	146
Effects of EVA and Long-term Exposure to Microgravity on Pulmonary Function (PuFF)	147
Renal Stone Risk During Space Flight: Assessment and Countermeasure Validation (Renal Stone)	149
Sleep-Wake Actigraphy and Light Exposure During Space Flight-Long (Sleep-Long)	150
Sleep-Wake Actigraphy and Light Exposure During Space Flight-Short (Sleep-Short)	151
Stability of Pharmacotherapeutic and Nutritional Compounds (Stability)	152
Subregional Assessment of Bone Loss in the Axial Skeleton in Long-term Space Flight (Subregional Bone)	154
Surface, Water, and Air Biocharacterization (SWAB) –A Comprehensive Characterization of Microorganisms and Allergens in Spacecraft	156
Organ Dose Measurement Using a Phantom Torso (Torso)	157
Effect of Microgravity on the Peripheral Subcutaneous Veno-arteriolar Reflex in Humans (Xenon-1)	159

Observing the Earth and Educational Activities

Amateur Radio on the International Space Station (ARISS)	161
Crew Earth Observations (CEO)	162
Crew Earth Observations-International Polar Year (CEO-IPY)	167
Commercial Generic Bioprocessing Apparatus Science Insert-01 (CSI-01)	169
Commercial Generic Bioprocessing Apparatus Science Insert-02 (CSI-02)	171
DreamTime (DreamTime)	172
Education Payload Operations (EPO)	173
Education Payload Operation–Demonstration (EPO-Demos)	174
Education Payload Operations-Educator (EPO-Educator)	175
Education Payload Operations-Kit C (EPO-Kit C)	176
Earth Knowledge Acquired by Middle School Students (EarthKAM)	177
Space Exposed Experiment Development for Students (Education-SEEDS)	179
Space Experiment Module (SEM)	180

Results from ISS Operations

Clinical Nutrition Assessment of ISS Astronauts (Clinical Nutrition Assessment)	183
Education-How Solar Cells Work (Education-Solar Cells)	185
Environmental Monitoring of the International Space Station (Environmental Monitoring)	186
International Space Station In-flight Education Downlinks (In-flight Education Downlinks)	188
International Space Station Acoustic Measurement Program (ISS Acoustics)	190
Periodic Fitness Evaluation with Oxygen Uptake Measurement (PFE-OUM)	192
Analysis of International Space Station Plasma Interaction (Plasma Interaction Model)	193
Saturday Morning Science (Science of Opportunity)	194
Soldering in Reduced Gravity Experiment, SDTO 17003U (SoRGE)	196
International Space Station Zero-Propellant Maneuver (ZPM) Demonstration	197
Environmental Monitoring of the International Space Station (Environmental Monitoring): Miscellaneous Results	199

Space Station Science Benefiting Life on Earth ... 201

Appendices

Appendix A: Scientific Research by Expedition

Expedition 0 (Sep 8 2000 – Nov 2 2000)	204
Expedition 1 (Nov 2 2000 – Mar 18 2001)	205
Expedition 2 (Mar 10 2001 – Aug 20 2001)	206
Expedition 3 (Aug 12 2001 – Dec 12 2001)	207
Expedition 4 (7 Dec 2001 – 15 Jun 2002)	209
Expedition 5 (Jun 7 2002 – Dec 2 2002)	211
Expedition 6 (Nov 25 2002 – May 3 2003)	213
Expedition 7 (Apr 28 2003 – Oct 27 2003)	214
Expedition 8 (Oct 20 2003 – Apr 29 2004)	215
Expedition 9 (Apr 21 2004 – Oct 23 2004)	217

Expedition 10 (Oct 15 2004 – Apr 25 2005) 218
Expedition 11 (Apr 16 2005 – Oct 10 2005) 219
Expedition 12 (Oct 3 2005 – Apr 8 2006) 220
Expedition 13 (Apr 1 2006 – Sep 28 2006) 221
Expedition 14 (Sep 20 2006 – Apr 21 2007) 223

Appendix B: Publications Resulting from Research Aboard the International Space Station 228

Appendix C: Acronyms and Abbreviations 245

Introduction

The International Space Station (ISS) celebrated 10 years of operations in November 2008. Today, it is more than a human outpost in low Earth orbit (LEO). It is also an international science laboratory hosting state-of-the-art scientific facilities that support fundamental and applied research across the range of physical and biological sciences.

The launch of the first ISS element in 1998, the Russian Zarya module, was a highly visible milestone for international cooperation in human exploration. Later, when the first international crew that included Bill Shepard, Sergei Krikalev, and Yuri Gidzenko, moved into the ISS to establish a continuous human presence in space, a new, global chapter in the history of human space flight was opened. As of this writing, 18 multinational crews comprising 52 astronauts and cosmonauts have called the ISS their home and workplace since November 2000. Dozens more have visited and assisted construction and science activities.

While the ISS did not support permanent human crews during the first 2 years of operations (November 1998 to November 2000), it hosted a few early science experiments months before the first international crew took up residence. Since that time—and simultaneous with the complicated task of ISS construction and overcoming impacts from the tragic *Columbia* accident—science returns from the ISS have been growing at a steady pace. From Expedition 0 through 15, 138 experiments have been operated on the ISS, supporting research for hundreds of ground-based investigators from the U.S. and International Partners. Many experiments are carried forward over several ISS increments, allowing for additional experimental runs and data collection.

This report focuses on the experimental results collected to date, including scientific publications from studies that are based on operational data. Today, NASA's priorities for research aboard the ISS center on understanding human health during long-duration missions, researching effective countermeasures for long-duration crewmembers, and researching and testing new technologies that can be used for future Exploration crews and spacecraft. Most research also supports new understandings, methods, or applications that are relevant to life on Earth, such as understanding effective protocols to protect against loss of bone density or better methods for producing stronger metal alloys. Experiment results have already been used in applications as diverse as the manufacture of solar cell and insulation materials for new spacecraft and the verification of complex numerical models for behavior of fluids in fuel tanks.

Figure 1. STS124e9982 — The International Space Station in June 2008. The inset is a close-up view of the three ISS laboratory facilities.

Over the first 10 years of operations, events shaped the ability and capacity of the ISS for performing space research, as well as the focus of the ISS research itself.

- The ISS has been under continuous assembly during this time period (fig. 1). The U.S. *Destiny* laboratory was deployed in early 2001. The European *Columbus* Module and Japanese *Kibo* Laboratory were both launched and mated to the ISS in 2008. Today, major research outfitting has grown to include 18 racks and facilities within the laboratory space (fig. 2)—a complement of several multipurpose ExPRESS [Expedite the Processing of Experiments to Space Station] racks; Human Research Facility racks; a versatile glovebox; racks supporting fluid physics, combustion physics, life sciences, and capabilities for deep freezing of samples—as well as facilities for externally mounted experiments [8].

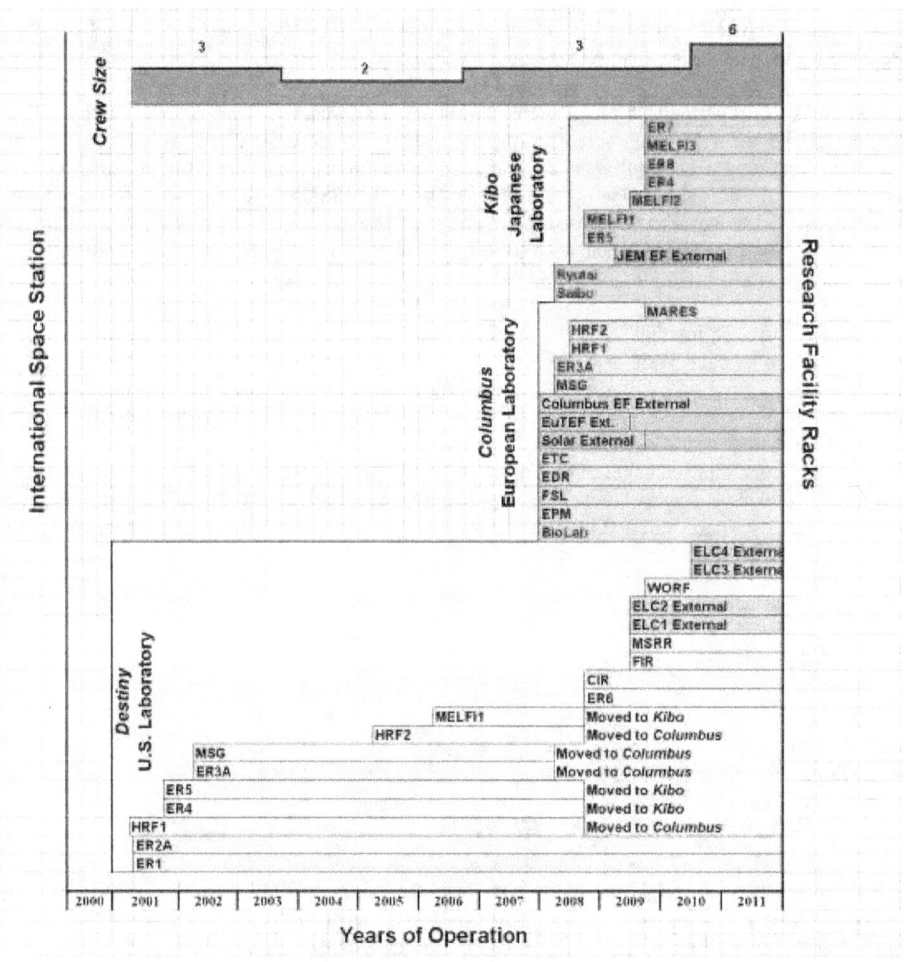

Figure 2. Graphic showing the build-up of ISS research facilities over time.

- The space shuttle fleet was grounded following the *Columbia* accident, halting ISS assembly. Prior to the *Columbia* accident, more than 6,600 kg (14,600 lbs) of research equipment and facilities had been brought to ISS. Between the accident and the return to flight of the Space Shuttle *Discovery* in Jul 2005, 75 kg (165 lbs) of research supplies had been brought up on Russian Progress and Soyuz vehicles. The crew of the ISS was also reduced from three to two, and the research program was drastically adjusted to accommodate these changes. Since return to flight in 2005, the research space has grown, and the crew size has increased back to three.

- Assembly operations have required real-time changes in plans. For example, the contamination of the starboard solar array rotary joint [SARJ] precluded full rotation of the starboard solar arrays and limited power for science and operations. At the same time, assembly operations have also supported research related that is to operations and important for future missions. For example, research on rates of micro-meteoroid and orbital debris (MMOD) strikes and studies on the extent of thermal flexing of new parts of the ISS structure are ongoing.
- The focus of NASA's ISS research has changed strategically to support the Vision for Space Exploration that was announced by the President on Jan 14, 2004. While still including some fundamental research in microgravity, emphasis has shifted to programs that are targeted at developing and testing new exploration technologies and reducing the risks to human explorers on missions to the moon, Mars, and beyond.
- The NASA Authorization Act of 2005 designated the ISS as a National Laboratory and secured a pathway for new experiments that are sponsored by other agencies or partnerships with the private sector. Details and current plans are provided by Uhran [11]. The National Lab Pathfinder program is now under way, with limited payload opportunities on each shuttle flight to the ISS.
- The anticipated shutdown of the Space Shuttle Program in 2010 has reprioritized the manifests for research equipment and operational supplies, and created pressure on investigators to develop new experimental protocols that require fewer sample returns.

At the 10-year point, the stage is set for increasing the scientific returns from the ISS. During the 2008 calendar year, the laboratory space and research facilities were tripled with the addition of the European Space Agency's (ESA's) *Columbus* and Japanese Aerospace Exploration Agency's *Kibo* scientific modules joining NASA's *Destiny* Laboratory. In 2009, the number of crewmembers will increase from three to six, greatly increasing the time that is available for research. The realization of the international scientific partnership provides new opportunities for scientific collaboration and broadens the research potential on the ISS. Engineers and scientists from around the world are working together to refine their operational relationships and build from their experiences in conducting early science to ensure maximum utilization of the expanded capabilities aboard ISS.

This NASA Technical Publication is intended to provide an archival record of the United States-sponsored ISS research that was accomplished through Expedition 15, both as part of formal investigations or "payloads" and from early scientific analysis of data collected as part of operating the station. By scientific theme, the investigations are collected as follows:

- Technology Development—studies and tests of new technologies for use in future Exploration missions. Areas of emphasis included spacecraft materials and systems, and characterization and control of the microgravity environment on ISS.
- Physical Sciences—studies of physics and chemistry in microgravity. Areas of emphasis included materials sciences experiments, including physical properties and phase transitions in polymers and colloids, fluid physics, and crystal growth experiments.
- Biological Sciences—studies of biology using microgravity conditions to gain insight into the effect of the space environment on living organisms. Areas of emphasis included cellular biology and biotechnology, and plant biology.
- Human Research for Exploration—human medical research to develop the knowledge that is needed to send humans on Exploration missions beyond Earth orbit. These studies focused on the effect of living in space on human health and countermeasures to reduce health risks that will be incurred by living in space in future. Areas of emphasis included physiological studies related to the effects of microgravity on bone and muscle, other physiological effects of space flight, psycho-social studies, and radiation studies.
- Observing the Earth and Education—these activities and investigations allowed students and the public to connect with the ISS mission; inspired students to excel in science, technology, engineering, and math; and shared the astronauts' unique view of the Earth system with scientists and the public. A detailed summary of all educational activities on ISS is also available [10].
- Results from ISS Operations—in addition to the formal, peer-reviewed scientific research and experiments, the ISS supports a large body of research using data from ISS operations, including routine medical monitoring of the crew and data that are collected on the ISS environment, both inside and outside of the ISS.

This report includes an executive summary of all of the research results. The investigations that were conducted during the first 15 Expeditions are presented topically by the research disciplines, and alphabetically within each topical section. Of the investigations that are summarized here, some are completed with results released, some are completed with preliminary results, and some remain ongoing. For each case we provide an overview of the research objectives and the results that have been returned to date. We also indicate whether additional activities are planned for future ISS missions at the time of writing.

For interested readers, the appendices provide a chronological listing of the investigations that were performed in each Expedition, and a full listing of research publications sorted by discipline and payload acronym.

In addition to specific scientific results, a number of reviews and updates on ISS science have been written ([2]–[6], Tate et al. 2006, [7], [9], [1]). Research that was conducted independently by the International Partners on ISS, especially the Russian Space Agency and ESA, is not included. Results from many of the investigations continue to be released. Continuously updated information on ISS research and results is available on the NASA portal at http://www.nasa.gov/mission_pages/station/science/index.html/.

REFERENCES

[1] Evans CA, Uri JJ, Robinson JA. Ten Years on the International Space Station: Science Research Takes Off. 59th International Astronautical Congress. Glasgow, Scotland. 2008, 29 Sep – 3 Oct; IAC-08-B3.4.2.

[2] Pellis NR, North RM. Recent NASA Research Accomplishments Aboard ISS. 54th International Astronautical Congress of the International Astronautical Federation, the IAA, and the International Institute of Space Law, Bremen, Germany. IAC-03-T.4.07, Sep 29–Oct 3, 2003.

[3] Rhatigan J. Research Accomplishments on the International Space Station. AIAA/ICAS International Air and Space Symposium and Exposition: The Next 100 Years, Dayton, Ohio. AIAA 2003-2703, 2003.

[4] Rhatigan JL, Robinson JA, Sawin CF. Exploration-Related Research on ISS: Connecting Science Results to Future Missions. NASA TP-2005-213166. 1–34, 2005. (Also published at the 44th AIAA Meeting, 2006-344, 2006.)

[5] Robinson JA, Rhatigan JL, Baumann DK. Recent Research Accomplishments on the International Space Station. Proceedings of the IEEE Aerospace Conference, Big Sky, Mont. Paper no. 310, Mar 5–12, 2005.

[6] Robinson JA., Thomas DA. NASA Utilization of the International Space Station and the Vision for Space Exploration, ISTS 2006-o-1-06, International Symposium on Space Technology and Science, Kanazawa, Japan, Jun 4–10, 2006, pp. 1–6.

[7] Robinson JA, Thumm TL, Thomas DA. NASA Utilization of the International Space Station and the Vision for Space Exploration. *Acta Astronautica*. 2007; 61(1-6): 176-184 (Presented at the International Astronautical Congress, Oct 2006 and the 45th AIAA Aerospace Sciences Meeting and Exhibit, Jan 2007: AIAA 2007-139).

[8] Robinson JA, Evans CA, Tate JM, Uri JJ. International Space Station Research—Accomplishments and Pathways for Exploration and Fundamental Research. 46th AIAA Aerospace Sciences Meeting and Exhibit. Reno, Nev. 7 – 10 Jan 2008; AIAA 2008-0799.

[9] Robinson JA, Thumm TL, Thomas DA. NASA Utilization of the International Space Station and the Vision for Space Exploration. *Acta Astronautica*. 2007, 61(1-6): 176-184 (Presented at the International Astronautical Congress, Oct 2006 and the 45th AIAA Aerospace Sciences Meeting and Exhibit, Jan 2007: AIAA 2007-139).

[10] Thomas DA, Robinson JA, Tate J, Thumm T. Inspiring the Next Generation: Student Experiments and Educational Activities on the International Space Station, 2000–2006. NASA TP-2006-213721. Lyndon B. Johnson Space Center, Houston, 108 pp., 2006.

[11] Uhran ML. Progress toward establishing a US National Laboratory on the International Space Station, International Astronautical Congress, Glasgow, Scotland, Sep 2008. IAC-08-B3.4.

STS124-E-9956 — With three fully equipped laboratory modules, the ISS research potential becomes reality. This full view of ISS was photographed by a crewmember on board Space Shuttle *Discovery* after the undocking of the two spacecraft in June 2008.

EXECUTIVE SUMMARY

INTERNATIONAL SPACE STATION RESEARCH ACCOMPLISHMENTS DURING THE ASSEMBLY YEARS: 2000–2008

The first 15 Expeditions aboard the International Space Station (ISS) have established the ISS as a laboratory that provides a unique opportunity for research capabilities. This paper reviews the science accomplishments that were achieved by the collaborative efforts of researchers, crewmembers, and the ISS Program.

Early space station science utilization, while an important part of crew activities, competed with ISS assembly and maintenance, anomaly resolution, and the re-planning efforts that were required after the *Columbia* accident. Simply learning to live on and operate the ISS consumed a large percentage of ISS program resources. Despite the environment of construction and system shakedowns that dominated the early years, important science activities were conducted.

In 2005, the ISS Program Science Office created an online experiment database that was aimed at tracking and communicating ISS scientific research [1] and published a summary of all of the U.S. experiments and their results through Expedition 10 [2]. Today, this database is continually updated. It captures ISS experiment summaries and results, and functions as a reference that is used for planning ISS science operations. An important aspect of the database is the inclusion of scientific publications from all ISS experiments as they become available. To date, more than 200 results publications and research summaries have been included in the database; the number continues to grow at a steady pace [3]. Additional science data and results have been discussed in scientific symposia, with publications pending. The scientific findings support future research and guide upcoming exploration activities.

NASA's research activities on the ISS span several scientific fields, including exploration technology development, microgravity research in the physical and biological sciences, human research, Earth science, and education. Data from ISS operations have been used in additional research on crew health, the ISS structure, and the ISS environment. Figure 1 is a graphical representation of the distribution of ISS experiments by discipline, based on the number of experiments that were performed through the first 15 Expeditions. Below, we examine and summarize the published research stemming from NASA-supported experiments in each scientific field.

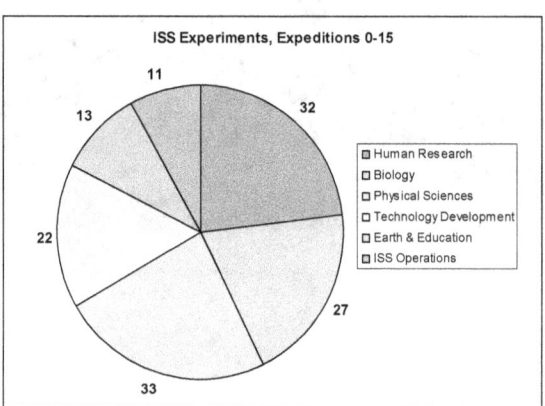

Figure 1. Distribution by discipline of NASA's ISS experiments for Expeditions 0 through 15.

International Space Station Research Results

Technology Development for Exploration

One of NASA's top priorities for research aboard the ISS is the development and testing of new technologies and materials that are being considered for future exploration missions. To date, 22 different technology demonstrations have been performed. These experiments include research characterizing the microgravity environment, monitoring the ISS environment both inside and outside the spacecraft, testing spacecraft materials, developing new spacecraft systems, and testing picosatellites and new satellite commanding and controls. We are tracking 34 scientific publications, and recognize that classified and proprietary proceedings include a much greater number of results documenting technology developments. Recent experiments range from combustion physics and soot production (important data for redesign of spacecraft smoke detectors [4]) to the successful demonstration of microfluidic technologies for rapidly detecting different contaminants such as bacteria and fungi. Other new technology experiments and stand-alone instrument packages monitor other air contaminants.

One of the most prolific series of investigations on the ISS is the Materials International Space Station Experiment (MISSE). MISSE (fig. 2) tests how spacecraft materials withstand the harsh space environment, including solar radiation, atomic oxygen (AO) erosion, thermal cycling, micrometeoroid and orbital debris impacts, and contamination from spacecraft. Hundreds of materials have been tested to date. MISSE results have been used to understand and calibrate how materials that are already in use on spacecraft degrade in the space environment [5] (for example, polymers used for insulation, and solar array materials) and predict the durability of new materials (e.g., the solar cell materials that are planned for the Commercial Orbital Transportation System). Samples that are currently being tested on the ISS include materials that are part of the design of the new Orion vehicle [6]. In addition to several recent and upcoming publications, MISSE investigators have devised methods for measuring and predicting AO erosion yields, and have assembled data that will provide important reference for designers, developers, and builders of future spacecraft and instruments. The MISSE team has also spearheaded innovative collaborations among industry, academia, NASA, and the Department of Defense (DOD), including hundreds of investigators.

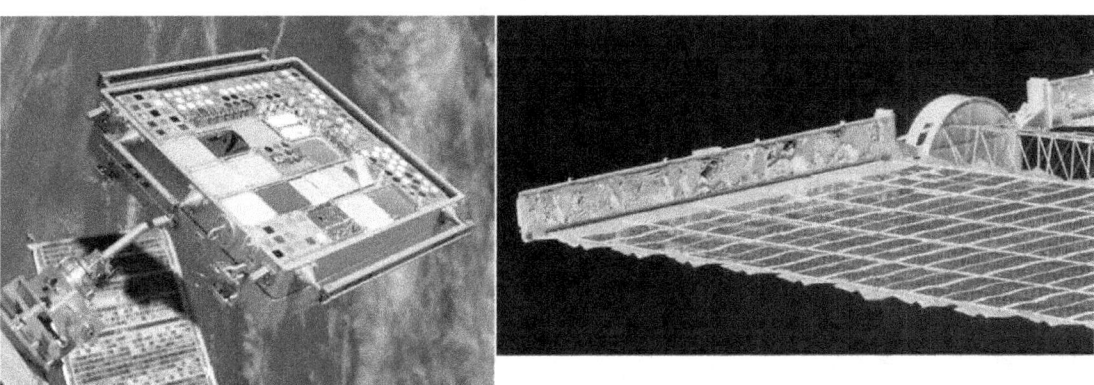

Figure 2. Left: MISSE 3 experiment on the ISS, exposing samples to the space environment. Right: AO damage to materials on an ISS solar array blanket box.

Physical Sciences in Microgravity
The research themes of the microgravity program in the physical sciences have shifted since 2000. Over the first 15 Expeditions, 33 physical science experiments were performed, yielding more than 40 publications. Early physical science experiments emphasized the growth of protein crystals in microgravity conditions and investigated fundamental properties in fluids as well as the specific behaviors of colloids. Today, a major research thread that is sponsored by NASA on the ISS focuses on the fundamental properties of colloids and complex fluids, part of the relatively new field of the physics of soft matter [7]. The fundamental properties of these materials are determined by the behaviors and properties of the particles and their tendency to assemble or cluster into particular structures, depending on conditions. Microgravity research enables the definition of phase boundaries, and responses to directional forces other than gravity and interactions without convection. Several experiments examine basic phase transitions (solid-liquid-gas transitions) in the absence of gravitationally induced complications like density stratification of materials.

While the ISS research addresses questions that are framed by fundamental physics and chemistry, the results often have very practical applications for future exploration technologies and on Earth. For example, a demonstration of capillary flow of fluids in microgravity (Capillary Flow Experiment (CFE), which is designed to record the flow of fluid in containers with different geometries) produced the first-space-validated numerical models describing fluid behavior [8]. The research to describe capillary flow in microgravity is important for the management of large volumes of fluids in space; results will be benchmarked and applied to the designs of fuel tanks, and other spacecraft fluid systems.

Another experiment, Investigating the Structure of Paramagnetic Aggregates from Colloidal Emulsion (InSPACE), provided new observations and unexplained structures and transitions of structures in magnetically controlled colloids called magnetorheologic fluids (fluids with suspended iron particles that cluster and form structures when subjected to a magnetic field, causing rapid changes in viscosity of the fluid). Microgravity allows for longer observations of the formation of magnetic particle aggregates, and their structures and kinetics [9]. Recently, the laboratory environment on the ISS facilitated the real-time re-planning effort to explore unexpected structures that were observed during InSPACE sample runs. Magnetically controlled materials have widespread applications in several industries including robotics, the automotive industry (e.g., suspension and damping systems), and civil engineering (e.g., bridges, earthquake-protection systems).

Microgravity physics experiments include other colloidal systems that examine the clustering properties of colloidal particles in microgravity. These experiments complement the fast-growing field of research in colloids and complex fluids, with many potential applications in both commercial and exploration sectors—from plastics to household cleaning and personal hygiene products to the manufacture of new high-technology materials.

Biological Sciences in Microgravity

Through Expedition 15, 27 different biological experiments were performed on the ISS. The early experiments were centered on testing new biotechnology tools (see table 1). Cellular biology and parameters of plant growth affected by microgravity comprised a large percentage of the research. Experiments on animal biology and microbiology were also conducted. To date, we have collected 25 publications reporting results from biological research on the ISS.

Table 1. Biotechnology support hardware tested as part of early ISS utilization

Facility	Purpose
Cellular Biotechnology Operations Support System (CBOSS)	Cell incubator; supports multiple experiments
Commercial Generic Bioprocessing Apparatus (CGBA)	Bioprocessor for microorganisms; programmable, accurate temperature control
Advanced Astroculture™ (AdvAsc)	Plant growth unit
Biomass Production System (BPS)	Microgravity plant growth environment with environmental control subsystems
Microencapsulation Electrostatic Processing System (MEPS)	Automated system used to produce liquid-filled micro-balloons
Plant Generic Bioprocessing Apparatus (PGBA)	Closed chamber plant growth facility with light, moisture, temperature, and gas control
Group Activation Pack-Fluid Processing Apparatus (GAP-FPA)	Test tube system for controlled, sequential mixing of two or three fluids in microgravity

One of the most exciting results reported from ISS research is the confirmation that common pathogens change and become more virulent during space flight [10]. The Microbe (Effect of Spaceflight on Microbial Gene Expression and Virulence) experiment was performed in Sep 2006; it examined changes in three microbial pathogens. Initial data from one of the microbes, *Salmonella typhimurium* (a leading cause of human gastroenteritis), showed that 167 genes were expressed differently in flight when compared with ground controls. The data indicated a response to the microgravity environment, including widespread alterations of gene expression that increased disease-causing potential. These results show great promise for both understanding mechanisms used by pathogens to spread disease and also designing ways to better protect humans in space [10]. With these new insights, similar experiments will continue on the ISS with related sets of pathogens. The original experiment was funded because of the human health risks for exploration missions; but because of the potential for applications to prevent disease on Earth, follow-on studies have been implemented as pathfinders for the use of the ISS as a National Laboratory.

Human Research Program

Research on the human body in space—how it reacts to microgravity and radiation—is a high priority for NASA's ISS science portfolio (fig. 2). The Human Research Program (HRP) experiments aboard the ISS build from the large body of work that has been collected since the early days of space program, including a robust set of experiments that was conducted on *Skylab*, *Mir*, and shorter-duration shuttle flights. Clinical evidence demonstrated important physiological changes in astronauts during space flight. The HRP, together with ISS Medical Operations, sponsored experiments that study different aspects of crew health, and efficacy of countermeasures for extended-duration stays in microgravity. Up through the 15th ISS expedition, 32 experiments focused on the human body, including research on bone and muscle loss, the vascular system, changes in immune response, radiation studies, and research on psycho-social aspects of living in the isolation of space. Several of the early experiments have led to new experiments, testing details of observations or pursuing new questions that were raised by early results. One or two new experiments are started nearly every expedition (fig. 3).

The HRP research now focuses on knowledge gaps in our understanding of the physiological changes that are observed during long-duration space flight and research that is aimed at ameliorating the greatest health risks [11]; today's experiments are designed to provide more detail in the complex changes in crew health. Today, an integrated set of parameters is monitored on ISS crewmembers. For example, Nutrition (Nutritional Status Assessment), which is a comprehensive in-flight study of human physiologic changes during long-duration space flight [12], and Integrated Immune (Validating Procedures for Monitoring Crewmember Immune Function) sample and analyze participant's blood, urine, and saliva before, during, and after space flight. These samples are used to study the changes that are related to functions such as bone metabolism, oxidative damage, and immune function. These studies are unique because of the information that they collect on the timing of changes during the course of a space mission. Another collaborative set of experiments measures and monitors body fluid shifts; electrocardiograms are collected to monitor the heart function and vascular health of the crew. The crewmembers periodically test their pulmonary function, and keep journals that are used to quantitatively analyze their response to isolation. Future research is also ensured. Extra biological samples are collected for the Repository investigation, which is a long-term archive of critical biological samples that are collected from ISS astronauts, for future analysis when new tools and methods can be used and new questions are posed.

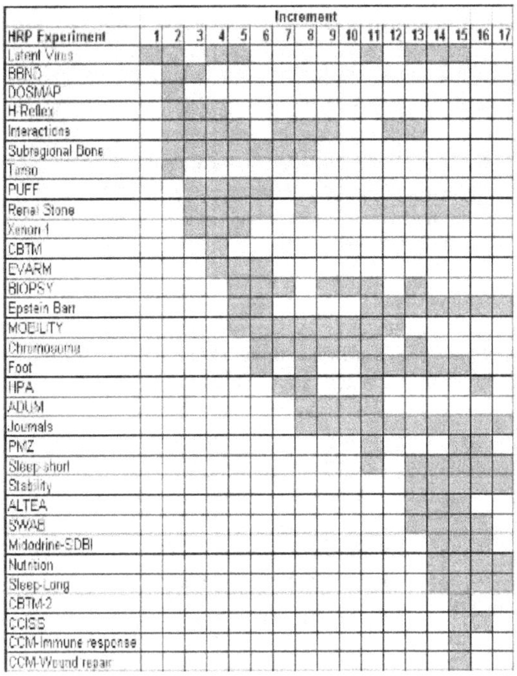

Results from the initial experiments on the ISS are just now being published; most studies require multiple subjects over several years to derive the necessary data. Nevertheless, we have already identified 43 scientific publications from research sponsored by the HRP that was performed on ISS. These results document, in increasing detail, locations and parameters of bone loss, links between bone and muscle loss, renal stone development, rates of recovery and changes in recovered bone mass, changes related to the immune system, associated profiles of other physiological or biochemical parameters, and roles of diet, drug countermeasures, and exercise. Compilations that include the collective results and collaborations of ground and space-based human research experiments have also been published (e.g., Cavanaugh and Rice 2007). Since many of the human research studies continue aboard the ISS, results will continue to flow in from the early experiments.

Figure 3. A listing of experiments by Expedition, sponsored by the HRP. The program has initiated new experiments nearly every Expedition.

Observing the Earth and Educational Activities

The crew continues to perform the Crew Earth Observations (CEO) investigation that supports a variety of Earth Science initiatives. Crew observations document urban growth, monitor changes along coastlines and long-term ecological research sites, record major events such as volcanic eruptions or hurricanes, and provide observations to support the International Polar Year. In a recent example, researchers working with the National Snow and Ice Data Center requested images of icebergs that broke from the Larsen ice shelf in Antarctica. The high-resolution ISS images that answered this request provided the first observations of ponded meltwater on the icebergs as they drifted into the South Atlantic Ocean (fig. 4). These data allowed scientists to use the icebergs as analogs of ice sheets and model the accelerated breakup of an ice shelf [13]. While Earth observation activities have less formal experiment protocols, the data are fully accessible to scientists around the world. ISS Earth observations have supported more than 20 publications and one patent, several Web-based articles, and a robust database that serves more than 325,000 images of Earth that were taken by astronauts on the ISS (http://eol.jsc.nasa.gov).

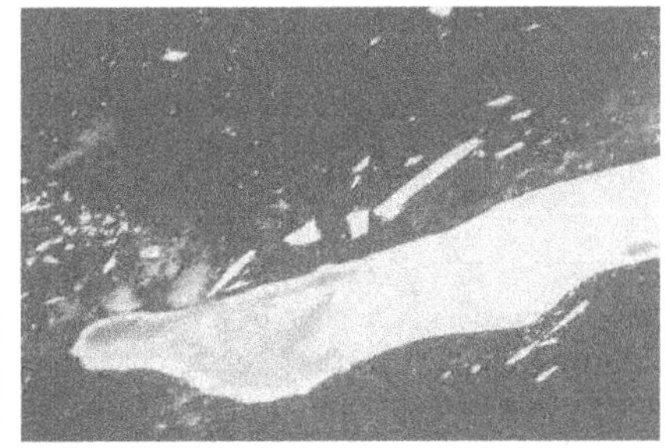

Figure 4. ISS008-E-12558 — ISS astronauts, working with scientists studying the breakup of Antarctic ice shelves, tracked Iceberg A39B near South Georgia Island as it collected ponded meltwater on the surface and disintegrated in the South Atlantic Ocean.

ISS education activities have touched millions of students around the world. For example, tens of thousands of students have participated in the EarthKAM experiment; even more have participated in crew conferences with schools through HAM radio [14]. In addition, crews continue to create short videos demonstrating elements of life in microgravity, profiling technologies involved in living off the planet, and providing a behind-the-scenes look at some of the science experiment hardware. Recent examples of these include a feature called "Toys in Space" [15] and a demonstration of Newton's Laws.

Science from International Space Station Operations
The formal experiments that are performed on the ISS, and their published results, represent part of the body of ISS research that is supported by NASA. For example, some ISS crewmembers initiate their own experiments and demonstrations of microgravity phenomena (e.g., Saturday Morning Science [16]). Ground studies in analog settings, experiments that refine techniques or instrumentation, and follow-on studies to the ISS experiments support almost all of the specific research projects that are performed on the ISS. Beyond the continued experimentation and analysis and reporting of peer-reviewed research, there is a large body of science (at least 30 publications) resulting from data that were collected from day-to-day operations of the ISS. These data and their analyses are critical for future exploration. They document daily parameters of the crew and the spacecraft, verify as-built configurations and responses of hardware, record changes over time in configurations, document the space environment (radiation, micrometeoroid and orbital debris (MMOD) flux, local contamination from outgassing and venting, daily wear due to thermal flexing), and the efficacy of countermeasures such as MMOD and radiation shielding.

International Space Station Research: Benefits to Life on Earth
Spin-offs and Patents
We continue to track innovations resulting from research in microgravity. Several recent patents and partnerships have demonstrated the back-to-Earth benefits of the public investment in ISS research. A few examples are provided here.

- The air-purifying technology (titanium dioxide (TiO_2)-based ethylene), which is employed in the plant growth chamber that is used in the AdvAsc experiment, was incorporated into an airborne pathogen scrubber that is effective against Anthrax spores [17].
- A researcher who is mining the public database of astronaut photographs of the Earth (http://eol.jsc.nasa.gov) for oblique views of large fans of sediments around river systems (fluvial fans) assembled and patented a compilation of the global distribution of these features. This work has implications for hydrocarbon exploration [18].
- An ISS investigator recently patented the Microparticle Analysis System and Method, which is an invention for a device that detects and analyzes microparticles [19]. This technology supports the chemical and pharmaceutical industries, and is one of a sequence of inventions that is related to technology development for experiments on the ISS and shuttle, including the MEPS experiment that demonstrated microencapsulation processing of drugs, a new and powerful method for delivering drugs to targeted locations. Also of note, the ISS demonstration of the microencapsulation of drugs has now been reproduced on the ground [20].
- The investigator for the Binary Colloidal Alloy Test -3 and 4: Critical Point (BCAT-3-4-CP) was facing a challenge using a microscope to observe the dynamic clustering behavior of his colloidal experiments: the moving clusters of colloidal beads would move out of the field of view. The investigator developed a target-locking technology for his microscope [21].
- The MISSE experiments have yielded dozens of important assessments that are directly applicable to current and future spacecraft materials. Additionally, spin-offs include the use of AO to remove organic content from surfaces for diverse applications. Examples include use of AO to restore valuable artwork, and to remove bacteria contaminants from surgical implants [23]. A recent patent was awarded to etch the surface of the optical fibers that are used for blood analyses to increase the surface area and enable rapid assessment of blood glucose levels [24].

Education: Benefits to the Next Generation of Space Explorers
Supporting science, technology, engineering, and mathematics (STEM) education is an important component of all ISS research; the estimated educational impact of the collection of ISS research is broad. In 2006, Thomas et al. [14] estimated that more than 30 million students and nearly 15,000 teachers have participated in ISS scientific activities or demonstrations. Specific educational activities are part of every ISS research complement, and include student-developed experiments, students performing classroom versions of ISS experiments and demonstrations, and stand-

alone educational demonstrations that can be used as teaching aids or resource materials. In one large coordinated effort, the first educator astronaut flew to the ISS in Dec 2006 and conducted classroom sessions throughout her mission.

In addition to these education activities, many research experiments are tied to U.S. academic institutions and involve student research from high school through advanced graduate degrees. Students may be listed as co-investigators, involved in hardware development or testing, or participate in critical parts of operations and data reduction. For example, two investigators for the MISSE-2 experiment enlisted high school students to collaborate with them on sample preparation and subsequent data analysis of their polymer experiment that was mounted on the MISSE-2 container for exposure to the space environment. This experiment, which is called the Polymer Erosion and Contamination Experiment (PEACE) polymers experiment, is now being examined to determine the effects of AO erosion [23,24]. PEACE polymers experiments were also flown as part of MISSE-5 and MISSE-6. Students who are involved in the MISSE PEACE polymer experiments have successfully competed in science fairs and demonstrations, and have gone on to science degrees from top universities. A team of three young women that participated in the PEACE polymer studies placed sixth in the prestigious National Siemen's Competition in Math, Science and Technology. For its work, the team received medals and $10,000 in scholarship money. (For more information, see http://www.hb.edu/html/about.php?id=1158.)

Many other experiments conduced on the ISS comprised critical research for undergraduate and graduate students. These educational venues are summarized in Thomas et al. 2006 [14].

Looking forward, using the ISS as an educational platform will be taken to a new level as the ISS National Laboratory develops. The America Competes Act (2007) provides direction to NASA to include STEM educational activities as a high priority. Educational activities are a key component of the National Lab legislation, and planning is under way with commercial, academic, and other U.S. agency partners to develop novel programs and experiments targeting student involvement [25].

SUPPORTING FUTURE EXPLORATION: SCIENCE AFTER THE FIRST 10 YEARS

Publications of ISS scientific results, which is one metric that is used to measure success of the research program, have shown steady increases in all scientific research areas on the ISS (fig. 5). To date, we have identified roughly 200 publications directly resulting from research on the ISS.

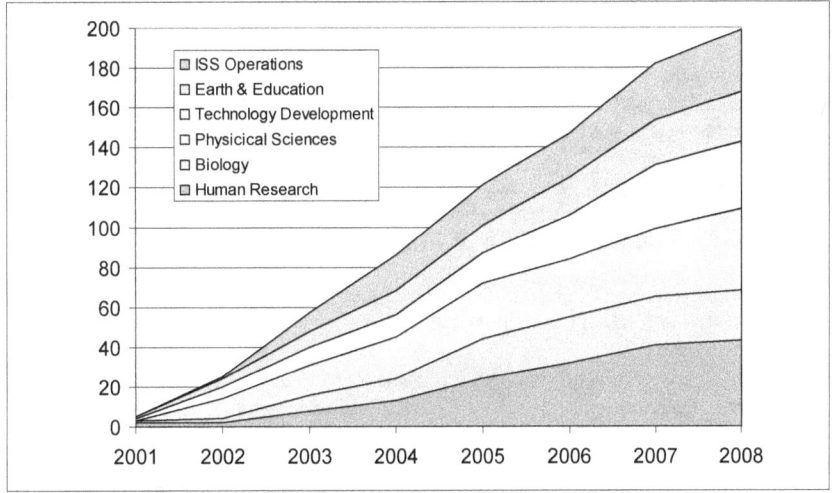

Figure 5. A compilation of the number of publications resulting from research aboard the ISS. Each discipline has produced a steady stream of published results, which is one metric for assessing scientific productivity.

Building from these successes, engineers and scientists from around the world are working together to refine their relationships and build from their experiences conducting early science to ensure maximum utilization of the expanded capabilities aboard ISS. These capabilities have already enabled scientists to replan their experiments in real time, and immediately respond to new observations and unexpected results by performing new experiment runs. Other research builds from questions arising from early ISS science results.

New experiments aboard the ISS include a broad range of science, as follows:

- Coordinated human research experiments that collaborate with International Partners' science objectives and activities, including shared baseline data collection and in-flight sampling, with the goal of understanding integrated causes and effects of changes in the human body.
- Research using new science racks, including the fluids integrated rack (FIR), the combustion integrated rack (CIR), and materials science research rack (MSRR), will enable new experiments exploring combustion, fluid behavior, and heat-dependent crystallization patterns in metal alloys.
- Exploration Technology Development will build from early experiments on materials exposure, smoke generation, liquid fuel management, and environmental monitoring.
- The new Window Observation Research Facility (WORF) will provide capabilities to support remote-sensing instruments, enabling Earth Science research that will, for example, document crop health and test the utility of blue-green bands for ocean research.

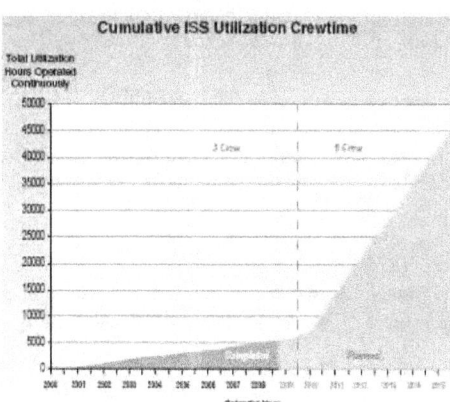

With a large body of published scientific results from the ISS, new science facilities, mature operational protocols for science operations, the realization of international scientific partnerships and new opportunities for scientific collaboration, and the increased crew size in the near future, the research from ISS is on the rise. Figure 6 illustrates the projected increase in crew time that is available for conducting scientific research. We anticipate that this graph also approximates the increased rate of scientific returns in the future.

Figure 6 Graph showing the projected increase in crew time available for research.

As ISS science activities and operations continue, scientific data that are derived from earlier experiments will be re-examined, mined, and assembled with new data and findings—perhaps data from other fields. We anticipate that successful ISS research will be used to seed new ideas and hypotheses to be tested on future missions.

REFERENCES

[1] National Aeronautics and Space Administration, International Space Station, ISS Experiment and Facility Information, http://www.nasa.gov/mission_pages/station/science/els_index2.html.

[2] Robinson JA, Rhatigan JL, Baumann DK, Tate J, Thumm T. International Space Station Research Summary Through Expedition 10. NASA TP-2006-213146. 2006: 1–142.

[3] National Aeronautics and Space Administration, ISS Experiment and Facility Results Publications, http://www.nasa.gov/mission_pages/station/science/experiments/Publications.html.

[4] Urban DL, Ruff GA, Brooker JE, Cleary T, Yang J, Mulholland G, Yuan Z-G. Spacecraft Fire Detection: Smoke Properties and Transport in Low-Gravity. 46th AIAA Aerospace Sciences Meeting and Exhibit, Reno, Nev. 2008: 7 – 10 Jan; AIAA 2008-806.

[5] de Groh KK, Banks BA, McCarthy CE, Rucker RN, Roberts LM, Berger LA. MISSE PEACE Polymers Atomic Oxygen Erosion Results. NASA TM-2006-214482 (2006).

[6] de Groh KK, Banks BA, Dever JA, Jaworske DA, Miller SK, Sechkar EA, Panko SR. NASA Glenn Research Center's Materials International Space Station Experiments (MISSE 1-7). NASA TM-2008-215482 (2008).

[7] McLeish TA. Tangled Tale of Topological Fluids, *Physics Today*, Aug 2008: 40–45.

[8] Weislogel MM, Jenson RM, Klatte J, Dreyer ME. The Capillary Flow Experiments aboard ISS: Moving Contact Line Experiments and Numerical Analysis. 46th AIAA Aerospace Sciences Meeting and Exhibit, Reno, Nev. 2008; 7–10 Jan; AIAA 2008-816.

[9] Vasquez PA, Furst EM, Agui J, Williams J, Pettit D, Lu E. Structural Transitions of Magnetoghreological Fluids in Microgravity. 46th AIAA Aerospace Sciences Meeting and Exhibit, Reno, Nev. 2008: 7–10 Jan; AIAA 2008-815.

[10] Wilson JW, Ott CM, Hoener zu Bentrup K, Ramamurthy R, Quick L, Porwollik S, Cheng P, McClelland M, Tsaprailise G, Radabaugh T, Hunt A, Fernandez D, Richter E, Shah M, Kilcoyne M, Joshi L, Nelman-Gonzalez M, Hing S, Parra M, Dumars P, Norwood K, Bober R, Devich J, Ruggles A, Goulart C, Rupert M, Stodieck L, Stafford P, Catella L, Schurr MJ, Buchanan K, Morici L, McCracken J, Allen P, Baker-Coleman C, Hammond T, Vogel J, Nelson R, Pierson DL, Stefanyshyn-Piper HM, Nickerson CA. Space flight alters bacterial gene expression and virulence and reveals a role for global regulator Hfq. Proceedings of the National Academy of Sciences of the United States of America. 2007: 104(41):16299–16304.

[11] Bioastronautics Roadmap. A Risk reduction strategy for Human Space Exploration. NASA SP-2004-6113 Washington D.C., National Aeronautics and Space Administration, 2005.

[12] Smith SM, Zwart SR. Nutrition issues for space exploration. *Acta Astronautica*. 2008; 63: 609–613.

[13] Scambos T, Sergienko O, Sargent A, MacAyeal D, Fastook J. ICES at profiles of tabular iceberg margins and iceberg breakups at low latitudes. *Geophys. Research Letters*, Vol 32, L23S09 2005.

[14] Thomas DA, Robinson JA, Tate J, Thumm T. Inspiring the Next Generation: Student Experiments and Educational Activities on the International Space Station, 2000–2006. NASA TP-2006-213721. 2006: 1–108.

[15] National Aeronautics and Space Administration Educational Product. International Toys in Space – Science on the Station. DVD. 2004: ED-2004-06-001-JSC.

[16] Love SG, Pettit DR. Fast, Repeatable Clumping of Solid Particles in Microgravity. *Lunar and Planetary Science* XXXV. 2004; 1119.

[17] AiroCide TiO_2, AdvAsc, Advanced Astroculture™ spin-off, http://www.sti.nasa.gov/tto/spinoff2002/er_5.html.

[18] M.J. Wilkinson, 2006, Method for Identifying Sedimentary bodies from images and its application to mineral exploration, USPTO Patent # 6,985,606. http://patft.uspto.gov/netacgi/nph-Parser?Sect1=PTO2&Sect2=HITOFF&p=1&u=%2Fnetahtml%2FPTO%2Fsearch-bool.html&r=1&f=G&l=50&co1=AND&d=PTXT&s1=6985606.PN.&OS=PN/6985606&RS=PN/6985606.

[19] Morison, DR, Microparticle analysis system and method, 2008 http://www.uspto.gov/web/patents/patog/week46/OG/html/1324-2/US07295309-20071113.html

[20] Morrison DR, Haddad RS, Ficht A. Microencapsulation of Drugs: New cancer therapies and improved drug delivery derived from microgravity research. Proceedings of the 40th Space Congress, Cape Canaveral, Fla. Apr, 2003.

[21] Lu P, TARC: Target-Locking Microscopy, http://www.physics.harvard.edu/~plu/research/PLuTARCU.

[22] National Aeronautics and Space Administration Spin-offs: Corrosive Gas Restores Artwork, Promises Myriad of Applications
http://www.sti.nasa.gov/spinoff/spinitem?title=Corrosive+Gas+Restores+Artwork%2C+Promises+Myriad+Applications+.

[23] Banks, B, Energetic Atomic and Ionic Oxygen Textured Optical Surfaces for Blood Glucose Monitoring, 2008.
http://patft.uspto.gov/netacgi/nph-Parser?Sect1=PTO2&Sect2=HITOFF&p=1&u=%2Fnetahtml%2FPTO%2Fsearch-bool.html&r=1&f=G&l=50&co1=AND&d=PTXT&s1=%22Energetic+Atomic+Ionic+Oxygen+Textured+Optical+Surfaces+Blood+Glucose+Monitoring%22.TI.&OS=TTL/.

[24] Stambler AH, Inoshita KE, Roberts LM, Barbagallo CE, de Groh KK, Banks BA. Ground-Laboratory to In-Space Atomic Oxygen Correlation for the PEACE Polymers, 9th International Conference on "Protection of Materials and Structures from Space Environment," May 20–23, 2008, Toronto, Canada.

[25] National Aeronautics and Space Administration (2008) An Opportunity to Educate: International Space Station National Laboratory NP-2008-03-503-HQ.

SUMMARY OF NASA EXPERIMENTS AND THEIR RESULTS

TECHNOLOGY DEVELOPMENT

Future exploration—the return to the moon and human exploration of Mars—presents many technological challenges. Studies on the ISS can test a variety of technologies, systems, and materials that will be needed for future Exploration missions.

Some of the technology development experiments have been so successful that the hardware has been transitioned to operational status. Other experimental results feed new technology developments.

CHARACTERIZING THE MICROGRAVITY ENVIRONMENT ON ISS

Many people think of ISS as a "zero-gravity" environment because the continuous free-fall as ISS orbits the Earth simulates the absence of gravity. However, tiny vibrational disturbances aboard station from aerodynamic drag, venting of air or water, movement of solar arrays and antennas, dockings, reboosts, and crew activity all exert forces on station. Experiments have monitored the microgravity environment and evaluated technical solutions to protect ISS experiments from unwanted forces.

ENVIRONMENTAL MONITORING ON ISS

The living space aboard the ISS is monitored continuously for contamination or other threats to the health of the crew. New technologies are being tested that rapidly detect and analyze both contaminants in the cabin air (gaseous, particulate, and microbial), in the water, and on the surfaces of the station. Because the ISS is a closed environment, early detection and analysis of contamination is critical. The job of accurate sampling and detection is made more difficult by microgravity: air does not circulate by convection, and geography inside the station can create zones that are difficult to monitor. Variable concentrations of contaminants can also provide misleading results.

SPACECRAFT ENVIRONMENTS

Understanding the space environment outside of the ISS living space is also critical to life aboard the station and future exploration missions. Quantifying radiation hazards, the effects of AO, large thermal cycles, and micrometeorites and orbital debris is a large area of research aboard the ISS..

SPACECRAFT MATERIALS AND SYSTEMS

ISS provides a testbed for a variety of spacecraft systems. Results of studies on station help to determine which materials are most resistant to the conditions of the space environment, provide insight into in-space repairs, test satellite control algorithms, and give information on the physical processes underlying systems as diverse as fire suppression and propellant tank design.

PICOSATELLITES AND CONTROL SYSTEMS

Better engineering autonomous operations in space, including free-flyers, and understanding control system behavior in microgravity will be increasingly important for future space missions—both crewed and uncrewed.

Technology Development Experiments Performed on the International Space Station, Grouped by Discipline

Characterizing the Microgravity Environment on ISS
ARIS-ICE (Active Rack Isolation System – ISS Characterization Experiment)
MAMS (Microgravity Acceleration Measurement System)
SAMS-II (Space Acceleration Measurement System-II)
Environmental Monitoring on ISS
ANITA (Analyzing Interferometer for Ambient Air)
LOCAD-PTS (Lab-on-a-Chip Application Development-Portable Test System)
Picosatellites and Control Systems
MACE-II (Middeck Active Control Experiment-II)
SPHERES (Synchronized Position Hold, Engage, Reorient, Experimental Satellites)
STP-H2-MEPSI (Space Test Program-H2-Microelectromechanical System-Based (MEMS) PICOSAT Inspector)
STP-H2-RAFT (Space Test Program-H2-Radar Fence Transponder)
Spacecraft Materials and Systems
EMCH (Elastic Memory Composite Hinge)
ISSI (In Space Soldering Experiment)
MISSE-1 and -2 (Materials International Space Station Experiment -1 and -2)
MISSE-3 and -4 (Materials International Space Station Experiment -3 and -4)
MISSE-5 (Materials International Space Station Experiment-5)
MISSE-6A and -6B (Materials International Space Station Experiment-6A and -6B)
Spacecraft Systems
DAFT (Dust and Aerosol Measurement Feasibility Test)
Maui (Maui Analysis of Upper Atmospheric Injections)
SAME (Smoke and Aerosol Measurement Experiment)
SNFM (Serial Network Flow Monitor)
Spacecraft Environments
RAMBO (Ram Burn Observations)
STP-H2-ANDE (Space Test Program-H2-Atmospheric Neutral Density Experiment)

Analyzing Interferometer for Ambient Air (ANITA)
Principal Investigator(s): Gijsbert Tan, European Space Research and Technology Center, Noordwijk, The Netherlands
Expeditions 15–17

Research Area Environmental Monitoring on ISS

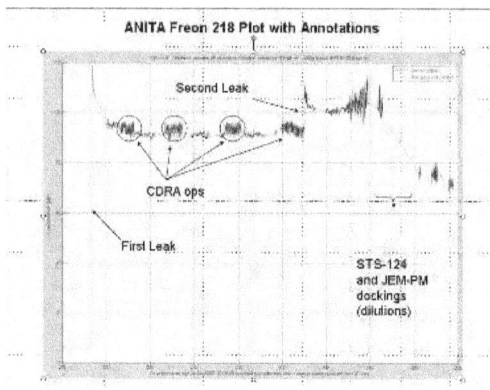

The Analyzing Interferometer for Ambient Air (ANITA) experiment is a trace gas monitoring system based on Fourier Transform Infrared (FTIR) technology. The initial flight of ANITA tested the accuracy and reliability of the FTIR technology as a potential next generation atmosphere trace gas monitoring system for the International Space Station (ISS).

ANITA data plot for Freon 218 demonstrating the ability to detect levels of contaminant, and operations of the carbon dioxide removal assembly (CDRA).

ANITA was calibrated to simultaneously monitor 32 gaseous contaminants (including formaldehyde, ammonia, and carbon monoxide (CO)) at low as parts per million levels in the ISS atmosphere. The hardware design—a quasi on-line, fast-time resolution gas analyzer—allowed air quality to be analyzed in near real time. For this experiment, the actual ANITA analysis results (contaminant identification and concentration) were not monitored by the crew, but were downlinked to the ground support team. In the future, ANITA analyses could be used by ISS crewmembers for detecting contaminants and immediate initiation of countermeasures, if required.

The ANITA calibration models that were developed prior to flight were based on a set of reference spectra of the 32 gaseous species that were considered possible contaminants. Earlier air quality analyses have determined that these trace gases may be present in the ISS cabin atmosphere. The experiment set-up allowed for rebuilding calibration models and resulting software changes to the in-flight calibration codes by the ANITA ground team if a contaminant is new or present at a different concentration level. A real-time modification of the operational parameters can be uplinked to the ISS.

Results
ANITA data, which were recently completed, are still being analyzed . Once final data collection is completed, the results will be available.

However, ANITA was used in mid 2008 to detect a Freon leak (Khladon 218) from the Russian air conditioner, and to monitor the timeline of Freon concentrations with CDRA operations and shuttle docking. The ANITA data helped to determine that the zeolite bed in the CDRA was not effective in scrubbing the Freon leak, but that diluting the ISS air after the docking with the shuttle substantially reduced the level of Freon.

The preliminary data analysis suggested that ANITA data were fairly accurate; the peak concentration that was measured by ANITA corresponds well to Russian estimates of mass released. Air was captured and returned to Earth for data analysis to corroborate ANITA results.

Publication(s)
There are no publications at this time.

ACTIVE RACK ISOLATION SYSTEM-ISS CHARACTERIZATION EXPERIMENT (ARIS-ICE)

Principal Investigator(s): Glenn S. Bushnell and Ian J. Fialho, The Boeing Company, Seattle, Wash.
Expeditions 2–4

Research Area Characterizing the Microgravity Environment on ISS

Constant microgravity conditions are essential for some ISS experiments. Sources of disturbance can include minor changes in acceleration, the movement of hardware such as the space station remote manipulator system (the robotic arm), or even normal crew activities, any of which can cause subtle vibrations to be transferred through station. The Active Rack Isolation System (ARIS) protects equipment by absorbing the shock of motion before it can affect an experiment.

ARIS is installed in two ExPRESS racks on ISS (ExPRESS Rack 2 and ExPRESS Rack 3), where it reduces vibrations using a combination of sensors and actuators. When the sensors detect a disturbance, the actuators counter the effect by sending a reactive force between the ExPRESS rack and the laboratory module, much as a shock absorber on an automobile would—except this "smart" shock absorber is finely tuned to react to, and cancel out, very minute vibrations. Accelerometer assemblies measure the disturbances and send data to the ARIS electronic control unit. The electronic control unit signals pushrods to press against the framework of station, stabilizing the rack. A microgravity rack barrier prevents accidental disturbances to the active ARIS rack. ARIS is designed to isolate all frequencies greater than 0.01 Hz, and is most effective in the 0.05- to 300-Hz range.

The ARIS-ISS Characterization Experiment (ICE) was a payload activity to characterize the ARIS on-orbit performance by monitoring the ambient vibration environment and by generating disturbances. The shaker unit provided a precise, measurable disturbance that simulated possible station vibrations. The other major component of ICE is the payload on-orbit processor, which executed characterization tests and acquired, synchronized, and processed ICE and Space Acceleration Measurement Systems-II (SAMS-II) data for downlink.

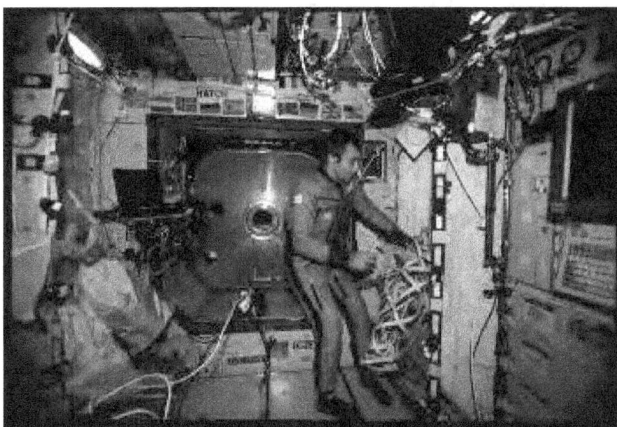

SS003327013 — Cosmonaut Vladimir Dezhurov performs ARIS-ICE hammer tests in the U.S. *Destiny* laboratory during Expedition 3.

RESULTS

ARIS-ICE operations were performed for over a year during ISS Expeditions 2–4. During that period more than 1,700 test runs were completed, ranging from short 1-second stability tests to 5-hour isolation characterization tests. Station vibrations were isolated to levels well below the science requirements of investigations in ExPRESS racks that were equipped with ARIS (Bushnell et al. 2002a). The ARIS-ICE command and data handling architecture helped to streamline the operational efficiency of the ARIS. The level of testing that was needed was greatly reduced by this system (Bushnell et al. 2002b). Quick-control design cycles were able to facilitate on-orbit payload operations, allowing ease of performing ARIS-ICE (Fialho et al. 2003).

The investigators determined, through a series of acceleration characterization experiments conducted during ARIS-ICE, that the ARIS facility provides the ability to predict and prevent the potentially damaging effect of station vibrations. It was also determined that sensitive experiments that are installed in ARIS would be isolated and protected from both vibrational and acceleration movements. These capabilities are critical to make ISS a unique, world-class research laboratory in microgravity.

PUBLICATION(S)

Bushnell GS, Fialho IJ, McDavid T, Allen JL, Quraishi N. Ground and On-Orbit Command and Data Handling Architectures for the Active Rack Isolation System Microgravity Flight Experiment. AIAA 53rd International Astronautical Congress, The World Space Congress, Houston. IAC-02-J.5.07, Oct 10–19, 2002a.

Bushnell GS, Fialho IJ, Allen JL, Quraishi N. Microgravity Flight Characterization of the International Space Station Active Rack Isolation System. AIAA Microgravity Measurements Group Meeting, The World Space Congress, Houston. Oct 10–11, 2002b.

Fialho IJ, Bushnell GS, Allen JL, Quraishi N. Taking H-infinity to the International Space Station: Design, Implementation and On-orbit Evaluation of Robust Controllers for Active Microgravity Isolation. AIAA Guidance, Navigation and Control Conference, Austin, Texas. Aug 2003.

JSC2001E34286 — Astronaut James Voss performing ARIS-ICE hammer operations in the U.S. *Destiny* laboratory on ExPRESS Rack 2.

DUST AND AEROSOL MEASUREMENT FEASIBILITY TEST (DAFT)
Principal Investigator(s): David Urban, NASA Glenn Research Center, Cleveland, Ohio
Expeditions 10, 13

Research Area Spacecraft Systems

Our understanding of how fires burn in a spacecraft has evolved since the fire detection equipment on space shuttle and space station was developed. One thing we have learned is that smoke particles that form in microgravity can be larger than those formed on Earth. Since smoke detectors are gauged to detect certain sizes of particles, this knowledge could help design more accurate smoke detectors for future spacecraft. The Smoke and Aerosol Measurement Experiment (SAME) is planned to gather particulate size information on ISS. The DAFT experiment, which was initiated after the *Columbia* accident, is to be used to obtain data in preparation for SAME using very little upmass.

Video screen shot of ISS commander and science officer Leroy Chiao performing DAFT operations on station during Expedition 10.

DAFT is intended to assess and characterize the distribution of particles in the air inside ISS to allow assessment of the suitability of current shuttle and ISS smoke detectors. This experiment was begun on Expedition 10 and completed during Expedition 13. DAFT is designed to test the effectiveness of the P-Trak Ultrafine Particle Counter, a device that counts ultra-fine dust particles in a microgravity environment. Most particle counters work by using a laser to record instances when the beam is interrupted; however, this method will not record ultra-fine particles that are much smaller than the wavelength of the light. P-Trak works by passing dust-laden air through a chamber of vaporous isopropyl alcohol. When a droplet of alcohol condenses over an ultra-fine dust particle, the particle becomes large enough to break the light beam and be counted. The alcohol is then recycled as it condenses on sidewalls and gravity pulls the alcohol back to the saturator. If the results are satisfactory, P-Trak will be used in SAME, which requires counts of particles ranging from 0.2 to 1 micron.

RESULTS
Preliminary results, which are based on DAFT sessions that were performed on station in 2005 and 2006, demonstrated that the P-Trak Ultrafine Particle Counter could be successfully built into the SAME payload, and established the optimal ranges for particle detection using this instrument. The data collected also indicate very low levels of particulate in the ISS environment relative to that previously measured on shuttle. This low particulate level is not surprising due to the small crew size on ISS (two or three vs. the typical seven-person crew on shuttle) and the high-efficiency particular accumulator (HEPA) filtration system of ISS. It is suspected that the particulate level will rise once station hosts a larger crew. In later experiment runs, the results indicated that the average size of the particles were larger on orbit than on Earth. These results could be due to the inefficiency of the large particle filter. Once the median particulate level throughout ISS is known, it can be used to design future smoke detectors that accurately distinguish normal dust from the presence of dangerous smoke particles.

A series of events (malfunction of the electronic data transmitted, crew time limits, and an alcohol wick that did not open) shortened the overall experiment, but adequate data were collected to label the demonstration successful.

PUBLICATION(S)
Urban D, Griffin D, Ruff G, Cleary T, Yang J, Mulholland G, Yuan Z. Detection of Smoke from Microgravity Fires. Proceedings of the International Conference on Environmental Systems. 2005-01-2930, 2005.

Elastic Memory Composite Hinge (EMCH)

Principal Investigator(s): Lieutenant Corey Duncan, Air Force Research Laboratory, Kirtland Air Force Base, N.M.
Expedition 15

Research Area Spacecraft Materials and Systems

The elastic memory composite hinge (EMCH) experiment provided test data on new materials that will further space hardware technology. This technology may eliminate the need for highly complex deployment mechanisms by providing a simpler, lightweight alternative to mechanical hinges. EMCH builds on the previous space shuttle experiment, lightweight flexible solar array hinge (LFSAH) that was flown on STS-93.

The six EMCH test articles that were tested on the ISS were approximately 10 × 2.5 × 2.5 cm in dimension and made of a unique resin and carbon fiber laminate that was developed at Composite Technology Development, Inc. End fixtures and metrology devices were developed to allow the EMCH test articles to be deployed through remote actuation, while their force-torque history and deployment accuracy are recorded automatically. A resistive heater was embedded in the elastic memory composite (EMC) laminates in the hinge to provide the heat that was necessary for actuation. The hinge was folded on the ground and deployed in space. A proximity sensor was attached to the end mass to determine the final position upon deployment. A second motion sensor was used to determine the deployment vs. time history. The test articles were re-settable to allow for multiple deployments.

ISS015E08404 — View of the EMCH experiment assembly in the U.S. *Destiny* laboratory. The photograph was taken during Expedition 15.

Results

EMCH was successfully completed on board ISS during Expedition 15. The investigation was returned to Earth for a complete analysis by the investigator team in 2007. However, the preliminary assessment indicated that the experiment demonstrated the robustness and reliability of the TEMBO® EMC hinge in the zero-gravity environment. This test campaign consisted of both nominal and off-nominal conditions, with the final series of tests presenting the most challenging conditions for on-orbit TEMBO® hinges. The successful completion of these tests indicates that the hinges meet the designed performance goals of a next-generation space-flight-qualified actuator. Additionally, the science that was gained from this experiment confirms the engineering assumptions that were used to design the TEMBO® EMC hinges, as well as other TEMBO® EMC deployable structures that are being developed for space applications.

Publication(s)

There are no publications at this time.

In-SPACE Soldering Experiment (ISSI)

Principal Investigator(s): Richard Grugel, NASA Marshall Space Flight Center, Huntsville, Ala.
Expeditions 7–10

Research Area Spacecraft Materials and Systems

Video screen shot of ISS science officer Mike Fincke using a soldering iron to perform the ISSI on station during Expedition 9.

The In-SPACE Soldering Experiment (ISSI) is another payload that was rapidly developed after the *Columbia* accident to provide a low-mass experiment using hardware already on board station. It was designed to promote understanding of joining techniques, shape equilibrium, wetting phenomena, and micro-structural development in space. Its primary objective was to better understand the effects and consequences of soldering in a microgravity environment such as that found on ISS. In Earth's gravity, soldering has a defined behavior and is reliant on gravity and convection to assist in solidification, joint shape, integrity, and microstructure. Unfortunately, on Earth detrimental gas bubbles (void spaces) are still found in the solder joint and at contact surfaces. These voids reduce the thermal and electrical conductivity and provide sites for crack initiation. Bubbles have less chance to escape in the reduced-gravity environment of space and, therefore, are likely to be more of a problem. To better understand this potential problem, a systematic series of soldering samples was designed to investigate and understand porosity development, surface wetting, and equilibrium shape formation. After the samples were heated on orbit, they were returned to Earth for property testing and metallographic examination.

Results

Five soldering sessions resulted in 86 samples. Experiment samples were returned to the investigator team in late 2005, and were evaluated both nondestructively and then destructively.

Real-time downlink video of the experiment yielded direct observation of the solder melting, equilibrium shape attainment by the liquid, and flux movement. The flux movement was particularly noteworthy because it was entirely unexpected. When the flux was released from the solder during heating, it formed a droplet that spun around the larger solder drop. This surprising movement is driven by thermocapillary flow that is induced by the temperature gradient. This type of behavior cannot be duplicated on Earth.

ISS009E14472 — Astronaut Edward M. (Mike) Fincke, Expedition 9 science officer and flight engineer, works on the ISSI in the U.S. *Destiny* laboratory.

Publication(s)

Grugel R, Cotton LJ, Segre PN, Ogle JA, Funkhouser G, Parris F, Murphy L, Gillies D, Hua F, Anilkumar AV. The In-Space Soldering Investigation (ISSI): Melting and Solidification Experiments Aboard the International Space Station. Proceedings of the 44th AIAA Aerospace Sciences Meeting and Exhibit, Reno, Nev. AIAA 2006-521, Jan 9–12, 2006.

Grugel RN, Luz P, Smith G, Spivey R, Jeter L, Gillies D, Hua F, Anilkumar AV. Materials research conducted aboard the International Space Station: Facilities overview, operational procedures, and experimental outcomes. *Acta Astronautica*. 2008 ;62;491–498. (*Also presented at the 57th International Astronautical Congress IAC-06-A2.2.10*)

LAB-ON-A-CHIP APPLICATION DEVELOPMENT-PORTABLE TEST SYSTEM (LOCAD-PTS)

Principal Investigator(s): Norman R. Wainwright, Ph.D., Charles River Endosafe, Charleston, S.C.
Expedition 15–18

Research Area Environment Monitoring on ISS

ISS014E18794 — Expedition 14 Flight Engineer Suni Williams uses the swabbing unit to collect samples that will be placed into the cartridges to be analyzed by the LOCAD reader.

The Lab-on-a-Chip Application Development-Portable Test System (LOCAD-PTS) is a handheld device that enables the crew to perform complex laboratory tests for the detection of a variety of biological and chemical target molecules on an interchangeable thumb-sized cartridge with a press of a button. Every thumb-sized plastic cartridge has four channels, and each channel contains a dried extract of horseshoe crab blood cells and colorless dye. In the presence of bacteria and fungi, the dried extract reacts strongly to turn the dye a green color. Therefore, the more green dye, the more microorganisms there are in the original sample.

Tests by LOCAD-PTS will become increasingly specific with the advent of new cartridges. Current cartridges target bacteria and fungi. New cartridges, which are to be launched on subsequent flights, will target bacteria only, followed by groups of bacteria, and eventually individual species or strains that pose a specific risk to crew health. Cartridges can also be adapted to detect chemical substances of concern to crew safety on the ISS (e.g., hydrazine, ammonia, and certain acids) and proteins in urine, saliva, and blood of astronauts to provide added information for medical diagnosis.

RESULTS

LOCAD-PTS was operated for the first time on ISS during Expeditions 14 and 15. Five separate sessions were performed over 6 weeks, with four separate swabs taken for analysis at each site. In Session 1, control test results showed that the instrument and procedures functioned nominally, with the positive control (a swab of the palm) giving a high reading (2.4 endotoxin units (EUs) per 25 cm^2) and the negative control (no swab) giving the lowest possible reading (less than 0.05 EU/25 cm^2). This session also enabled the LOCAD-PTS team to identify an improvement in the procedures (implemented during subsequent sessions) that would help remove air bubbles during the dispensing step and make it easier for the crew to dispense consistent droplet volumes. The consistent readings that were obtained in Session 2 (and remaining sessions) indicated that this procedure change had been successful.

ISS014E18822 — Astronaut Suni Williams, Expedition 14 flight engineer, works with the LOCAD-PTS. Williams is placing the sample mixed with water from the swabbing unit into the LOCAD-PTS cartridge.

Sessions 1 and 2 revealed that the smooth and flat Node 1 surfaces were relatively clean, an average of 0.1 EU/25 cm^2, approximately 25 times cleaner than the positive control. The fabric surface of the temporary sleep station (TeSS) in the ISS that was analyzed in Session 3 gave consistently elevated readings, but the readings were still fairly low as compared with a typical office desk. The audio terminal unit (ATU) and air supply diffuser (ASD)—which were analyzed in Sessions 4 and 5, respectively—gave more variable readings, correlating with the increased variability of these sites in terms of surface materials and air flow between the four swab areas.

As the number of cartridge types on ISS increases, LOCAD-PTS is set to have an extended array of applications: from monitoring the cabin environment for other biological and chemical contaminants to monitoring blood and saliva of the crew to support medical diagnostics. Looking ahead, it is hoped that this type of rapid and portable technology has the potential to be implemented on future human lunar missions to monitor the spread of Earth-derived biological material on the lunar surface following landing. This will be important preparation for the human exploration of Mars, where a major scientific goal will be the search for life and differentiation of that signal from biological material brought there by crewmembers and their spacecraft.

Currently, the technology is being used to assess fluids that are used in pharmaceutical processing. The technology has been used to swab the Mars exploration rovers (MERs), for planetary protection, and to assess microbial contamination in the NASA Extreme Environment Mission Operations (NEEMO) project. This technology will provide quick medical diagnostics in clinical applications. It will also provide environmental testing capabilities that may serve homeland security.

PUBLICATION(S)
There are no publications at this time.

MIDDECK ACTIVE CONTROL EXPERIMENT-II (MACE-II)

Principal Investigator(s): R. Rory Ninneman, Air Force Research Laboratory (AFRL), Kirtland Air Force Base, N.M.
Expeditions 1, 2

Research Area Picosatellites and Control Systems

The Middeck Active Control Experiment-II (MACE-II) was designed to allow engineers to design future spacecraft and facilities with lightweight, inexpensive structures and materials without sacrificing the stability that is demanded by sensitive payloads. MACE-II, the first hands-on experiment on board station, consists of two basic parts that are designed to detect and compensate for vibrations. The multi-body platform (MBP) test article, which is the structure undergoing tests, has four 1-inch-diameter struts that are connected to five nodes. It is loosely tethered in the aisle between racks during operations and is stowed between operations. The entire platform has 20 separate sensors that monitor vibration. The experiment support module (ESM) is a self-contained computer with a power interface to the ExPRESS rack and an umbilical connection to the MBP.

During experiments, scientists used a gimbal on the MBP to create a disturbance at one end of the platform. The ESM detected these movements and, using an adaptable set of algorithms, calculated the opposing forces to be applied at the opposite gimbal, thereby stabilizing the platform. The algorithms could be adapted to changes due to moving parts, variations in temperature, and normal wear and tear on mechanical systems.

A collaborating team at the Massachusetts Institute of Technology (MIT) planned to study how control systems such as that used for MACE-II can be applied to hardware and systems that change over time, such as telescopes, antennas, and robotic arms that must be moved to perform specific duties.

RESULTS

MACE-II provided data autonomously (no human intervention or prior knowledge of the system), decreasing the effects of vibration in moving structures in space. Algorithms were developed to control mechanical systems in real time using only information from on-board sensors and actuators to respond to changes in the system. The system was able to reduce unwanted vibrations without human intervention once it was turned on. These algorithms were able to "adapt" whenever they sensed changes in vibration or the loss of a sensor or actuator.

ISS002E6721 —Susan Helms, Expedition 2 flight\ engineer, works with MACE-II, which is shown "floating" in the microgravity environment of station.

Fourteen test protocols were completed during Expedition 1, and an additional 62 test protocols were completed during Expedition 2. The MACE-II unit, which was returned on shuttle flight STS-105, successfully completed all its experiment objectives that were associated with the Air Force Research Laboratory (AFRL) Science Team while on station. On orbit they demonstrated a decrease in vibration by a factor of 10 while the system was under control. They then showed that their system could adapt to failure of a primary actuator on the system and still decrease vibration by a factor of six.

However, due to data downlink constraints, the MIT Science Team was unable to meet its science objectives. The MIT team required downlink of specific on-orbit tests to build its control algorithms. By the time data were provided to the university, there was insufficient time to uplink the commands to run the critical experiments.

PUBLICATION(S)

Davis L. Economical and Reliable Adaptive Disturbance Cancellation. AFRL-VS-TR-2002-1118 Vol. I, AFRL-VS-TR-2002-1118 Vol. II Pt. 1, AFRL-VS-TR-2002-1118 Vol. II Pt. 2, AFRL-VS-TR-2002-1118 Vol. II Pt. 3, Sep 2002. (*DOD clearance is required to view this paper*)

Ninneman R, Founds D, Davis L, Greeley S, King J. Middeck Active Control Experiment Reflight (MACE II) Program: Adventures in Space. AIAA Space 2003 Conference and Exhibition, Long Beach, Calif. AIAA 2003-6243, 2003.

MICROGRAVITY ACCELERATION MEASUREMENT SYSTEM (MAMS) AND SPACE ACCELERATION MEASUREMENT SYSTEM II (SAMS-II), TWO INVESTIGATIONS

Principal Investigator(s):   Richard DeLombard, NASA Glenn Research Center, Cleveland, Ohio
Expeditions 2–12, ongoing

Research Area Characterizing the Microgravity Environment on ISS

Apparent weightlessness is created as the station circles and falls around the Earth; its continuous free-fall simulates the absence of gravity. A number of scientific investigations on station rely on the absence of gravity for successful completion. However, tiny disturbances aboard ISS, including the reboosts that are required to maintain the station's orbit, mimic the effects of gravity. The SAMS-II and Microgravity Acceleration Measurement System (MAMS) experiments were designed to characterize the microgravity environment on the ISS, and verify whether microgravity requirements are being met. Both SAMS-II and MAMS instruments were deployed in Apr 2001, and continue collecting data for specific experiments.

The microgravity acceleration environment includes quasi-steady-state (frequencies less than natural frequency of the ISS, varying over relatively long times), vibratory accelerations (oscillatory, greater than or equal to the natural ISS frequency), and transients accelerations (short-duration and larger amplitude vibrations).

Vibrational disturbances occur within the frequency range of 0.01 to 300 Hz. SAMS-II measures vibrations from vehicle acceleration, systems operations, crew movements, and thermal expansion and contraction. The SAMS instrument, one of the first U.S. experiments to be operated on the *Mir* space station, was identical to the instrument that was used to characterize the microgravity environment on the space shuttle. The second-generation SAMS (SAMS-II) was deployed on the ISS to characterize the microgravity environment of the ISS during construction in racks near experiments that required complete characterizations of accelerations that were experienced during experiment runs.

The MAMS complements the SAMS-II data by recording accelerations that are caused by aerodynamic drag and ISS movements, which are caused by small attitude adjustments, gravity gradient, and the venting of water. These quasi-steady-state accelerations occur in the frequency range below 1 Hz.

ISS006E42571 — This view features a reboost of station. Ground controllers at Mission Control Moscow ignited the thrusters of a Progress rocket that was docked to the *Zvezda* service module. This 14-minute firing raised the average altitude of ISS by about 3 km. Periodic reboosts, vehicle dockings and other maneuvers create small forces that may influence experimental results. MAM and SAM-II provide critical data for documenting the microgravity environment and interpreting results.

MAMS consists of a low-frequency triaxial accelerometer, the miniature electro-static accelerometer (MESA), a high-frequency accelerometer, the high-resolution accelerometer package (HiRAP), and associated computer, power, and signal processing subsystems that are contained within a double middeck locker enclosure. SAMS-II has multiple remote triaxial sensor (RTS) systems that are used to monitor individual experiments. Each RTS is capable of measuring between 0.01 Hz to beyond 300 Hz of vibration, which is also known as g-jitter.

RESULTS

Both MAMS and SAMS-II experiments have been monitoring the microgravity environment and how it changes during the construction of the ISS, and during certain payload operations to verify exact microgravity conditions. MAMS data have been analyzed to examine the quasi-steady regime on station with a frequency that is below 0.01 Hz. These are related to flight attitude, aerodynamic drag, gravity gradient and rotational effects, venting of air or water, and appendage movement, such as that of the solar arrays and antennas. Analysts determined that the movement of the Ku-band antenna was the source of the unusual characteristics in the quasi-steady data that were collected by MAMS. (A Ku-band antenna is used to transmit payload science data and video.) The correlation was made after comparing the data with real-time observations from ISS (DeLombard et al., 2002, 2004).

A special study using MAMS data was performed by ISS science officer Don Pettit during Expedition 6 as a part of Saturday Science. Pettit examined the motion of air bubbles in water to see how it correlated with quasi-steady accelerations. (Quasi-steady accelerations are vibrations that are at or below a frequency of 0.01 Hz for a period greater than 100 seconds.)

SAMS-II has been used in microgravity and non-microgravity modes of ISS operations to measure vibratory acceleration disturbances. Data have been collected to document vibrations resulting from experiment hardware (e.g., fans, blowers, and pumps) and crew activity, and effectiveness of the ARIS to reduce the effect of the vibrational environment on the ISS. These data are critical for interpreting results from experiments on the ISS that assume a uniform microgravity environment.

As the ISS nears completion, the microgravity environment changes. During early increments—at least through Increment 8 (2004)—the ISS was not meeting the documented microgravity requirements. For certain acceleration regimes, the XPOP attitude ((X axis perpendicular to the orbit plane) appeared to provide a better environment (Jules et al. 2004).

MAMS is currently being activated intermittently to meet operational requests for data during major mission events such as dockings by Soyuz and Progress vehicles. SAMS-II is also still on orbit.

PUBLICATION(S)

DeLombard R, Kelly EM, Foster, K, Hrovat K, McPherson, KM; Schafer, CP. Microgravity Acceleration Environment of the International Space Station, AIAA 2001-5113

Del Basso S, Laible M, O'Keefe E, Steelman A, Scheer S, Thampi S. Capitalization of Early ISS Data for Assembly Complete Microgravity Performance. Proceedings of the 40th AIAA Aerospace Sciences Meeting and Exhibit, Reno, Nev. AIAA 2002-606, Jan 14–17, 2002.

DeLombard R, Hrovat K, Kelly EM, McPherson K, Jules K. An Overview of the Microgravity Environment of the International Space Station Under Construction. AIAA 2002-0608.

Jules K, McPherson K, Hrovat K, Kelly E, Reckart T. A Status on the Characterization of the Microgravity Environment of the International Space Station. 54th International Astronautical Congress, 29 Sep to 3 Oct 2003, Bremen, Germany, IAC-03-J.6.01.

Jules K, Hrovat K, Kelly EM. The Microgravity Environment of the International Space Station during the Buildup Period: Increments 2 to 8. 55th Int. Astronautical Congress 2004 Vancouver, Canada, IAC-04-J.6.01.

DeLombard R, Kelly EM, Hrovat K, McPherson. Microgravity Environment of the International Space Station. AIAA 2004-125.

DeLombard R, Kelly EM, Hrovat K, Nelson ES, Pettit DR. Motion of Air Bubbles in Water Subjected to Microgravity Accelerations. Proceedings of the 43rd AIAA Aerospace Sciences Meeting and Exhibit, Reno, Nev. AIAA 2005-722, Jan 10–13, 2005.

DeLombard R, Hrovat K, Kelly EM, Humphreys B. Interpreting the International Space Station Microgravity Environment. Proceedings of the 43rd AIAA Aerospace Sciences Meeting and Exhibit, Reno, Nev. AIAA 2005-0727, Jan 10–13, 2005.

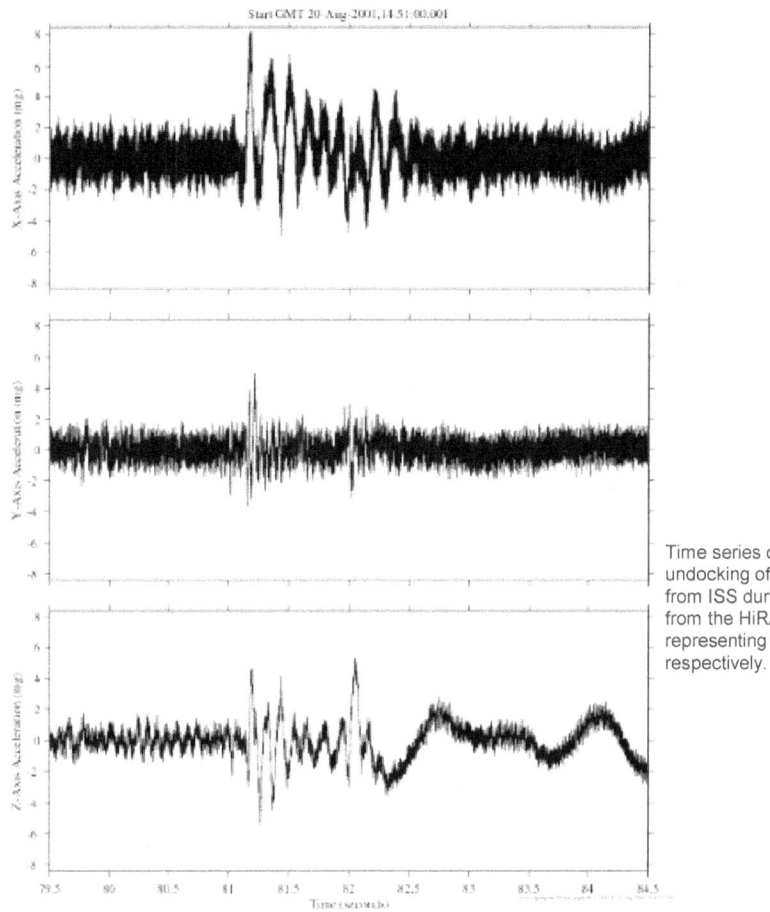

Time series of vibrations produced by the undocking of shuttle flight STS-105 (7A.1) from ISS during Expedition 2. Data are from the HiRAP sensor with graphs representing the X, Y, and Z axes, respectively.

MAUI ANALYSIS OF UPPER ATMOSPHERIC INJECTIONS (Maui)

Principal Investigator(s): Rainer A. Dressler, Ph.D., Hanscom Air Force Base, Lexington, Mass.
Expeditions 13–16, 18

Research Area Spacecraft Systems

The Maui analysis of upper atmospheric injections (Maui) experiment observes the space shuttle engine exhaust plumes from the Maui Space Surveillance Site (MSSS) in Hawaii. Observations occur when the shuttle fires its engines at night or twilight when passing over the MSSS. Spectrally filtered images and spectra of the radiation resulting from exhaust-atmosphere interactions are taken by the optical telescope and all-sky imagers. The data should determine the chemical and physical mechanisms that are associated with the interaction between the chemical species in engine exhaust and the space environment. Improved models of this interaction will result in enhanced space event characterization as well as the determination of sensor requirements for effective plume and contamination analysis of other spacecraft.

View of Orbital Maneuvering System (OMS) engine burn to boost the orbit of the space shuttle for rendezvous with the *Mir* space station. Image courtesy of NASA Johnson Space Center.

RESULTS
Data are still being collected for this experiment.

PUBLICATIONS(S)
There are no publications at this time.

MATERIALS INTERNATIONAL SPACE STATION EXPERIMENT-1 AND -2 (MISSE-1 AND -2)

Principal Investigator(s):   William H. Kinard, NASA Langley Research Center, Hampton, Va.
Expeditions 3–11, ongoing

Research Area Spacecraft Materials and Systems

Researchers from the private and public sector prepared a wide range of samples for the first externally mounted experiment on ISS. MISSE-1 and -2 were testbeds for more than 400 materials and coatings samples, testing their survivability under the corrosive effects of the space environment; including MMOD strikes, AO attack, intense ultraviolet radiation from the sun, and extreme temperature swings. Results will provide a better understanding of the durability of various materials in this environment. Many of the materials may have applications in the design of future spacecraft.

ISS009E22435 — MISSE on ISS, over cloud tops.

Both MISSE-1 and -2 were deployed in Aug 2001 on Expedition 3 and were planned for a 1-year exposure. Due to the delays that were incurred following the *Columbia* accident, they were not retrieved until 4 years later during ISS Expedition 11 in Aug 2005. Follow-on samples testing new materials, new technologies and with increasingly complex missions were part of subsequent MISSE experiments mounted on the station (MISSE-5, MISSE-3, MISSE-4, and MISSE-6).

RESULTS

In late 2005, 35 investigators who were taking part in MISSE-1 and -2 traveled to NASA Langley Research Center to inspect their samples and prepare them for return to their respective laboratories for further analysis. Researchers who took part in this investigation have interests in polymers, thermal control coatings, nano-composites, radiation shielding, environmental monitors, and marking processes that are designed to label parts that will be exposed to the space environment. The primary data from MISSE will be obtained by comparing the preflight laboratory characterization of the test specimens with postflight laboratory characterizations made after the specimens are retrieved.

Some particulate contamination was observed. Optical property changes in thermal control materials were also observed. Several materials did well in the harsh environment. Lack of widespread molecular contamination on MISSE gives confidence in using station for future material studies. A number of results are anticipated to be released over the next few years. While the samples are still under investigation, researchers indicate that over 100 micrometeoroid and space debris strikes were found. Many polymer film samples were completely eroded by AO, but some samples survived and are undergoing analysis.

Many of the experiments provide space-validated results for ground-based experiments , such as the durability of materials to withstand AO (deGroh et al. 2008)..Because AO erosion is the primary weathering force to spacecraft materials, and true space environmental conditions are difficult to replicate on Earth, MISSE provides a valuable test platform that enables methods for correlating and extrapolating ground results. Snyder et al. (2006) discuss results from MISSE-2 testing of AO erosion of silicon oxide (SiOx)-coated Kapton compared to uncoated Kapton (the AO erosion profile of Kapton is well documented). They calculated mass loss. De Groh et al. (2006) analyzed 41 different PEACE polymers, with the objective to determine the AO erosion yield for a variety of materials—such as Kevlar, polyethylene, Lucite, Kapton, and Teflon—that are used in spacecraft and exposed to the space environment. The erosion yield data are immediately applicable to spacecraft designs. Although the length of exposure

ISS009E22432 – Close-up of MISSE with Earth backdrop.

was four times longer than planned, the sample preparation method of stacking many thin layers allowed for meaningful data even after 4 years: not all of the material had eroded away. Samples masses were weighed after flight and compared with preflight masses to calculated erosion yields.

Other investigators studied specific polymers with various compositional additives or coatings to test resistance to AO erosion.(e.g., Tomczak et al. 2007, Juhl et al. 2007). One approach (Tomczak et al. 2007) is to embed polymers with nano-sized silicon dioxide (SiO_2). With AO exposure, a silica passivation layer is formed.

In addition to testing various materials, experiments were also set up to measure the geometry of AO scattering from oxidized aluminum surfaces.

MISSE-1 and -2 materials, and insights into AO erosion, have resulted in several patents and spin-offs, including cleaning artwork, etching parts to be used in human grafts, developing new methodologies for testing blood sugar, and more. Because MISSE assembles partners across industry and the DOD, in addition to NASA scientists and academic partners, many of the results are proprietary.

PUBLICATIONS

Finckenor M. The Materials on International Space Station Experiment (MISSE): First Results from MSFC Investigations, Proceedings of the 44th AIAA Aerospace Sciences Meeting and Exhibit, Reno, Nev. AIAA 2006-472, Jan 9–12, 2006.

Snyder A, Banks BA, Waters DL. Undercutting Studies of Protected Kapton H Exposed to In-Space and Ground-Based Atomic Oxygen. NASA TM-2006-214387, Aug 2006.

de Groh K, Banks B. MISSE PEACE Polymers Atomic Oxygen Erosion Results. NASA TM-2006-214482 (2006).

Harvey GA, Kinard WH. MISSE 1 and 2 Tray Temperature measurements. Proceedings of MISSE Post Retrieval Conference and the 2006 National Space & Missile Materials Symposium

de Groh K , Banks B. Materials International Space Station Experiment (MISSE) Polymers Degradation. 9th International Conference on "Protection of Materials and Structures from Space Environment," May 20–23, 2008, Toronto, Canada.

de Groh K, Banks B, Arielle H, Stambler LM. Roberts KE, Barbagallo I, Barbagallo CE. Ground-Laboratory to In-Space Atomic Oxygen Correlation for the PEACE Polymers, 9th International Conference on "Protection of Materials and Structures from Space Environment," May 20–23, 2008, Toronto, Canada.

Tomczak SJ, Marchant D, Mabry JM, Vij V, Yandek GR, Minton TK, Brunsvold AL, Wright ME, Petteys BJ, Guenthner AJ. Studies of POSS-Polyimides Flown on MISSE-1. 25–29 Jun 2007; Keystone, Colo.; 2007 National Space & Missile Materials Symposium.

Juhl SB, Akinlemibola B, Kasten L, Vaia R. Durability of Poly(Caprolactam) (Nylon 6) and Poly(Caprolactam) Nanocomposites in Low Earth Orbit. 25–29 Jun 2007; Keystone, Colo.; 2007 National Space & Missile Materials Symposium.

Rice N, Shepp A, Haghighat R, Connell J. Durable TOR Polymers on MISSE. 25–29 Jun 2007; Keystone, Colo.; 2007 National Space & Missile Materials Symposium.

Watson KA, Ghose S, Lillehei PT, Smith Jr. JG, Connell JW, Effect of LEO Exposure on Aromatic Polymers Containing Phenylphosphine Oxide Groups, SAMPE Proceedings Vol. 52, Jun 6 2007, Baltimore, Md.

MATERIALS ON THE INTERNATIONAL SPACE STATION EXPERIMENT-3 AND -4 (MISSE-3 AND -4)

Principal Investigator(s): William H. Kinard, Ph.D., NASA Langley Research Center, Hampton, Va.
Expeditions 13–15

Research Area Spacecraft Materials and Systems

MISSE-3 and -4 were successfully deployed in Aug 2006 and retrieved in Aug 2007. Approximately 875 specimens of various materials were contained in suitcase-like cases called passive experiment containers (PECs). These specimens were exposed to the harsh environment of microgravity to observe the effects that AO (single-oxygen molecules) and ultraviolet light have on materials.

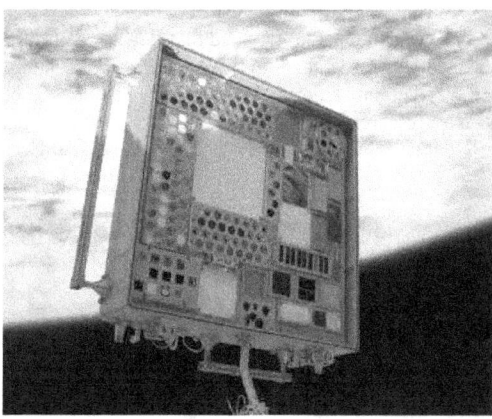

ISS015E19092 — This image of tray 2 of the MISSE-3 PEC was taken in Jul 2007. At this point, MISSE-3 has been exposed to the space environment for approximately 11 months.

The specimens include a variety of materials (e.g., paint and protective coatings) that will be used on future spacecrafts such as satellites. Environmental monitors recorded the thermal cycling (the change in temperature) that the experiment was subjected to while on orbit. New material that might be used in the next generation of extravehicular activity (EVA) suits was tested to examine how the material reacts to the harsh space environment.

As part of an education outreach program, three million basil seeds were placed in containers located underneath the sample trays on MISSE-3 and -4 PECs. These seeds were returned to Earth as part of the STS-118/13A.1 mission in which Astronaut Barbara Morgan initiated the grown cycle of basil seeds inside the ISS, The seeds were sent to school children for them to plant and observe the differences between seeds exposed to space and seeds that have remained on Earth.

RESULTS

Preliminary assessments, which include results from previous MISSE-flown materials, suggest that the contamination control for the station—the method for tracking whether scientific instruments, windows, radiators and other hardware are staying clean from contaminants such as dust, dirt, or hair—is working.

The earlier experiments showed that samples of the glass that were used in station windows were better than 90% clear, and samples of the same white thermal coatings that were used on station radiators looked like new even after 4 years in space. Full analyses of the MISSE-3 and -4 materials are under way.

PUBLICATIONS(S)

There are no publications at this time.

MATERIALS INTERNATIONAL SPACE STATION EXPERIMENT-5 (MISSE-5)

Principal Investigator(s): Robert Walters, Ph.D., Naval Research Laboratory, Washington, D.C. (District of Columbia); William H. Kinard, Ph.D., NASA Langley Research Center, Hampton, Va. Expeditions 11–13

Research Area Spacecraft Materials and Systems

MISSE-5 was an external payload that flew on board the ISS from Aug 2005 until Sep 2006, thus providing an opportunity for researchers to test a wide range of samples in the low Earth orbit (LEO) environment. MISSE-5, which was a collaboration between NASA Langley Research Center, Glenn Research Center, Ohio State University, the Naval Research Laboratory, and the U.S. Naval Academy, consisted of three experiments: Prototype Communications Satellite-2 (PCSat-2), Forward Technology Solar Cell Experiment (FTSCE), and the Thin Film Materials Experiment.

S114E7352 — Views of MISSE-5 mounted on the ISS P6 truss during Expedition 11. PcSat-2 is protected by a golden thermal blanket with flexible material samples attached.

PCSat-2 was a communication system that was sponsored by the U. S. Naval Academy. It had two objectives: (1) to test the Amateur Satellite Service off-the-shelf solution for telemetry command and control; and (2) to provide a communication system for the FTSCE. PCSat-2 was able to transmit solar cell data for FTCSE using the Amateur Satellite Service with a 145.825 uplink and a 435.275 ± 10 kHz Doppler downlink.

FTSCE characterized the durability and electrical output of 39 advanced solar cell samples that could be used on future space exploration vehicles. Several types of solar cell technologies were tested: triple junction InGaP/GaAS/Ge; thin film amorphous Si and $CuIn(Ga)Se_2$; and single-junction GaAs cells. It is known that solar cells degrade over time when they are exposed to the space environment. FTSCE used their on-board instrumentation to measure the performance and downlink the data to Earth through the PCSat-2.

The Thin Film Material Experiment consists of 254 thin film samples that were attached to the thermal blanket protecting the PCSat-2 hardware. The samples are exposed directly to the space environment to evaluate the degradation of the materials over time. These materials range from testing polymer coatings to solar array blanket material to paints that are used on spacecraft and many others. An additional aspect of the Thin Film Materials Experiment is the educational component. Of the 254 samples, 49 are part of a collaboration between the Glenn Research Center and the Hathaway Brown School in Cleveland, Ohio.

The technology testing that occurred during the MISSE-5 investigation provided the data that are necessary to develop the new space exploration vehicles, satellites, and communication systems that will take us to the moon, Mars, and beyond.

RESULTS

The MISSE team has reported many results from the MISSE-5 experiment. Because MISSE assembles partners across industry and the DOD in addition to NASA scientists and academic partners, many of the results are proprietary.

MISSE-5 also tested new approaches for technology tests, including innovative collaborations that leverage the costs and assembly environment of flight article preparation (piggy-backed thin films that are stitched onto a thermal blanket) (Kinard et al. 2007).

A wide range of solar cell technologies were tested with favorable results (Walters et al. 2006). In some cases, the samples that were flown were orders of magnitude larger than previous samples.

In addition to the solar cell experiment, MISSE-5 tested a wide variety of materials. Some focused on new polymers with additives to slow the AO erosion process. Others tested both new and old thermal control materials that are used in multilayer insulation (MLI) blankets (Finckenor et al. 2007). The data are compared with other flight experiments and analyzed for solar absorptance, contamination, and other attributes.

PUBLICATIONS(S)

Walters RJ, Garner JC, Lam SN, Vazquez JA, Braun WR, Ruth RE, Warne JH, Lorentzen JR, Messenger SR, CDR Bruninga R, Jenkins PP, Flatico JM, Wilt DM, Piszczor MF, Greer LC, Krasowski MJ. Forward Technology Solar Cell Experiment First On-Orbit Data, 19th Space Photovoltaic Research and Technology Conference, (2007) NASA CP-2007-214494, 79–94.

Krasowski M, Greer L, Flatico J, Jenkins P, Spina D. Big Science, Small-budget Space Experiment Package aka MISSE-5: A Hardware and Software perspective. 19th Space Photovoltaic Research and Technology Conference, (2007) NASA CP-2007-214494, 95–117.

Kinard WH. Materials Experiment Flown on MISSE 5. 25–29 Jun 2007; Keystone, Colo.; 2007 National Space & Missile Materials Symposium.

Finckenor M, Zweiner JM, Pippin G, Thermal Control Materials on MISSE-5 with Comparison to Earlier Flight Data. 25–29 Jun 2007; Keystone, Colo.; 2007 National Space & Missile Materials Symposium.

Simburgera EJ, Matsumotoa JH, Giantsa TW, Garcia III A, Liua S, Rawalb SP, Perry AR, Marshall CH, Linc JK, Scarborough SE, Curtis HB, Kerslake TW, Peterson TT. Development of a thin film solar cell interconnect for the PowerSphere concept, Materials Science and Engineering, Volume 116, Issue 3, 15 Feb 2005, pp. 321–325.

Walters RJ, Garner JC, Lam SN, Vasquez JA, Braun WR, Ruth RE, Warner JH, Lorentzen JR, Messenger SR, CDR Bruninga R, Jenkins PP, Flatico JM, Wilt DM, Piszczor MF, Greer LC, Krasowski MJ. Materials on the International Space Station Experiment-5, Forward Technology Solar Cell Experiment: First On-Orbit Data, Conference Record of the 2006 IEEE 4th World Conference on Photovoltaic Energy Conversion, May 2006; Volume 2, pp. 1951–1954.

Wilt DM, Clark EB, Ringel SA, Andre CL, Smith MA, Scheiman DA, Jenkins PP, Maurer WF, Fitzgerald EA, Walters RJ, et al.. LEO Flight Testing of GaAs on Si Solar Cells Aboard MISSE 5. 19th European Photovoltaic Solar Energy Conference and Exhibition, Paris, France, 2004.

Pippin G, deGroh K, Finckenor M, Minton T. Post-Flight Analysis of Selected Fluorocarbon and Other Thin Film Polymer Specimens Flown on MISSE-5. 25–29 Jun 2007; Keystone, Colo.; 2007 National Space & Missile Materials Symposium.

deGroh K, Finckenor M, Minton T, Brunsvold A, Pippin G. Post-Flight Analysis of Selected Fluorocarbon and Other Thin Film Polymer Specimens Flown on MISSE-5. 25–29 Jun 2007; Keystone, Colo.; 2007 National Space & Missile Materials Symposium.

Ram Burn Observations (RAMBO)

Principal Investigator(s): William L. Dimpfl, Ph.D., Aerospace Corporation, Los Angeles, Calif.
Expeditions 13–16

Research Area Spacecraft Environments

The Ram Burn Observations (RAMBO) experiment uses a satellite to observe the spectral characteristics and direction of the movement of plumes that are created from shuttle OMS burns in LEO. The engine burns create high-temperature, high-velocity molecular collisions between the chemical species that are in the engine exhaust (e.g., water and CO) and AO. Three specific objectives are listed below.

STS00718778 — This image shows the Glow experiment documentation of OMS/Reaction Control System (RCS) pods and vertical stabilizer from STS-007.

1. **Determination of the distribution of internal states of CO that are excited through collisions with AO at hyperthermal collision velocities.** Past experiments have shown that CO is efficiently excited to high internal (vibrational and rotational) energy states in collisions with AO at high-temperature (hyperthermal) energies. The theory is that such atom-molecule collisions are generally inefficient at transferring collision energy to internal states (the state at which an object is in regards to its internal properties). The AO + CO system represents an interesting scientific anomaly. The anomaly is attributed to a chemical interaction in the AO + CO system, which allows the exchange of AO atoms to take place. A rigorous theoretical treatment of this system has been developed, and predictions of the energy transfer have been calculated. RAMBO measurements are a method by which the theoretical understanding can be validated over the range of collision velocities from 4 to 11 km/s by observing CO radiant emission that is excited by collisions between CO in the plume and AO atoms in the atmosphere.

2. **Determine the total scattering cross section for atomic and molecular species at hyperthermal energies.** Analysis of the space shuttle orbiter engine plumes while in orbit have indicated that models for molecular scattering that are based on laboratory interactions of flame temperatures up to 2,000 through 4,000 K are not accurate in the hyperthermal regime that is experienced by orbiting spacecraft. Analysis of infrared plume radiance resulting from atmosphere-plume collisions that induce molecular excitation in the range of 4 to 11 km/s from the RAMBO experiment will help quantify total molecular scattering cross sections in the hyperthermal regime.

3. **Determining the rate constants for hydroxide producing reaction at hyperthermal energies.** Emission in high-altitude rocket plumes in the region from 3 to 4 microns has generally been attributed to emission from water. Better understanding has indicated that there is also significant emission in that region from internally excited hydroxide (OH) radicals that are formed through the reaction of atmospheric AO and plume water and molecular hydrogen. Rate constants that have been determined in the laboratory for these reactions are generally limited to the thermal regime of energies below 1 electron volt (eV).

High-altitude plumes involve energies that extend into the hyperthermal regime up to about 10 eV. Extrapolation of laboratory-measured rate constants into the hyperthermal regime provides modeling data, but extrapolated values may have a significant error. RAMBO experiments sample OH emission that is produced directly from the relevant reactions at hyperthermal energies, and are being used to establish valid hyperthermal rate constants.

RESULTS
This experiment continues on the ISS. Results are pending, but are important for better constraining models for the high-temperature, high-velocity atomic and molecular collisions that are induced by spacecraft operations.

PUBLICATIONS(S)
There are no publications at this time.

SMOKE AND AEROSOL MEASUREMENT EXPERIMENT (SAME)
Principal Investigator(s):   David Urban, Ph.D., NASA Glenn Research Center, Cleveland, Ohio
Expedition 15

Research Area Spacecraft Systems

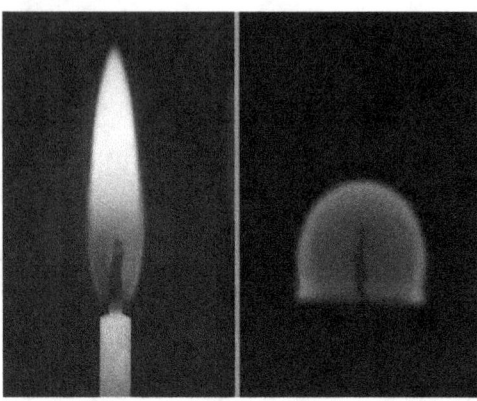

A candle flame in Earth's gravity (left) and microgravity (right) shows difference in the processes of combustion in microgravity. Image courtesy of NASA Johnson Space Center.

Spacecraft smoke detectors must detect different types of smoke. For example, hydrocarbon fuels typically produce soot and plastics produce droplets of recondensed polymer fragments while paper and silicone rubber produce smoke that is comprised of liquid droplets of recondensed pyrolysis products. Each of these materials produces a different type of smoke, with particles of various sizes and properties.

SAME will assess the size and distribution of the smoke particles that are produced by the different types of material that are found on spacecraft such as Teflon, Kapton, cellulose, and silicone rubber. It will evaluate the performance of the ionization smoke detectors, which are used on space shuttles; evaluate the performance of the photoelectric smoke detectors, which are used on the ISS; and collect data for which a numerical formula can be developed and used to predict smoke droplet growth and to evaluate alternative smoke detection devices on future spacecraft.

The experimental design and practical application of the data will be complimented by the development of a numerical code to predict the smoke droplet growth as a function of the fuel pyrolysis rate, the thermodynamic properties of pyrolysis vapor, and the flow environment. SAME also has the capability to evaluate other fire detection/particulate sensing devices for the test materials. The results will provide statistics of the smoke particulate size distribution for a range of smoke-generation conditions and measurement of a readily modeled reference for validation of smoke growth models.

RESULTS
Building from the earlier Dust and Aerosol Measurement Feasibility Test (DAFT) experiment that successfully tested a particulate detector, SAME tested the size of smoke aerosols of spacecraft materials such as Kapton and Teflon.

Smoke particulate that is produced in low gravity by SAME was found to be typically 50% larger in count mean diameter than similar conditions in normal gravity. The particle sizes were all below 300 nm, suggesting that discriminating smoke from spacecraft dust possibly could be achieved by detecting in the sub-micrometer range (Urban et al. 2008). These results have significant implications for the design of smoke detection systems for current and future spacecraft.

The experiment also modeled the smoke transport in the U.S. Laboratory using Environmental Control and Life Support System (ECLSS) data. Numerical modeling of smoke transport predicted that actual detection times can be quite long and strongly dependent on detector location and the inside geometry of obstructions that block cabin air flow (Urban et al. 2008).

PUBLICATION(S)
Urban DL, Ruff GA, Brooker JE, Cleary T, Yang J, Mulholland G, Yuan Z-G. Spacecraft Fire Detection: Smoke Properties and Transport in Low-Gravity. 46th AIAA Aerospace Sciences Meeting and Exhibit, Reno, Nev. 2008, 7 – 10 Jan; AIAA 2008-806.

SERIAL NETWORK FLOW MONITOR (SNFM)

Principal Investigator(s):   Carl Konkel, The Boeing Company, Houston
Expeditions 9–11

Research Area Spacecraft Systems

The serial network flow monitor (SNFM) is a commercial off-the-shelf (COTS) software package that monitors packet traffic through the payload Ethernet local area networks (LANs) on board station. The SNFM experiment characterized the network equivalent of data traffic jams on board ISS. The SNFM team targeted historical problem areas including the SAMS-II communication issues, data transmissions from ISS to the ground teams, and multiple users on the network at the same time. By looking at how various users interact with each other on the network, conflicts can be identified and work can begin on solutions.

RESULTS
SNFM data are still being analyzed, and will provide "lessons learned" for ongoing network operations on space station and future spacecraft systems.

PUBLICATIONS(S)
There are no publications at this time.

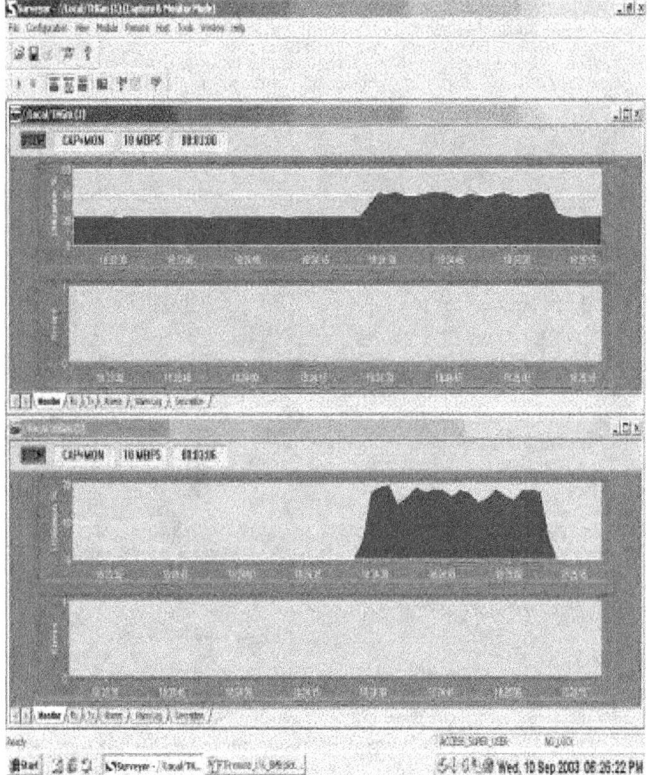

Computer screen capture image provides a graphic example of network load monitoring.

SYNCHRONIZED POSITION HOLD, ENGAGE, REORIENT, EXPERIMENTAL SATELLITES (SPHERES)

Principal Investigator(s): David Miller, MIT, Cambridge, Mass.
Expedition 8, 13–15, ongoing

Research Area Picosatellites and Control Systems

The Synchronized Position Hold, Engage, Reorient, Experimental Satellites (SPHERES) experiment is a testbed for formation flying by satellites—the theories and calculations that coordinate the motion of multiple bodies maneuvering in microgravity. To achieve this inside the ISS cabin, bowling-ball-sized spheres perform various maneuvers (or protocols), with one to three spheres operating simultaneously. The SPHERES experiment tests relative attitude control and station-keeping between satellites, re-targeting and image plane filling maneuvers, collision avoidance and fuel balancing algorithms, and an array of geometry estimators that are used in various missions.

SPHERES consists of three self-contained satellites, all of which are 18-sided polyhedrons that are 0.2 m in diameter and weigh 3.5 kg. Each satellite contains an internal propulsion system, power, avionics, software, communications, and metrology subsystems. The propulsion system uses carbon dioxide (CO_2), which is expelled through the thrusters. SPHERES satellites are powered by AA batteries. The metrology subsystem provides real-time position and attitude information. To simulate ground station-keeping, a laptop will be used to transmit navigational data and formation-flying algorithms. Once these data are uploaded, the satellites will perform autonomously and hold the formation until a new command is given.

SPHERES is an ongoing demonstration. Each session tests progressively more complex two- and three-body maneuvers that include docking (to fixed, moving, and tumbling targets), formation flying, and searching for "lost" satellites.

RESULTS

During Expedition 8, several interference tests were conducted to characterize the effects of lights and other sources of electromagnetic radiation. The "Beacon-Beacon Test" (SPHERES-BBT) used one beacon (with mount) and one beacon tester to demonstrate the functionality of the ultrasound/infrared positioning system of the SPHERES experiment. This portion of the investigation was a risk mitigation experiment for SPHERES to determine whether any sources of infrared radiation or ultrasonic waves exist in the work area that may interfere with SPHERES operations. Infrared interference from a general lighting assembly and a laptop in the Node was produced during the SPHERES-BBT. Based on these results, SPHERES test sessions use the newer laptop and the general lighting assembly will be dimmed by 25% in the workspace during SPHERES operations.

SPHERES test sessions that were begun during Expedition 13 demonstrated maneuvers that would lead to the docking of a satellite to a beacon, tested algorithms that automatically determine the mass properties of the satellites, and performed initial two-sphere satellite tests. During these sessions, the docking maneuvers were completed successfully and updated to the estimator (calculation of the satellite's location with respect to the beacon), which helped to achieve successful approaches. The two satellite tests were successful; these tests demonstrated the functionality of the new satellite and the ability of the satellites to maintain both angular formation (i.e., they point in the same direction all the time) as well as position formation (if one translates the other does, too).

Other SPHERES test sessions that were performed during Expedition 13 demonstrated the functionality of the Global Metrology System (during all previous sessions the satellites calculated their position with respect to a single beacon or with respect to other satellites, but not with respect to the U.S. Laboratory frame). This session set up the complete metrology system so that the satellites could know their location and pointing with respect to the U.S. Laboratory. Tests included collection of large amounts of data to process on the ground to test for the functionality of all the transmitters and receivers in the system as well as the presence of any noise (such as reflections on unexpected objects). The test also demonstrated the functionality of the astronaut interface to indicate to the satellite the location of the beacons in the U.S. Laboratory. All of the necessary data were collected. Secondary tests included successful demonstration of two satellites docking (controlled contact between the two satellites). These tests will ultimately enable scientists to assemble large space structures and make autonomous resupply of consumables and upgrades a reality.

Additional one- and two-satellite tests continued during Expedition 14; three-satellite tests began during Expedition 15 and continue. Expeditions 14 and 15 tested more complex docking maneuvers and formation flying. The SPHERES team also includes a "Guest Scientist" Program that enables remote scientists to have the ability to design and code new algorithms to test.

PUBLICATIONS(S)
There are no publications at this time.

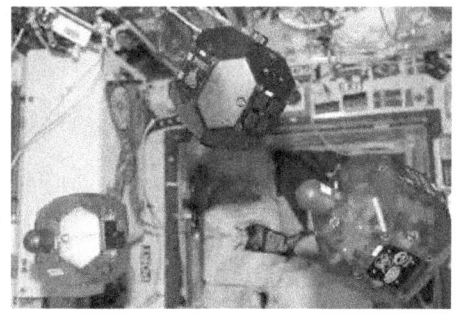

ISS014E17874 — Three satellites fly in formation as part of the SPHERES investigation. This image was taken during Expedition 14 in the *Destiny* laboratory module.

ISS013E68304 — Astronaut Jeffrey N. Williams, Expedition 13 NASA space station science officer and flight engineer, does a check of the SPHERES satellites in the *Destiny* laboratory of the ISS.

Space Test Program-H2 (STP-H2): Three Investigations

Principal Investigator(s): Andrew Nicholas, Naval Research Laboratory, Washington, D.C. (ANDE)
James Keeney, Ph.D., Kirtland Air Force Base, Albuquerque, N.M. (MEPSI)
United States Naval Academy, Annapolis, Md. (RAFT)
Expedition 14

Research Area Picosatellites and Control Systems, Spacecraft Environments

The complement of Space Test Program-H2 (STP-H2) investigations was deployed from STS-116/12A.1 on Dec 20, 2006.

S116E07836 — Shown is one of the ANDE satellites floating free from the internal cargo unit (ICU) after deployment. The second satellite did not leave its half of the ICU until 4 hours later.

Atmospheric Neutral Density Experiment (ANDE)

ANDE consists of two spherical microsatellites: the Mock ANDE Active (MAA) spacecraft and the Fence Calibration (FCal) spacecraft. These satellites were launched from the space shuttle cargo bay into a circular orbit just below the ISS altitude. The main objective of the ANDE mission is to measure the total atmospheric density and composition between 100 and 400 km. Density data that are gathered will be used to better characterize the parameters that are used to calculate the drag coefficient of a satellite and improve orbit determination calculations of resident space objects.

Both satellites will be tracked by the Satellite Laser Ranging (SLR) system and the U.S. Space Surveillance Network (SSN). These satellites have similar dimensions, but are constructed of different materials and have different masses. Because of the difference in mass, the satellites will drift apart over time. Observing the satellites' position will provide a study on spatial and temporal variations in atmospheric drag associated with geomagnetic activity. The FCal sphere will also be used to perform calibrations for the U.S. Radar Fence.

Microelectromechanical System (MEMS)-Based Picosat Inspector (MEPSI)

The Microelectromechanical System (MEMS)-based Picosat Inspector (MEPSI) experiment series was designed to demonstrate the concept of an on-board intelligent hardware agent, "InfoBot," that can be used to assist satellite operations. It is designed to enhance satellite command and control operations by providing active on-board imaging capability to assess spacecraft damage from human-made or environmental threats, monitor satellite early orbit testing operations, and augment servicing operations. MEPSI was developed through a series of four preflight missions, each of increasing complexity and each improving overall satellite performance over the previous version. In Dec 2002, MEPSI completed its third development mission with a successful launch from the Space Shuttle *Endeavour*, which was its first shuttle mission. Improvements from the 2002 version were included in this payload, which was deployed in Dec 2006 from STS-116.

Radar Fence Transponder (RAFT)

The radar fence transponder (RAFT) mission is a student experiment from the U.S. Naval Academy that used picosatellites to test the Space Surveillance Radar Fence and experimental communications transponders. More specifically, RAFT was designed to provide the Navy Space Surveillance System (NSSS) radar fence with a means by which to determine the limits of a constellation of picosatellites that would be otherwise undetectable to the radar fence, and to enable NSSS to independently calibrate its transmit and receive beams using signals from RAFT. This was accomplished with two picosatellites (RAFT1 and MARScom): one that actively transmitted and received signals, and one with a passively augmented radar cross section. Additionally, RAFT provided experimental communications transponders for the Amateur Satellite Service, the Navy Military Affiliate Radio System, and the Naval Academy's Yard Patrol Craft.

RESULTS

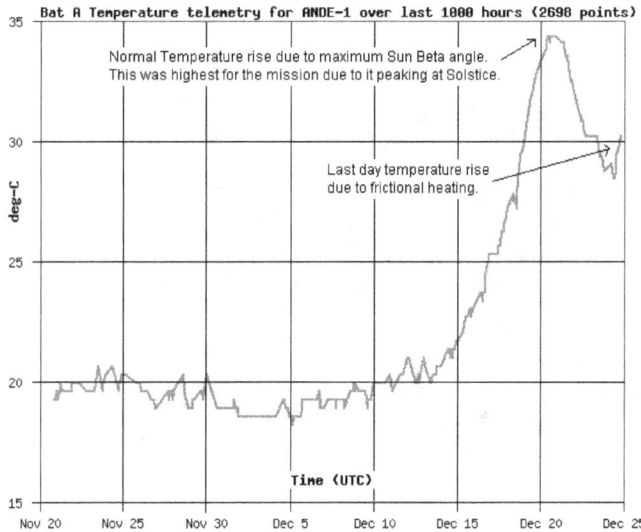

ANDE
The ANDE satellite re-entered the Earth's atmosphere on Dec 25, 2007, more than 1 year after deployment. Its orbital decay was tracked by the Maui Laser Ranging Tracking Station. Because the satellite carried packet radio communications systems operating in the Amateur Satellite Service, HAM radio volunteer ground stations were critical for telemetry feeds that included temperature and battery life.

MEPSI
This image of the Space Shuttle *Discovery* was taken by cameras that were located in MEPSI. This was taken shortly after deployment on Dec 20, 2006.

RAFT
RAFT was deployed on Dec 20, 2006 and de-orbited on 30 May 2007 after 5 months in space. The deployment resulted in an applied torque to the satellites (see image sequence, left). Several subsystems on the satellite (solar panel and thermal performance) were monitored. Volunteer ground stations were used to track the satellites.

The sequence of images shows the two RAFT satellite cubes being deployed from the space shuttle (STS-116). The bottom photo shows the onset of tumbling of the satellites.

Below is a Digipan display of a detection of the radar fence from the UC Irvine Ground Station on the May 27 2007. The Doppler starts high and then proceeds downward at an ever-increasing rate as it is approaching the center of the pass.

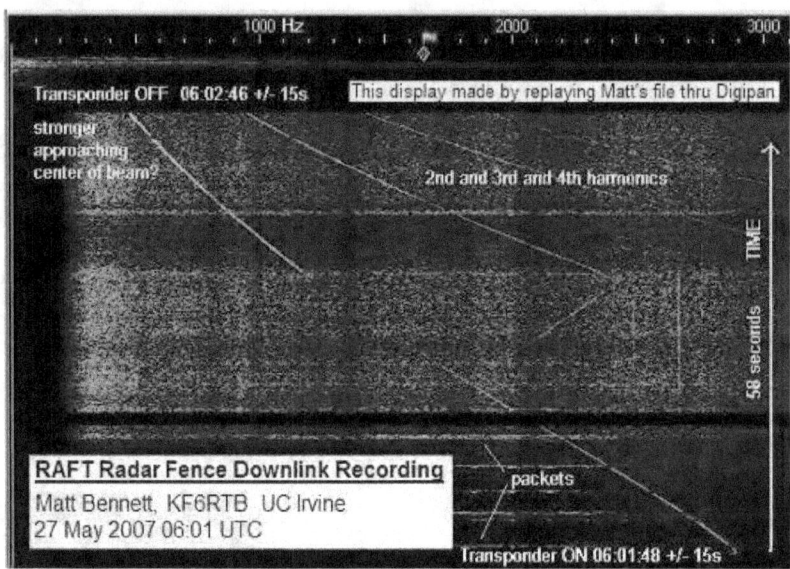

PUBLICATION(S)
There are no publications at this time.

PHYSICAL SCIENCES IN MICROGRAVITY

Much of our understanding of physics is based on the inclusion of gravity in fundamental equations. Using a laboratory environment found nowhere else, the U.S. Destiny laboratory on space station provides the only place to study long-term physical effects in the absence of gravity, without the complications of gravity-related processes such as convection and sedimentation. This unique microgravity environment allows different physical properties to dominate systems, and these have been harnessed for a wide variety of investigations in the physical sciences.

The ISS Physical Science Experiments portfolio includes experiments in different sub-fields of physics and material science. Several experiments are broadly classified as fluid physics, which test phenomena such as capillary flow, the mixing of miscible fluids, and bubble dynamics. Other experiments test concepts in the relatively new field of the physics of soft matter and colloidal systems. Yet another class of experiments focuses on crystal growth, especially protein crystal growth in microgravity. Growing crystals (both protein crystals and zeolite crystals) in space, free from the gravitational effects of sedimentation and convection, provides an opportunity to grow crystals that are larger or more pure than crystals that are grown on Earth. Crystallization experiments on ISS have examined proteins, viruses, and other macrobiological molecules to better understand their structure and function for maintaining human health and fighting disease.

FLUID PHYSICS

The current generation of fluid experiments focuses on colloidal and liquid crystal systems. The Light Microscopy Module will be used for some of these experiments. Colloidal investigations will study small colloidal particles that are used to model atomic systems and engineer new systems without sedimentation and particle jamming. Liquid crystal investigations will exploit the unique characteristics of freely suspended liquid crystal films. Gravity-driven sedimentation produces inhomogeneities in film thickness, precluding the formation of large, homogeneous, boundary-free films in Earth's gravity.

MATERIALS SCIENCE

Experiments in microgravity remove the effects of gravity-related processes such as sedimentation and convection. This allows testing of physical hypotheses that cannot be tested in any other environment.

PROTEIN CRYSTAL GROWTH

Experiments obtain high-quality crystal for ground-based research. Study of protein crystals is essential for visualizing proteins and developing new drugs and agricultural products.

Physical Sciences Experiments Performed on the International Space Station, Grouped by Discipline

Fluid Physics
CFE (Capillary Flow Experiment)
FMVM (Fluid Merging Viscosity Measurement)
MFMG (Miscible Fluids in Microgravity)
Materials Science
BCAT-3-BA (Binary Colloidal Alloy Test-3: Binary Alloys)
BCAT-3-SC (Binary Colloidal Alloy Test-3: Surface Crystallization)
BCAT-3-4-CP (Binary Colloidal Alloy Test-3 and -4: Critical Point)
CSLM-2 (Coarsening in Solid Liquid Mixtures-2)
EXPPCS (ExPRESS Physics of Colloids in Space)
Foam (Viscous Liquid Foam - Bulk Metallic Glass)
InSPACE (Investigating the Structure of Paramagnetic Aggregates from Colloidal Emulsions)
PFMI (Toward Understanding Pore Formation and Mobility During Controlled Directional Solidification in a Microgravity Environment)
SUBSA (Solidification Using a Baffle in Sealed Ampoules)
ZCG (Zeolite Crystal Growth)
Protein Crystal Growth
APCF-Camelids (Advanced Protein Crystallization Facility - Extraordinary Structural Features of Antibodies from Camelids)
APCF-Crystal Growth (Advanced Protein Crystallization Facility - Solution Flows and Molecular Disorder of Protein Crystals: Growth of High Quality Crystals, Motions of Lumazin Crystals and Growth of Ferritin Crystals)
APCF-Crystal Quality (Advanced Protein Crystallization Facility - Effect of Different Growth Conditions on the Quality of Thaumatin and Aspartyl-tRNA Synthetase Crystals Grown in Microgravity)
APCF-Lipoprotein (Advanced Protein Crystallization Facility - Crystallization of Human Low Density Lipoprotein (LDL) Subfractions in Microgravity)
APCF-Lysozyme (Advanced Protein Crystallization Facility - Testing New Trends in Microgravity Protein Crystallization)
APCF-Octarellins (Advanced Protein Crystallization Facility - Crystallization of the Next Generation of Octarellins)
APCF-PPG10 (Advanced Protein Crystallization Facility - Protein Crystallization in Microgravity, Collagen Model (X-Y-Gly) Polypeptides - the case of (Pro-Pro-Gly) 10)
APCF-Rhodopsin (Advanced Protein Crystallization Facility - Crystallization of Rhodopsin in Microgravity)
CPCG-H (Commercial Protein Crystal Growth - High Density)
DCPCG (Dynamically Controlled Protein Crystal Growth)
PCG-EGN (Protein Crystal Growth-Enhanced Gaseous Nitrogen Dewar)
PCG-STES-IDQC (Protein Crystal Growth-Single Locker Thermal Enclosure System-Improved Diffraction Quality of Crystals)
PCG-STES-IMP (Protein Crystal Growth-Single Locker Thermal Enclosure System-Crystallization of the Integral Membrane Protein Using Microgravity)
PCG-STES-MM (Protein Crystal Growth-Single Locker Thermal Enclosure System-Synchrotron Based Mosaicity Measurements of Crystal Quality and Theoretical Modeling)
PCG-STES-MMTP (Protein Crystal Growth-Single Locker Thermal Enclosure System-Crystallization of the Mitochondrial Metabolite Transport Proteins)
PCG-STES-MS (Protein Crystal Growth-Single Locker Thermal Enclosure System - Crystal Growth Model System for Material Science)
PCG-STES-RDP (Protein Crystal Growth-Single Locker Thermal Enclosure System-Engineering a Ribozyme for Diffraction Properties)
PCG-STES-RGE (Protein Crystal Growth-Single Locker Thermal Enclosure System-Regulation of Gene Expression)
PCG-STES-SA (Protein Crystal Growth-Single Locker Thermal Enclosure System-Science and Applications of Facility Hardware for Protein Crystal Growth)
PCG-STES-VEKS (Protein Crystal Growth-Single Locker Thermal Enclosure System-Vapor Equilibrium Kinetics Studies)

ADVANCED PROTEIN CRYSTALLIZATION FACILITY (APCF), EIGHT INVESTIGATIONS

Payload developer: Italian Space Agency (ASI) for the European Space Agency
Expedition 3

Principal Investigator(s): Richard Giegé, Centre National de la Recherche Scientifique (CNRS), Strasbourg, France; Effect of Different Growth Conditions on the Quality of Thaumatin and Aspartyl-tRNA Synthetase Crystals Grown in Microgravity (APCF-Crystal Quality)

Manfred W. Baumstark, University of Freiburg, Germany; Crystallization of Human Low Density Lipoprotein (LDL) Subfractions in Microgravity (APCF-Lipoprotein)

Willem J de Grip, University of Nijmegen, Netherlands; Crystallization of Rhodopsin in Microgravity

Joseph. Martial, University of Liege, France; Crystallization of the Next Generation of Octarellins (APCF-Ocatrellins)

Fermin Otalora, University of Granada, Spain; Testing New Trends in Microgravity Protein Crystallization (APCF-Lysozyme)

Sevil Weinkauf, Technical University Munich, Germany; Solution Flows and Molecular Disorder of Protein Crystals: Growth of High Quality Crystals, Motions of Lumazine Crystals, and Growth of Ferritin Crystals (APCF-Crystal Growth)

Lode Wyns, Free University Brussels, Belgium; Extraordinary Structural Features of Antibodies from Camelids (APCF-Camelids)

Adriana Zagari, University of Naples, Italy; Protein Crystallization in Microgravity, Collagen Model (X-Y-Gly) Polypeptides: the case of (Pro-Pro-Gly)^{10} (APCF-PPG10)

Research Area Protein Crystal Growth

S105E5161 — Astronauts Frederick W. Sturckow and Scott J. Horowitz of the shuttle STS-105 crew pose with the APCF by the middeck locker during Expedition 3.

Understanding proteins is basic to understanding the processes of living things. While we know the chemical formula of proteins, learning the chemical structure of these macromolecules is more difficult. Mapping the three-dimensional structure of proteins, deoxyribonucleic acid (DNA), ribonucleic acid (RNA), carbohydrates, and viruses provides information concerning their functions and behavior. This knowledge is fundamental to the emerging field of rational drug design, replacing the trial-and-error method of drug development. Microgravity provides a unique environment for growing crystals— an environment that is free of the gravitational properties that can crush the delicate structures of crystals. Currently, several test facilities are used to grow crystals.

The Advanced Protein Crystallization Facility (APCF) can support three crystal-growth methods: liquid-liquid diffusion, vapor diffusion, and dialysis. In the vapor diffusion method, a crystal forms in a protein solution as a precipitant draws moisture in a surrounding reservoir. In the dialysis method, salt draws moisture away from the protein solution via a membrane separating the two and forming crystals. Only the vapor diffusion and dialysis methods were used for the APCF suite of experiments.

RESULTS

Initial analysis of the crystals that were returned from station supports the findings of earlier APCF flights: comparative crystallographic analysis indicates that space-grown crystals are superior in every way to control-group crystals that are grown on Earth under identical conditions. Crystals that are grown in microgravity generally have improved morphology, larger volume, higher diffraction limit, and are better ordered compared to Earth-grown crystals. Researchers reported that the electron-density maps that were calculated from diffraction data contained considerably more detail, allowing them to produce more accurate three-dimensional models.

APCF-Camelids, APCF-Crystal Quality, APCF-Growth, APCF-Lysozyme, APCF-Octarellins and APCF-PPG10 all produced excellent-quality crystals that had better resolution and other optical properties than those grown on Earth. APCF-Lipoprotein successfully produced crystals, but they did not achieve the expected level of resolution. APCF-Rhodopsin had slight technical problems that prevented the formation of suitable crystals.

A thaumatin crystal, grown in microgravity during Expedition 3, displays interference patterns in polarized light.

Although many of the investigators have not completed their analysis and modeling, early published results have come out for crystals of (Pro-Pro-Gly)[10] or (PPG[10]). PPG[10] is a collagen protein that is found in many tissues. This collagen is particularly concentrated in the skin, joints, and bones. Video that was collected during Expedition 3 showed the small movements within the crystallizing solutions. A direct correlation between crystal motion and acceleration from events on station (such as docking, venting, and crew movement) was determined for the first time. The PPG[10] crystals were independently studied by X-ray diffraction in various labs; the best resolution attained for microgravity-grown crystals from ISS was 1.5Å, which was superior to the 1.7Å obtained on the ground. The teams of APCF scientists are combining data from previous space flights, the ground, and the station to get the best possible information on protein structures for applications in pharmaceutical and physiological research.

PUBLICATION(S)

Berisio R, Vitagliano L, Vergara A, Sorrentino G, Mazzarella L, Zagari A. Crystallization of the collagen-like polypeptide (PPG)[10] aboard the International Space Station. 2. Comparison of crystal quality by X-ray diffraction. *Acta Crystallography*, Section D Biological Crystallography. 58:1695–1699, 2002.

Castagnolo D, Piccolo C, Carotenuto L, Vegara A, Zagari A. Crystallization of the collagen-like polypeptide (PPG)[10] aboard the International Space Station. 3. Analysis of residual acceleration-induced motion. *Acta Crystallographica*. Section D, Biological Crystallography. 59(pt4):773–776, 2003.

Lorber B. The crystallization of biological macromolecules under microgravity: a way to more accurate three-dimensional structures? *Biochimica et Biophysica Acta*. 1599(1–2):1–8, 2002.

Vergara A, Corvino E, Sorrentino G, Piccolo C, Tortora A, Caritenuto L, Mazzarella L, Zagari A. Crystallization of the collagen-like polypeptide (PPG)[10] aboard the International Space Station. 1. Video observation. *Acta Crystallography*, Section D Biological Crystallography. 58:1690–1694, 2002.

Vergara A, Lorber B, Zagari A, Giege R. Physical aspects of protein crystal growth investigated with the Advanced Protein Crystallization Facility in reduced gravity environments. *Acta Crystallographica*, Section D. 59:2–15, 2002.

Vergara A, Lorber B, Sauter C, Giege R, Zagari A. Lessons from crystals grown in the Advanced Protein Crystallization Facility for conventional crystallization applied to structural biology. *Biophysical Chemistry*. 118:102–112, 2005.

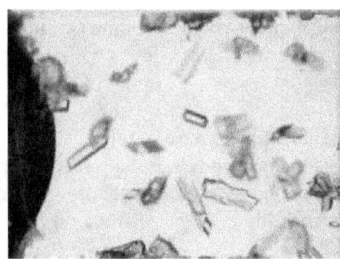

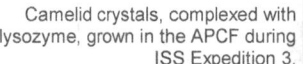

Camelid crystals, complexed with lysozyme, grown in the APCF during ISS Expedition 3.

Binary Colloidal Alloy Test-3 (BCAT-3), Three Investigations

Principal Investigator(s): Peter N. Pusey, Ph.D., University of Edinburgh, United Kingdom
Andrew Schofield, Ph.D., University of Edinburgh, United Kingdom; Binary Alloys (BCAT-3-BA, Expedition 8)

Arjun Yodh, Ph.D., University of Pennsylvania, University Park, Penn.; Jian Zhang, University of Pennsylvania, University Park, Penn.; Surface Crystallization (BCAT-3-SC, Expeditions 8–12)

David A. Weitz, Ph.D., Harvard University, Cambridge, Mass.; Peter Lu, Ph.D., Harvard University, Cambridge, Mass; Critical Point (BCAT-3-4-CP, Expeditions 8–10, 12, 13, 16, 17)

Research Area Materials Sciences

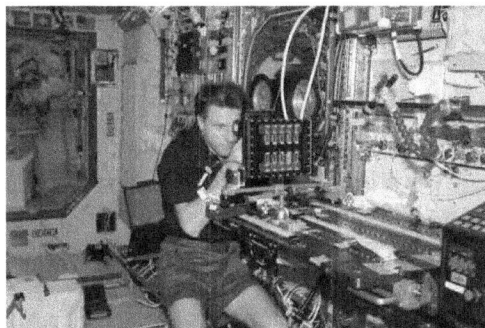

ISS008E20610 — Astronaut C. Michael Foale, Expedition 8 commander and NASA ISS science officer, uses a digital still camera to photograph a slow growth sample module (SGSM) for the BCAT-3 experiment. The SGSM is on a mounting bracket attached to the maintenance work area table set up in the U.S. *Destiny* laboratory.

The BCAT-3 hardware supported three investigations in which ISS crews photographed samples of colloidal particles (tiny nanoscale spheres suspended in liquid) to document liquid/gas phase changes, growth of binary crystals, and the formation of colloidal crystals that were confined to a surface. Colloids are small enough that in a microgravity environment without sedimentation and convection, they behave much as atoms. By controlling aspects of colloidal mixtures, they can be used to model all sorts of phenomena.

The BCAT-3 suite of payloads consists of 10 small samples of colloid alloys in which the microscopic colloid particles are mixed together into a liquid. These samples are contained within a small case that is the size of a school textbook. At the start of an experiment run, all samples are shaken to completely remix the colloid samples, using a magnet to run a magnetic stir bar up and down through the samples—much in the same way that salad dressing must be shaken to remix oil and vinegar. After the samples are mixed, they are periodically photographed using a digital camera until the colloid and liquid components of those samples have separated or the polymers have formed crystals. The samples can be remixed to repeat the experiment.

BCAT-3-BA (Pusey)

The binary alloy (BCAT-3-BA) investigations examine colloids, which are technologically interesting because they are the right size to manipulate light. For example, opal is a "photonic" crystal containing tiny spheres of silica and hydrous silica that differentially refract light, resulting in flashes of color under a directed light beam. The ability to better control the movement of light is a major technological goal, with many potential applications (e.g., fiber optics, other optical systems such as switches, displays). Crystal structures that are built from only one type of colloidal particle, e.g., the arrangement of colloidal silica spheres in an opal, are well understood, but their optical properties are limited. More useful photonic crystals can be built from two different types of spheres that are mixed together, yielding a binary alloy. Depending on the sizes and abundances of the different colloidal particles, the resulting structures and their optical properties are varied. Microgravity is crucial to the binary crystal experiment, facilitating the growth of crystals that are far larger than those that are created on the surface of the Earth.

8J2Z9926.DCR — BCAT-3 sample 10, taken May 25, 2004 at 18:57:07 during ISS Expedition 9, 2 months following mixing. The pink region is the phase separation region, which shows a possible indication of surface crystallization. The black asterisk indicates possible bulk crystal nucleation.

RESULTS
Unfortunately, the BCAT-3-BA sample dried out before crystallizing could occur on board the ISS.

BCAT-3-SC (YODH)
During the surface crystallization (BCAT-3-SC) experiment, astronauts photographed samples of suspended colloidal particles in a liquid that contains a small amount of polymer to document the formation of colloidal crystals. Under the right circumstances, these suspended particles will prefer to crystallize on the container surfaces rather than in the sample volume. Three BCAT-3-SC samples will study the formation of colloidal crystals that are confined to a surface, allowing comparison with bulk three-dimensional crystallization, to begin testing out how geometry affects crystallization itself. Results will help scientists develop fundamental physics concepts that were previously hindered by the effects of gravity. Ordered arrays of these micron-sized particles might be ideal for switching and controlling light.

RESULTS
Imagery for the BCAT-3-SC was collected over several increments and is still undergoing analysis.

BCAT-3-4-CP (WEITZ)
BCAT-3-4-CP samples were designed to help determine the phase separation rates and better define the phase diagram of a model critical fluid system. The colloidal mixtures were designed to model liquid/gas phase changes, specifically, to model the behavior of boiling water in a colloidal-polymer system, where the phases that are analogous to liquid and gas can be seen as two different colors. In an ordinary pot of boiling water, bubbles of water vapor coalesce on the bottom of the pot, growing until they detach and float to the surface where they escape into the atmosphere. At boiling temperature water exists simultaneously in two distinct phases—liquid and gas. The boiling water would look different in the absence of gravity, when the vapor no longer floats to the top. The phase separation behavior changes with increasing pressure: seal the pot, as in a pressure cooker, and the boiling temperature rises. Continue the pressure increase and the mixture will reach its critical point, a unique pressure and temperature value where the properties of liquid and gas merge. Just above the critical point is the supercritical regime where there are no distinct phases but, rather, a homogeneous supercritical fluid.

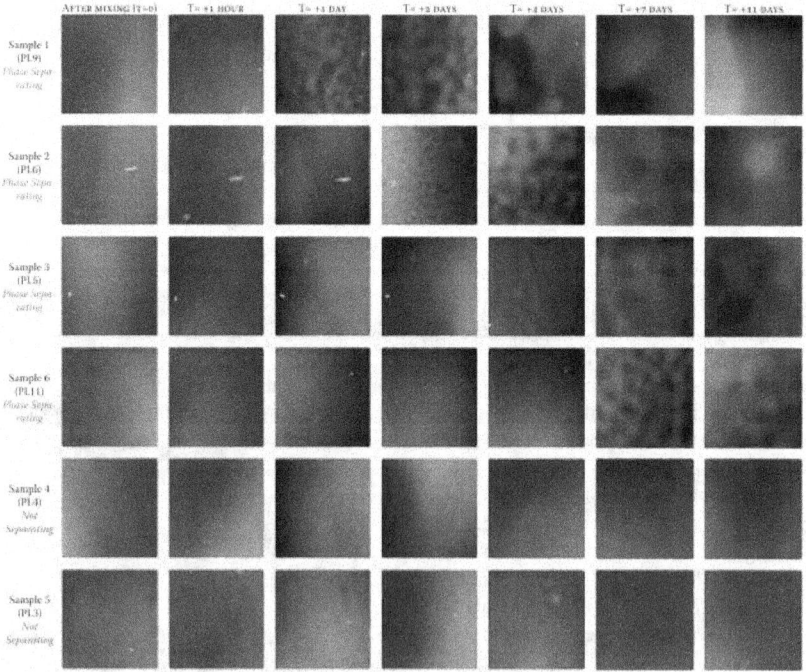

Images of BCAT-3-4-CP: Six samples of colloid/polymer mixtures separating at different times.

Supercritical fluids are technologically important because they uniquely combine the properties of liquids and gases, flowing easily (as gases) yet still having tremendous power to transport dissolved materials and thermal energy (like liquids). Understanding critical phenomena through ground-based experiments have been limited by gravity, which invariably causes the denser liquid phase to fall to the bottom of any container, precluding direct observation of phase separation alone. BCAT-3-4-CP is the first experiment to systematically attempt to precisely locate the critical point and visualize the behavior around it. This work also addresses the physics of complex fluids, such as colloidal suspensions, emulsions, or polymer solutions, and uses colloids as model systems with large particles and relatively slow, observable rates to characterize particle behavior near the critical point. It also allows a better understanding of phase behavior in colloid-polymer system, with many practical Earth and space applications—colloid-polymer solutions are used in many household and personal care products.

RESULTS

The BCAT-3-4-CP samples have yielded some surprising results: The samples that separated on Earth also separated in space; the samples that did not separate on Earth also did not separate in microgravity, indicating that the tendency for phase separation is not linked to gravity. However, for those samples that separated into different phases, the rates of separation for the microgravity samples were very different from the same mixtures on Earth. Those samples with concentrations that are closest to the critical point took weeks to months in microgravity, compared with a couple of days on Earth (Lu et al. 2007).

These dynamic data will help determine the boundary conditions for future models of critical behavior. Present observations also include a determination of the shape of the interface and which part of the sample wets the cell. The long-term observation of which samples phase separates will allow precise determination of the critical point of this colloidal mixture, and will allow inference of the fundamental physics underlying critical point behavior.

PUBLICATION(S)

Lu PJ, Weitz DA, Foale MC, Fincke M, Chiao LN, Meyer WV, Owens JC, Hoffmann MI, Sicker RJ, Rogers R, Havenhill MA, Anzalone SM, Yee H. Microgravity Phase Separation near the Critical Point in Attractive Colloids. 45th AIAA Aerospace Sciences Meeting and Exhibit. Reno, Nev. Jan 8–11, 2007; AIAA 2007-1152.

Capillary Flow Experiment (CFE)

Principal Investigator(s): Mark M. Weislogel, Ph.D., Portland State University, Portland, Ore.
Expeditions 9, 12–16

Research Area Fluid Physics

CFEs are a suite of fluid physics experiments whose purpose is to investigate capillary flows and phenomena in low gravity. The CFE data that is to be obtained will be crucial to future space exploration because they provide a foundation for physical models of fluids management in microgravity, including fuel tanks and cryogen storage systems (e.g., water recycling) and materials processing in the liquid state. NASA's current plans for Exploration missions assume the use of larger liquid propellant masses than have ever flown before. Under low-gravity conditions, capillary forces can be exploited to control fluid orientation so that such large mission-critical systems perform predictably.

The handheld experiments aim to provide results of critical interest to the capillary flow community that cannot be achieved in ground-based tests; for example, dynamic effects that are associated with a moving contact boundary condition, capillary-driven flow in interior corner networks, and critical wetting phenomena in complex geometries. The knowledge that is gained will help spacecraft fluid systems designers increase system reliability, decrease system mass, and reduce overall system complexity.

CFE encompasses three sub-experiments: CFE-Contact Line (CFE-CL), CFE-Interior Corner Flow (CFE-ICF), and CFE-Vane Gap (CFE-VG), with two unique experimental apparatuses per experiment for perfectly wetting and partially wetting fluids. There are multiple tests per experiment. Each of the experiments employs conditions and test cell dimensions that cannot be achieved in ground-based experiments. All of the units use similar fluid injection hardware, which is made of Lucite, have simple and similarly sized test chambers, and rely solely on video for highly quantitative data. Silicone oil is used as the fluid. The differences between units are primarily fluid properties, wetting conditions (determined by the coating inside the test chamber), and test cell cross section.

During each experimental run a crewmember disturbed the fluids by tapping the container, moving the vanes, etc. By digitizing and quantitatively analyzing video data of the resulting oscillations, natural frequencies and damping rates are determined. Using these data, the effects of partial wetting, the lag before contact angle changes, and fluid properties such as surface tension and viscosity will be quantified. Transient flow rates, stability limits, and coalescence time scales will be measured.

ISS009E23445 — ISS science officer Edward M. (Mike) Fincke is pictured next to CFE-CL2 in the U.S. *Destiny* laboratory aboard ISS during Expedition 9.

Contact Line

CFE-CL investigates the properties of the contact line (the boundary between the liquid and the solid surface of the container). The contact line controls the interface shape, stability, and dynamics of capillary systems in low gravity.

Results

The first contact line apparatus with perfect wetting properties, CFE-CL2 (pictured above with astronaut Mike Fincke), was tested during ISS Expeditions 9, 12, and 13. The partial wetting configuration, CFE-CL1, was performed during Expeditions 13 and 16. Scores of individual primary science tests and additional "extra science" tests were performed and downlinked, with commentary, via video by the crew. Full analyses of the CFE-CL data are under way using both the low-resolution video that was downlinked from ISS during real-time operations, and the higher-resolution video data from the returned tapes. The different tests explored techniques for imparting disturbances and contact line responses in "push," "slide," and "axial" modes, as well as droplet effects. Experts are debating whether current fluid dynamics models can accurately predict fluid behavior at contact lines in microgravity. Numerical analysts from the United States and Europe were given data (hardware descriptions and fluid properties of the CFE-CL systems) with which to create computer predictions of the experimental properties. Current fluid dynamic models will be compared to experimental results to determine the fidelity of current models and the need for future research (Weislogel 2005). Preliminary results on the first space-validated numerical models that predict contact line behavior are reported in Weislogel et al. (2008). For sets of initial boundary conditions, the model predictions were excellent.

Interior Corner Flow
CFE-ICF studies capillary flow in interior corners. Structured inside tanks providing interior corners are used in the design of fuel tanks so that the fuel will always flow to the outlet of the tank in the absence of gravity. The equations governing the process are known but, to date, have not been solved analytically because of a lack of experimental data identifying the appropriate boundary conditions for the flow problem. Experimental results will guide the analysis by providing the necessary boundary conditions as a function of container cross section and fill fraction. The benchmarked theory can then be used to improve propellant management aboard spacecraft.

RESULTS
Like the other CFE set-ups, the CFE-ICF apparatus provides benchmark data for validating numerical models predicting behavior of fluids in complex containers. In particular, this experiment set-up provided imagery of the bubble migrating from the vertex of the tapered container to the base, demonstrating spontaneous capillary flow within the container that appeared to be similar to the flow that was predicted by models (Weislogel et al. 2007).

Two geometries were tested. Both provided similar results; when the meniscus location is plotted against time (both dimensionless), the migration rates increase from dry, wet, and open loop tests. In general, migration rates increase for dry tests. The validated conclusion from the two geometries tested is that specific container geometry can control liquid flow.

Vane Gap
CFE-VG studies capillary flow when there is a gap between interior corners, such as in the gap formed by an interior vane and tank wall of a large propellant storage tank or the near intersection of vanes in a tank with a complex vane network.

RESULTS
The CFE-VG consists of a cylindrical container with and an elliptical cross section and a central vane that can be rotated within the cylinder. The vane is offset from the cylinder center so that the gaps between the cylinder walls are unequal. This experiment was designed to better understand the sensitivity of fluid behavior in low-gravity conditions to container shape, and to illustrate how very small shifts in container geometry can result in significant shifts in fluid distribution. The findings from this research have direct application to passive management of fluids aboard spacecraft, with potential for pumping large amounts of liquid by container configuration. The CFE-VG tests indicate that the degree of wetting is dependent on vane angle and the wetting properties of the fluid, but also on other geometric aspects of the container, vane, and vane gap (Weislogel et al. 2008). In more detail, the tests show that for the perfect-wetting fluid, three fluid configurations result: wetting along the small gap between the vane and the cylinder, wetting along the large gap, and a large shift in fluid distribution (bulk shift) that redistributes much of the fluid to one side of the container. The tests results show that very small rotations of the vane can cause bulk shifts of the fluid. In the partial-wetting configuration, the bulk shifting does not occur, but wetting along either the small or the large gaps does occur (Chen et al. 2008).

PUBLICATION(S)
Weislogel MM, Thomas EA, Graf JC. A Novel Device Addressing Design Challenges for Passive Fluid Phase Separations Aboard Spacecraft. *Journal of Microgravity Science and Technology*. ISSN 0938-0108 (Print) 1875-0494 (Online), DOI 10.1007/s12217-008-9091-7, Aug 6, 2008.

Weislogel MM, Chen Y, Bolleddula D. A Better Non-Dimensionalization Scheme for Slender Flows: The Laplacian Operator Scaling Method. *The Physics of Fluids*. 2008; 20(9):093602.

Weislogel, MM, Jenson, R, Chen, Y. The Capillary Flow Experiments aboard the International Space Station: Status, IAC-07-A2.6.02, 2007.

Weislogel MM. Preliminary Results from the Capillary Flow Experiment Aboard ISS: The Moving Contact Line Boundary Condition. Proceedings of the 43rd AIAA Aerospace Sciences Meeting and Exhibit, Reno, Nev. AIAA 2005-1439, Jan 10–13, 2005.

Weislogel MM, Jenson R, Klatte J, Dreyer M. Interim Results from the Capillary Flow Experiment Aboard ISS; the Moving Contact Line Boundary Condition. 45th AIAA Aerospace Sciences Meeting and Exhibit. Reno, Nev. AIAA 2007-747, Jan 8 – 11, 2007.

Weislogel MM, Jenson RM, Klatte J, Dreyer ME. The Capillary Flow Experiments aboard ISS: Moving Contact Line Experiments and Numerical Analysis. 46th AIAA Aerospace Sciences Meeting and Exhibit, Reno, Nev. AIAA 2008-816; 7 – 10 Jan 2008.

Chen Y, Jenson RM, Weislogel MM, Collicott SH. Capillary Wetting Analysis of the CFE-Vane Gap Geometry. 46th AIAA Aerospace Sciences Meeting and Exhibit, Reno, Nev. 2008, 7 – 10 Jan; AIAA 2008-817.

Weislogel MM, Jenson RM, Chen Y, Collicott SH, Williams S. Geometry pumping on Spacecraft (The CFE-Vane Gap Experiment on ISS), ISPS, 2007.

Video screen shot of the CFE contact line-2 set-up and initial condition on board ISS during Increment 9. The light blue liquid is in two narrow tubes, which are shown side by side, forming a meniscus in each tube.

Video screen shot of CFE CL-2 bubble test 1 on board ISS during Increment 9.

COMMERCIAL GENERIC PROTEIN CRYSTAL GROWTH-HIGH-DENSITY (CPCG-H)

Principal Investigator(s): Lawrence DeLucas, Center for Biophysical Sciences and Engineering, University of Alabama, Birmingham, Ala.
Expeditions 2, 4

Research Area Protein Crystal Growth

Proteins provide the building blocks of our bodies. Some proteins make it possible for red blood cells to carry oxygen while other proteins help transmit nerve impulses that allow us to see, hear, smell, and touch. Still other proteins play crucial roles in causing diseases. Pharmaceutical companies may be able to develop new or improved drugs to fight those diseases once the exact structure of the proteins are known.

The goal of the Commercial Protein Crystal Growth-High-density (CPCG-H) payload is to grow high-quality crystals of selected proteins so that their molecular structures can be studied. On Earth, gravity often has a negative impact on growing protein crystals. In microgravity, however, gravitational disturbances are removed, thus allowing some crystals to grow in a more regular and perfect form. During ISS Expeditions 2 and 4, CPCG-H was outfitted with High-Density Protein Crystal Growth (HDPCG) hardware. HDPCG was a vapor-diffusion facility that could process as many as 1,008 individual protein samples. The entire HDPCG assembly had four trays that held 252 protein crystal growth blocks, each consisting of six chambers. The chambers had a protein reservoir, a precipitant reservoir, and an optically clear access cap. The chambers were designed to reduce sedimentation problems and to produce highly uniform, single crystals.

The primary proteins that were involved in the testing of the CPCG-H hardware during ISS Expeditions 2 and 4 were mistletoe lectin-I (ML-I, *Thermus flavus* 5S RNA, brefeldin A-ADP ribosylated substrate (BARS, and a triple mutant myoglobin (Mb-YQR). ML-I is a ribosome inactivating protein that can stop protein biosynthesis (creation of proteins) in cells, and is also a major component of drugs that are used to treat cancer. Although the study of *Thermus flavus* 5S RNA has been ongoing for well over 30 years, the exact function of the protein remains obscure. Scientists believe that the crystallization of different domains of this protein may reveal functional properties. BARS is an enzyme that is involved in membrane fission, catalyzing the formation of phosphatidic acid by transfer. Mb-YQR was studied to assess the functional role of packing defects in proteins. The elucidation of these protein structures will provide valuable insight into the role of these proteins for application in the pharmaceutical industry.

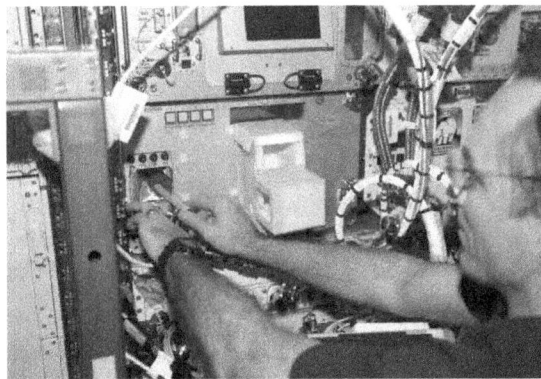

ISS004E10827 — Astronaut Carl E. Walz, Expedition 4 flight engineer, works in the U.S. *Destiny* laboratory installing the CRIM-M. CRIM-M provides the temperature control that is critical to the success of CPCG-H experiments.

RESULTS

Preliminary analysis indicated that at least 65% of the macromolecules that were flown in the CPCG-H experiments produced diffraction-sized crystals. X-ray diffraction studies of these crystals were conducted, and the data were used to determine and refine the three-dimensional structures of these macromolecules. Three benchmark proteins, ML-I, *Thermus flavus* 5S RNA, and BARS, were flown to validate the performance of the hardware. Diffraction-quality crystals, which were obtained from all of these proteins, yielded X-ray diffraction data that are comparable to those previously collected on Earth-grown crystals. Since the structure of each of the benchmark proteins is known to high resolution, these results indicate that the new HDPCG assembly worked very well, successfully producing high-quality crystals of the benchmark proteins.

Synchrotron diffraction data that were collected from the space crystals of the BARS protein were comparable in resolution but more intense and showed significantly less mosaicity than data from Earth-grown crystals. This indicates that the space-grown crystals had a higher order at the molecular level, and the X-ray diffraction data from the space crystals produced a more complete data set. These results contributed significantly to the structural study of BARS (Nardini et al. 2002).

ML-I is an enzyme that has the ability to inactivate ribosomes and inhibit cell replication. It is a target for new cancer treatments. Crystals of the protein that is attached to adenine (one of five building blocks of DNA or RNA) were flown; these crystals yielded X-ray data to 1.9 Å. These data were used to refine the structure of the complex and were especially valuable in refining the active site conformation (Krauspenhaar et al. 2002).

Perhaps the most exciting results from the macromolecular crystallization experiments that were conducted in the CPCG-H hardware were obtained from the *Thermus flavus* 5S rRNA [ribosomal ribonucleic acid] experiments. These experiments involved a synthetic RNA duplex of 5S rRNA, which is a model system for the study of the binding of ribosomal RNA to proteins. Crystallization under microgravity provided crystals that were of significantly higher quality than those grown in one-g. The space crystals diffracted to a maximum resolution of 2.6 Å in contrast to the best Earth-grown crystals, which diffracted to a maximum resolution of 2.9 Å. The improved X-ray data facilitated the completion of the structure of the RNA segment (Vallazza et al. 2002).

To understand the true function of a protein, the structure must be determined. The model of the structure must be accurate to allow scientists to create compounds that bind to the protein. The elucidation of the protein structure is of major importance with complex proteins (proteins that have significant folding). The three-dimensional structure of the triple mutant protein Mb-YQR was solved by growing the protein on ISS during Expeditions 2 and 4. Following return to Earth, three-dimensional models were created of the Mb-YQR proteins that were grown in space using X-ray crystallography techniques (Miele et al. 2004).

Structural studies of microgravity-grown crystals have provided important information for the development of new drugs. For example, previous studies that were conducted using crystals that were grown on shuttle flights have been used in the design of inhibitors, which may serve as broad-spectrum antibiotics. The CPCG-H payload offers a great increase in the amount of space that is available for protein crystal growth, enhancing the space station's research capabilities and commercial potential.

PUBLICATION(S)
Krauspenhaar R, Rypniewski W, Kalkura N, Moore K, DeLucas L, Stoeva S, Mikhailov A, Voelter W, Betzel C. Crystallisation under microgravity of mistletoe lectin-I from Viscum album with adenine monophosphate and the crystal structure at 1.9Å resolution. *Acta Crystallographica*. Section D, Biological Crystallography. 58:1704–1707, 2002.

Miele AE, Federici L, Sciara G, Draghi F, Brunori M, Vallone B. Analysis of the effect of microgravity on protein crystal quality: the case of a myoglobin triple mutant. *Acta Crystallographica*. Section D, Biological Crystallography. D59:928–988, 2004.

Nardini M, Spano S, Cericola C, Pesce A, Damonte G, Luini A, Corda D, Bolognesi M. Crystallization and preliminary X-ray diffraction analysis of brefeldin A-ADP ribosylated substrate (BARS). *Acta Crystallographica*. Section D, Biological Crystallography. 58:1068–1070, 2002.

Vallazza M, Banumathi S, Perbandt M, Moore K, DeLucas L, Betzel C, Erdmann V. Crystallization and Structure Analysis of *Thermus flavus* 5S rRNA helix B. *Acta Crystallographica*. Section D, Biological Crystallography. 58:1700–1703, 2002.

COARSENING IN SOLID LIQUID MIXTURES-2 (CSLM-2)
Principal Investigator(s):   Peter Voorhees, Northwestern University, Evanston, Ill.
Expeditions 7, 16

Research Area Materials Science

Coarsening is an increase in the size of grains in a metal, usually during heating at elevated temperatures. The process occurs in nearly any two-phase mixture, including industrial metallic alloys that are used in products such as dental fillings and turbine blades. It is important in industry because grain size and size distribution affects the strength of metal alloys. The objective of the Coarsening in Solid Liquid Mixtures-2 (CSLM-2) experiment was to assess coarsening mechanisms in a two phase metal alloy to better define the physical-chemical processes and validate models that are used for alloy manufacture on Earth. By operating on ISS, material transport phenomena such as gravitational settling and convection are eliminated, allowing researchers to focus on the effects of coarsening alone.

ISS007E10472 — Front view of the CSLM-2 hardware following setup in the microgravity sciences glovebox (MSG) before operation on Expedition 7.

During cooling of an alloy of several metals, the different constituent metals may solidify at different rates, and solid particles may experience competitive particle growth—a process that is called Ostwald ripening. During Ostwald ripening, coarsening occurs with large particles growing at the expense of smaller ones in a matrix, with the large particles stealing atoms from the smaller ones. Eventually, the sample consists of a few large particles that are crowded around a few remaining small particles. Materials that contain a few large particles rather than many small particles are structurally weaker.

In gravity, sedimentation will draw denser particles to the bottom and light particles to the top of a sample. As a result, the sample not only coarsens, it coarsens unevenly, thereby producing areas in the material that are much weaker than other areas in the material.

In CSLM-2, samples of a two-phase mixture of lead (Pb) and tin (Sn) were heated to the eutectic point (the lowest point of melting) for the mixture that was inside a large, cylindrical sample chamber inside the MSG. The melting resulted in Sn particles that were suspended in a liquid Pb-Sn matrix. The sample was then allowed to coarsen isothermally for a set period of time. After a sample was completed, pressurized water was pumped into the chamber to quench the sample, cooling it for removal. This system can quench samples from 185°C (365°F) (the temperature that is required to initiate coarsening in tin-lead [Sn-Pb] samples) to 120°C (248°F) in only 6 seconds. CSLM-2 is a follow-on to an earlier CSLM experiment that was performed on the space shuttle. The CSLM-2 samples test a wider range of volume fractions of Sn particles and longer coarsening time than the earlier CSLM experiment.

RESULTS
Sample return from the CSLM-2 experiment was delayed due to the *Columbia* accident. Initial samples were returned on STS-114 in 2005. Researchers had assumed that the data were compromised because the samples were left on board for years and not kept under refrigeration. To their delight and surprise, the samples were suitable

for analysis in spite of the extended storage at room temperature. Additional runs were performed recently, testing the effect of larger and smaller volume fractions of Sn particles and longer times for coarsening at isothermal temperatures. Investigators are now analyzing the microstructure of the samples. This will allow them to obtain new spatial distribution information on coarsening that is impossible to obtain using Earth-processed samples.

PUBLICATION(S)
Kammer D, Genau A, Voorhees PW, Duval WM, Hawersaat RW, Hickman JM, Lorik T, Hall DG, Frey CA. Coarsening In Solid-Liquid Mixtures: A Reflight. 46th AIAA Aerospace Sciences Meeting and Exhibit, Reno, Nev. 2008, 7 – 10 Jan; AIAA 2008-813.

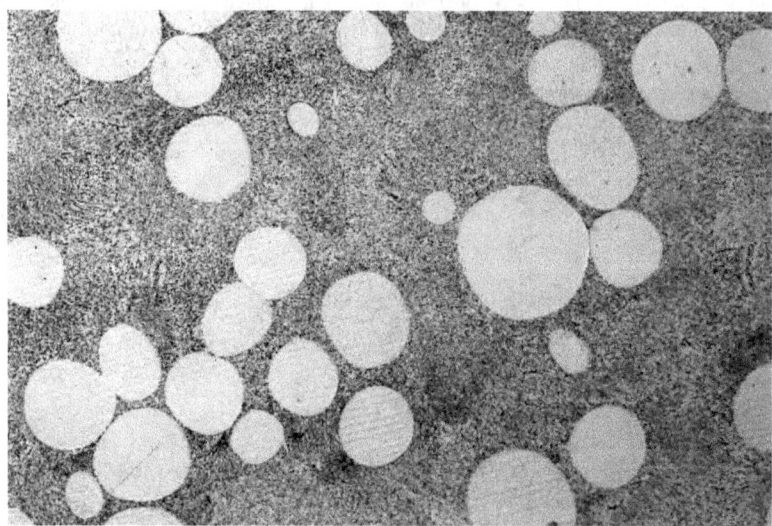

CSLM-2 sample showing Sn particles in matrix.

DYNAMICALLY CONTROLLED PROTEIN CRYSTAL GROWTH (DCPCG)
Principal Investigator(s): Lawrence DeLucas, University of Alabama, Birmingham, Ala.
Expedition 3

Research Area Protein Crystal Growth

Researchers have found that it is possible to grow high-quality protein crystals in the weightlessness of LEO, where gravitational forces will not distort or destroy a crystal's delicate structure. When crystals are returned to Earth, their structure is examined by sending X rays through them and using the resulting data to create computer-based models.

The goal of the Dynamically Controlled Protein Crystal Growth (DCPCG) experiment was to control and improve the crystallization process by dynamically controlling the elements that influence crystal growth. Current growth methods provide little or no control over growth rate and separation of the nucleation and growth phases. The DCPCG system provided researchers real-time control of the diffusion process (supersaturation) through control of the protein concentration. It also determined the differences in vapor diffusion rates (the speed at which the liquid surrounding a protein solution evaporates, leaving behind a protein crystal) between experiments that were conducted in microgravity and similar experiments that were conducted on Earth. DCPCG quantified the basic differences between crystal growth on Earth and in space, differences in growth rate and in the way crystals moved and organized in the two environments, thereby allowing researchers to assess in detail the best systems with which to grow high-quality crystals and how to optimize those systems.

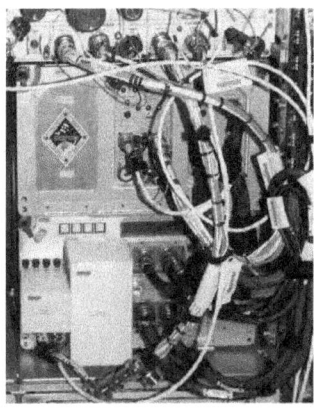

ISS003E8171 — Close-up image of DCPCG hardware, the V-locker, and the C-locker in ExPRESS Rack 1 on ISS Expedition 3.

Four different proteins were flown in DCPCG during Expedition 3. These proteins were glucose isomerase (an enzyme that catalyzes the conversion of glucose to fructose), equine serum albumin (a blood plasma protein that is produced in the liver and forms a large proportion of all plasma proteins), VEE capsid (target protein in the development of antiviral drugs to fight Venezuelan equine encephalitis (VEE)), and a chaperone protein that catalyzes the correct folding of newly synthesized proteins. DCPCG consists of a vapor locker (V-locker) that is connected to a Command and Data Management Locker (C-locker) that is installed in ExPRESS Rack 1. The V-locker contained 38 growth chambers surrounded by a closed-loop nitrogen management subsystem. Dry nitrogen flowing through the subsystem caused liquid to evaporate from the growth chambers. An in-line moisture sensor provided feedback as to how much of the liquid had evaporated. A static light-scattering sensor allowed researchers to modify the rate of evaporation, giving them far more control over the crystal growth process than is afforded by other methods. The C-locker housed the electronics and data ports for the experiment. A large portion of the C-locker is the ancillary equipment area (AEA) drawer, which contains a selection of tools and equipment for the experiment: a CD-ROM, spare flash disks, connector covers, and a tool for activating the experiment hardware.

RESULTS
DCPCG was the first flight test of an apparatus that was designed to control the crystal growth process by controlling the rate of evaporation. The apparatus worked on orbit, and crystals were grown for the test proteins; however, the investigators determined that the growth could have been better. The same apparatus was used in extensive testing on the ground. Researchers tested a selection of protein solutions, including insulin (a hormone that is produced by the pancreas to regulate the metabolism and use of sugar), serum albumin, and lysozyme (an enzyme that attacks bacteria) and found that a slower evaporation rate yielded better results than a more rapid evaporation rate. While the results of the ground tests were published, the DCPCG experiment investigators did not seek to publish any structures from crystals grown in orbit.

PUBLICATION(S)
Collingsworth OD, Bray L, Christopher GK. Crystal growth via computer controlled vapor diffusion. *Journal of Crystal Growth*. 219:283–289, 2000. (*Results from ground controls grown in parallel with ISS samples*)

ExPRESS Physics of Colloids in Space (EXPPCS)

Principal Investigator(s): David A. Weitz, Harvard University, Cambridge, Mass.; Peter N. Pusey, University of Edinburgh, Edinburgh, Scotland
Expeditions 2–4

Research Area Materials Science

Colloids can be defined as fluids with other particles dispersed in them, particularly particles of size between 1 nanometer and 1 micrometer. Since colloids have widespread uses in nature and industry, understanding of the underlying physics that controls their behavior is important. Under the proper conditions, colloidal particles can self-assemble to form ordered arrays, or crystals. On Earth, the ordering of these particles is mostly directed by gravitational effects, sedimentation, and buoyancy. Self-assembly does not occur. Thus, the weightlessness of LEO is an important element in the study of colloids.

Physics of Colloids in Space (PCS) focused on the growth, dynamics, and basic physical properties of four classes of colloids: binary colloidal crystals, colloid-polymer mixtures, fractal gels, and glass. These were studied using static light scattering (for size or positions of the colloids or structures formed), dynamic light scattering (to measure motions of particles or structures), rheological (flow) measurement, and still imaging.

Results

The results are discussed by class of colloid material that was studied. Analyses are still under way.

Binary colloidal crystals: These alloy samples are dispersions of two differently sized particles in an index-matching fluid. Two samples were studied: an AB_{13} crystal structure and an AB_6 crystal structure. Due to a hardware failure late in Expedition 4, the AB_6 experiment was not completed.

Unexpected "power law" growth behavior that is still under investigation was observed in the AB_{13} crystal structure sample.

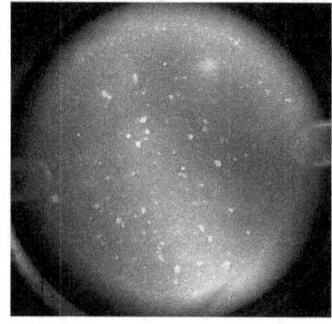

One of the first images from the PCS on ISS. During Expedition 2, sample AB_6 was illuminated with white light to produce this image. The colored regions result from refraction of white light by the sample and sample cell, splitting the light into its component colors.

Colloid-polymer mixtures: These mixtures induce a weak attractive interaction that allows precise tuning of the phase behavior of the mixtures, and approximates the phase separation below the critical point of a gas-liquid mixture. The phase behavior is controlled by the concentration of the colloid, the concentration of the polymer, and the relative size of the colloid and the polymer. The results form the ISS experiments studied the spinodal decomposition, or phase separation near the critical point, unencumbered by density differences of the phases. The growth of the phase separation was studied using both light scattering and imaging. Without gravity, the phase separation took 30 times longer than on Earth. The sample was mixed, then phase separation began, gradually coarsening until the container walls interacted with the mixture (at 42 hrs) and the colloid-rich phase wet the container wall, completely coating it after 60 hrs. Because the results follow very similar time evolution as a shallow quench of a binary liquid, they provide insight into the importance of the length scale of colloidal gels; separation depends more on coarsening rates than initial colloid size (Bailey et al. 2007).

Colloid-polymer gels: This sample was expected to be in a fluid-cluster state, but unexpectedly formed a solid gel. The elastic modulus, which was estimated using the experiment's rheology capabilities, will be compared to ground samples. "Aging" characteristics of this gel were found to be similar to those formed on Earth.

Colloid-polymer critical point: Immediately after mixing, the colloid-polymer critical point sample began to separate into two phases—one that resembled a gas and one that resembled a liquid, except that the particles were colloids and not atoms. The colloid-poor regions (the colloidal "gas" phase) grew bigger until, finally, complete phase separation was achieved and there was just one region of each—a colloid-rich phase and a colloid-poor phase. None of this behavior can be observed in the sample on Earth because sedimentation would cause the colloids to fall to the bottom of the cell faster than the de-mixing process could occur. Knowledge that was gained from these runs was used to develop the BCAT-3 that later operated on ISS.

Fractal gels: Fractal gels may form when charged colloids have their electrostatic repulsions screened out by the addition of a salt solution, permitting aggregation. These can be formed at very low volume fractions and form highly tenuous aggregates that exhibit a remarkable scaling property—their structure appears the same on all length scales up to a cluster size, and so can be described as a fractal. It was thought that the samples that were studied (colloidal polystyrene and silica gel) would, in the absence of sedimentation effects, ultimately form a continuous network of fractal aggregate; the polystyrene fractal sample never fully gelled as expected, however. Initial indications are that the volume fraction that was tested was too low. Large fractal clusters did nevertheless grow (larger than they do on Earth), allowing measurement of the internal vibration modes of these structures. The silica gel is thought to have gelled, and is currently being evaluated.

Colloidal glass: These samples are still under evaluation. Comparison to samples that were formed in one-g in the laboratory were needed to understand whether the crystallization observed was due to poor mixing or was a true microgravity phenomena.

PUBLICATION(S)

Bailey AE, Poon WC, Christianson RJ, Schofield AB, Gasser U, Prasad V, Manley S, Segre PN, Cipelletti L, Meyer WV, Doherty MP, Sankaran S, Jankovsky AL, Shiley WL, Bowen JP, Eggers JC, Kurta C, Lorik T Jr, Pusey PN, Weitz DA. Spinodal decomposition in a model colloid-polymer mixture in microgravity. *Physical Review Letters*. 99(20):205701, 2007.

Manley S, Davidovitch B, Davies NR, Cipelletti L, Bailey AE, Christianson RJ, Gasser U, Prasad V, Segre PN, Dohert MP, Sankaran S, Jankovsky AL, Shiley B, Bowen J, Eggers J, Kurta C, Lorik T, Weitz DA. Time-Dependent Strength of Colloidal Gels. *Physical Review Letters*. 95(4):048302(4), 2005.

Manley S, Cipelletti L, Trappe V, Bailey AE, Christianson RJ, Gasser U, Prasad V, Segre PN, Doherty MP, Sankaran S, Jankovsky AL, Shiley B, Bowen J, Eggers J, Kurta C, Lorik T, Weitz DA. Limits to Gelation in Colloidal Aggregation. *Physical Review Letters*. 93(10):108302, 2004.

Weitz D, Bailey A, Manley A, Prasad V, Christianson R, Sankaran S, Doherty M, Jankovsky A, Lorik T, Shiley W, Bowen J, Kurta C, Eggers J, Gasser U, Serge P, Cipelletti L, Schofield A, Pusey P. Results from the Physics of Colloids Experiment on ISS. NASA TM-2002-212011: IAC-02-J.6.04, 2002.

Doherty MP, Bailey AE, Jankovsky AL, Lorik T. Physics of Colloids in Space: Flight Hardware Operations on ISS. AIAA 2002-0762, 40th Aerospace Sciences meeting, Jan 14–17, 2002, Reno Nev.

Fluid Merging Viscosity Measurement (FMVM)

Principal Investigator(s): Edwin Ethridge, NASA Marshall Space Flight Center, Huntsville, Ala.
Expeditions 9, 11

Research Area Fluid Physics

ISS009E14560 — Mike Fincke, Expedition 9 science officer, is shown deploying one of the drops of liquid onto Nomex thread during FMVM testing.

The Fluid Merging Viscosity Measurement (FMVM) experiment was developed rapidly after the *Columbia* accident to provide a low-mass experiment using hardware already on board ISS. The purpose of FMVM is to measure the rate of coalescence of two highly viscous liquid drops and correlate the results with liquid viscosity and surface tension. The FMVM experiment will verify a new method for measuring the viscosity of highly viscous liquids by measuring the time it takes for two spheres of liquid to coalescence into a single spherical drop, where the time constant is proportional to the viscosity. The results are applied to calculating crystallization parameters of a liquid above and below the glass melting point, including an extremely broad range of viscosities of the liquid.

If this new method of measuring viscosity is validated, it could provide a method by which to measure viscosities of materials that cannot currently be measured. An example of this is liquid (molten) glass that crystallizes while cooling from liquid to solid. The viscosity in most of the crystallization range cannot be measured using current, Earth-based technology in spite of the fact that this is the most interesting range for the study of crystallization.

To obtain accurate data for precise models, it is best to measure viscosity in liquid that is free-floating and uncontained. The station's microgravity environment is an excellent testbed for this procedure because drops float freely in low gravity. The simple test apparatus was constructed of materials that were already on board ISS, and sample liquids, representing a range of viscosities, were deployed to station on a Russian Progress vehicle. These liquids consisted of glycerin, silicone oil (high and low viscosities), honey, honey mixed with water, corn syrup, and corn syrup mixed with water.

Results

Preliminary results were based on downlink video. The original data tapes were returned to Earth on shuttle flight STS-114/LF-1 and are still undergoing analysis. Glycerin proved difficult to deploy as planned (the viscosity was too low and/or the surface tension was too high), but the two different silicone oil viscosity calibration liquids were deployed easily, and five coalescence experiments were recorded with the two liquids. A range of drop sizes from 0.5 cc to 4 cc was coalesced. Since the honey had crystallized, the ISS food warmer was used to eliminate crystallization and the honey experiments were completed. Two corn syrup liquids formed a stiff skin. Real-time discussions with the astronaut resulted in a quick change to the procedure. New drops were deployed rapidly so as to avoid skin formation. Several successful coalescence runs were obtained with the corn syrup.

Coalescence of two unequal drops at early time.

Same as opposite figure, but at a later time.

Preliminary results from data analysis indicate a very good agreement with the predicted coalescence time. The experiments demonstrate that when the surface tension of a liquid is known, the coefficient of viscosity for that liquid can be determined by the contact radius speed (Antar et al. 2007). These data can be fit to numerical results to calculate the viscosity, thus validating the model for this new viscosity measurement method.

PUBLICATION(S)

Ethridge E, Kaukler W, Antar B. Preliminary Results of the Fluid Merging Viscosity Measurement Space Station Experiment, Proceedings of the 44th AIAA Aerospace Sciences Meeting and Exhibit, Reno, Nev. AIAA 2006-1142, Jan 9–12, 2006.

Antar BN, Ethridge E, Lehman D. Fluid Merging Viscosity Measurement (FMVM) Experiment on the International Space Station. 45th AIAA Aerospace Sciences Meeting and Exhibit. Reno, Nev. Jan 8 – 11, 2007; AIAA 2007-1151.

Viscous Liquid Foam-Bulk Metallic Glass (Foam)

Principal Investigator(s): William Johnson and Chris Veazy, California Institute of Technology, Pasadena, Calif.
Expedition 9

Research Area Materials Science

Bulk metallic glasses are a special class of metallic materials that are created by rapid solidification that causes them to form glass-like structures that are light but very strong. This experiment investigated the formation and structure of foams that are made from bulk metallic glass. Because the effects of buoyancy are minimized in space, more uniform foam structures with unique properties can be produced. These new materials have potential applications for use in future moon or Mars space structures (due to their high strength and low weight) as well as for potential shielding against micrometeorites and space debris impacts on spacecraft.

Results

Three planned runs for the Foam experiment were successfully completed on station during Expedition 9. Samples, which were returned to Earth in late Aug 2005, have been analyzed and reported (Veazy et al. 2008). The experiment was designed to test the hypothesis that amorphous metals exhibit foam-making qualities on the ground that mimic metallic foam textures that are made in microgravity conditions. The amorphous metals, when softened or melted, have a super-cooled state that has a very high viscosity—ideal conditions for foam processing. Foam that is made from a $Pd_{40}Ni_{40}P_{20}$ glass-forming metallic liquid was prepared both on the ground and aboard ISS. Pellets of the material that contained 1-atmosphere bubbles were sealed in ampoules. The ampoules were made to thread into a soldering iron tip for heating aboard the ISS. The samples were heated at 360°C (680°F) for 5 minutes, enabling foam creation before it was allowed to cool. The ground samples contained textures that were similar to those that were produced in microgravity—equally distributed bubbles dominated by surface tension forces; the bubbles did not experience sedimentation (floating).

ISS009E1479 — Expedition 9 flight engineer and science officer Edward M. (Mike) Fincke) performs the Foam experiment in the maintenance work area of space station.

These types of foams have great potential for future exploration applications because of their great strength and light weight. In particular, such foams may make very effective shields against micrometeorite and orbital debris strikes.

Publication(s)

Veazy C; Demetriou MD; Schroers J; Hanan JC; Dunning LA; Kaukler WF; Johnson WL. Foaming of Amorphous Metals Approached the Limit of Microgravity Foaming, *J. Adv. Materials*, v40, no 1 (2008) 7–11.

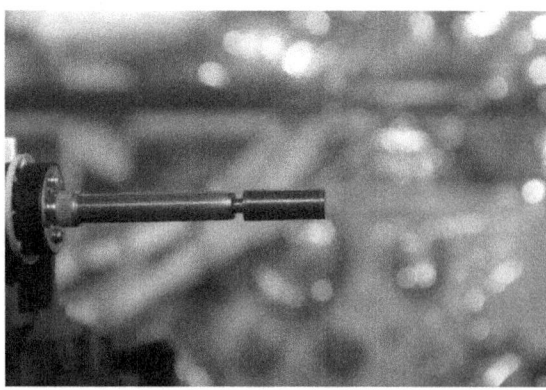

ISS009E14593 — Foam apparatus showing ampoules threaded into a soldering iron on board ISS. Samples were heated at 680°F (360°C) for 5 minutes, thus enabling foam creation before the sample was cooled.

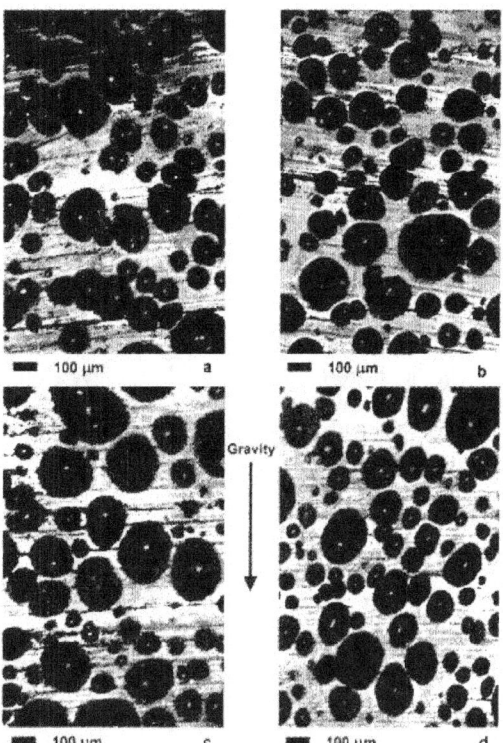

Close-up optical micrographs of foams that were processed at 680°F (360°C) for 5 minutes under microgravity (a and b) and under gravity (c and d). The gravity vector points down.

INVESTIGATING THE STRUCTURE OF PARAMAGNETIC AGGREGATES FROM COLLOIDAL EMULSIONS (InSPACE)

Principal Investigator(s): Alice P. Gast, Massachusetts Institute of Technology (MIT), Cambridge, Mass.
Expeditions 6, 7, 12, 13

Research Area Materials Science

Magnetorheological (MR) fluids are colloidal suspensions of magnetic particles whose properties can be controlled by magnetic fields. These fluids are classified as "smart materials" that transition to a solid-like state by the formation and cross-linking of microstructures in the presence of a magnetic field. The InSPACE experiment addresses the fundamental physics of liquid-solid phase changes in these fluids. On Earth these materials are used for vibration-dampening systems that can be turned on or off, in the automotive and construction industries, and for exploration technologies such as robots and smart space suits. InSPACE collects video imagery of fine structures of MR fluids in a pulsed (alternating on and off) magnetic field. The experiment requires a microgravity environment because the magnetic structures settle out in gravity, precluding the observation of steady-state structures. This study will help researchers understand the competing forces that govern the final shape of the structures.

The InSPACE coil assembly holds a Helmholtz coil assembly containing sealed vials of MR fluid, the camera/lens assemblies, and the power control. The coil assembly is attached to the "floor" of the MSG. The magnetic fields are applied to the various samples, and the operation of the experiment is monitored via video.

ISS006E41778 — During Expedition 6, flight engineer Donald R. Pettit works with the InSPACE samples in the MSG in the U.S. *Destiny* laboratory.

RESULTS

InSPACE was performed in the MSG during Expeditions 6, 7, and 13. Nine tests were performed for each Helmholtz coil for a total of 27 experimental runs. The collected data were processed, enabling a quantitative assessment of the structural data, including aggregate sizes and shapes. These are key parameters for defining the aggregate kinetics, and are used to test theoretical models of the microstructures. Furthermore, understanding the complex properties of the fluids and the interaction of the microparticles will enable the development of more sophisticated methods for controlling and use of these fluids. Results suggest that InSPACE runs did not achieve steady-state structures. However, intriguing data suggesting the onset of instability at low frequency were collected. Both of these phenomena will be further addressed in InSPACE 2.

PUBLICATIONS

Vasquez PA, Furst EM, Agui J, Williams J, Pettit D, Lu E. Structural Transitions of Magnetoghreological Fluids in Microgravity. 46th AIAA Aerospace Sciences Meeting and Exhibit, Reno, Nev. 2008, 7 – 10 Jan; AIAA 2008-815.

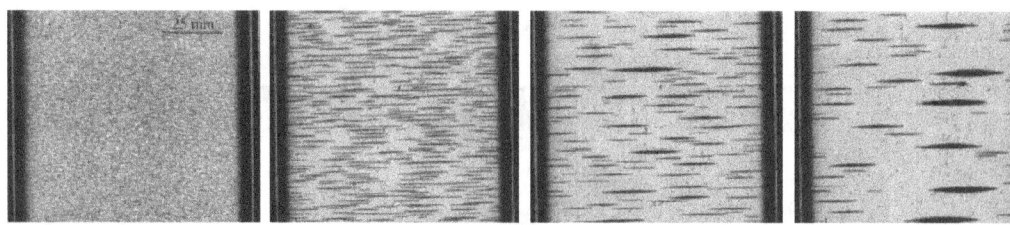

Structure evolution in a magnetorheological fluid over time while an alternating magnetic field is applied. The far left image shows the fluid after one second of exposure to a high-frequency-pulsed magnetic field. The suspended particles form a strong network. The images to the right show the fluid after three minutes, 15 minutes, and one hour of exposure. The particles have formed aggregates that offer little structural support and are in the lowest energy state.

Miscible Fluids in Microgravity (MFMG)

Principal Investigator(s):   John Pojman, University of Southern Mississippi, Hattiesburg, Miss.
Expeditions 8–11

Research Area Fluid Physics

The goal of the Miscible Fluids in Microgravity (MFMG) experiment is to test the fluid dynamics between two miscible (or mixable) liquids in a microgravity environment that is provided by ISS. Earth-based applications of this work may help researchers better understand the processing of miscible polymers.

This experiment was designed to demonstrate the existence of surface-tension-induced convection in miscible fluids. The effects of surface tension on the interaction between two miscible liquids here on Earth are masked by the effects of gravity and density . On Earth, miscible liquids effectively combine into one relatively homogenous (or equally distributed) solution. In microgravity, miscible liquids may behave completely differently, potentially taking on properties that are more akin to immiscible (or non-mixable) liquids. The role of interfacial surface tension in miscible liquids was proposed by the Dutch mathematician Korteweg over 100 years ago. Testing Korteweg's hypothesis is challenging on Earth because the force of gravity overwhelms surface tension, but the microgravity environment on station provides an ideal opportunity to do so.

Edward M. (Mike) Fincke, Expedition 9 science officer and flight engineer, uses syringes to inject honey into water to test the mixing ability of the two liquids for the MFMG experiment.

This experiment originated from a call for simple experiments requiring little upmass following the grounding of the shuttle fleet after the loss of *Columbia*. The MFMG experiment was proposed as a simple study of miscible fluids that was limited to the use of ordinary items that were already on board ISS (unused syringes, water, honey, Ziploc bags, a still camera, and a video camera). In the isothermal experiment (where diffusive forces predominate in microgravity), a stream of either honey or water was introduced into a syringe of the opposite fluid to observe the transient behavior of the miscible fluids. In the thermal experiment, a temperature gradient was created across the syringe holding one of the fluids, and a second fluid was introduced at ambient temperature.

Korteweg's theory predicts that miscible fluids will demonstrate interfacial tension transiently until diffusion prevails. Under normal gravity, the stream of honey would break apart under its own weight and surface tension would cause the fluid to have as little surface area as possible for a given volume. The droplets that form as the stream breaks apart would have less surface area than a cylinder (stream) of the same volume—an effect known as Rayleigh instability. The experiment was designed to determine whether the stream exhibits the Rayleigh instability characteristic of immiscible fluids.

Results

Isothermal results: Four sessions were performed with no observation of Korteweg's prediction of the Rayleigh instability. It was found that the honey did not break into small drops, neither did it change its shape when injected. The behavior that was exhibited was that of simple diffusion, which is seen on Earth when mixing two miscible fluids.

Thermal results: Two sessions of MFMG were performed with a thermal gradient introduced. The stream migrated towards the warmer side of the temperature gradient, which may indicate the presence of Korteweg's predicted behavior. Further sessions with thermal gradients are still being analyzed.

ISS011E07771 — Close up view of the MFMG syringe containing honey that is injected with tinted water during Expedition 11.

Despite the negative results—the authors concluded that the surface tension stresses were insufficient to overcome the large difference in viscosity of the honey-water system—the data were used to calculate the square gradient parameter for the water-honey system.

PUBLICATION(S)

Pojman JA. Miscible Fluids in Microgravity (MFMG): A Zero-Upmass Experiment on the International Space Station. Proceedings of the 43rd AIAA Aerospace Sciences Meeting and Exhibit, Reno, Nev. AIAA 2005-718, Jan 10–13, 2005.

Pojman JA, Bessonov N, Volpert V. "Miscible Fluids in Microgravity (MFMG): A Zero-Upmass Investigation on the International Space Station," *Microgravity Sci. Tech.* 2007, XIX, 33–41.

Protein Crystal Growth-Enhanced Gaseous Nitrogen (PCG-EGN)

Principal Investigator(s): Alexander McPherson, University of California, Irvine, Calif.
Expeditions 0 (prior to human occupation of ISS), 1, 2, 4

Research Area Protein Crystal Growth

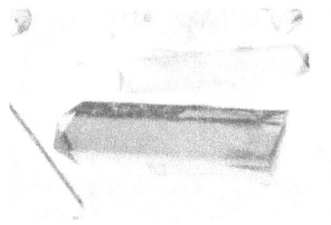

Crystal of trypsin that was grown in microgravity during ISS Expedition 4.

The microgravity environment on board the ISS is relatively free from the effects of sedimentation and convection and can provide an exceptional environment for crystal growth. Uniform, large crystals are key for determining the structure of protein and other large biological molecules.

The primary purpose of this experiment was to provide a simple trial platform for the production of a large number of crystals of various biological macromolecules. A second objective was an education program that was called "Student Access to Space" in which students participated in preparing some of the samples that were flown on orbit and learned about crystallization, the methods of analysis of crystals, and the impact of studies of crystals on advancing biotechnology, medicine, and basic research in structural biology. Through the Student Access to Space program, more than 500 samples were mixed by middle and high school students across the United States.

Protein Crystal Growth-Enhanced Gaseous Nitrogen (PCG-EGN) samples were brought to station frozen in liquid nitrogen in a Dewar (a stainless-steel and aluminum container assembly that is similar to a Thermos bottle) at −196°C (−321°F) in sealed plastic capillary tubes. On board ISS, the nitrogen warmed and boiled off, turning into a gas, and the samples began to thaw. After 8 days, when the samples had reached the station ambient temperature of 22°C (71.6°F), crystals began to form.

Results

Successful crystallization rates were as follows: Expedition 0 (prior to permanent human occupation of ISS), 10 of 24 proteins and viruses; Expedition 1, four of 23 proteins and viruses; Expedition 2, six of eight proteins and both viruses; Expedition 4, three of nine proteins and zero of two viruses. Major crystals obtained included Bence-Jones protein, Bromegrass Mosaic Virus, canavalin, lysozyme, pea lectin, thaumatin, trypsin, and 4a-hydroxy-tetrahydropterin dehydratase (DcoH). Overall the rate of successful crystallizations was not as high as expected. Although many of the crystals that were produced were no better than those that were obtained in the ground laboratory, there were still some significant structural results.

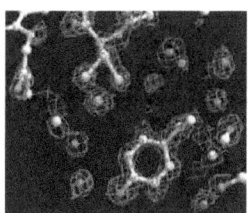

Electron density map of thaumatin crystal grown on ISS during Expedition 2.

When compared to their Earth-grown counterparts, the space-grown thaumatin crystals diffracted to a higher resolution, and some crystals showed as much as 40% more intensity during the diffraction process. This resulted in a more accurate protein structure model (electron density map) being produced from the space-grown crystal data. The pea lectin crystals also diffracted to a higher resolution than their Earth-grown counterparts. Data from the space-grown crystals were the best obtained, giving rise to the highest resolution structure for pea lectin. A refinement for the structural model of pea lectin is in progress. DcoH crystals that were grown on Expedition 1 also appeared to be of better quality than those that were grown on Earth.

Student investigations across the four Expeditions were successful in crystallizing a number of proteins. Although many of the crystals did not appear to be better than previously analyzed crystals, some of the crystals from Expedition 2 were used for microscopic observation and X-ray examination.

Publication(s)

Barnes CL, Snell EH, Kundrot CE. Thaumatin crystallization aboard the International Space Station using liquid-liquid diffusion in the Enhanced Gaseous Nitrogen Dewar (EGN). *Acta Crystallographica*. Section D, Biological Crystallography. 58(Pt 5):751–760, 2002.

Ciszak E, Hammons AS, Hong YS. Use of Capillaries for Macromolecular Crystallization in a Cryogenic Dewar. *Crystal Growth & Design*. 2(3):235–238, 2002.

PROTEIN CRYSTAL GROWTH-SINGLE LOCKER THERMAL ENCLOSURE SYSTEM (PCG-STES), NINE INVESTIGATIONS

Payload Developer: NASA Marshall Space Flight Center, Huntsville, Ala.
Expeditions 2, 4–6, samples held on board and returned during Expedition 11

Principal Investigator(s): Daniel Carter, New Century Pharmaceuticals, Inc., Huntsville, Ala., Protein Crystal Growth Facility-based Hardware: Science and Applications (PCG-STES-SA, Expeditions 2, 4–6)

Aniruddha Achari (Technical Investigator), Raytheon, Huntsville, Ala., Vapor Equilibration Kinetic Studies (PCG-STES-VEKS, Expeditions 2, 4–6)

Craig Kundrot, Marshall Space Flight Center, Huntsville, Ala., Optimizing the Use of Microgravity to Improve the Diffraction Quality of Problematic Biomacromolecular Crystals (PCG-STES-IDQC, Expeditions 2, 4–6)

Gloria Borgstahl, University of Toledo, Toledo, Ohio, and the University of Nebraska Medical Center, Omaha, Neb., Searching for the Best Protein Crystals: Synchrotron Based Mosaicity Measurements of Crystal Quality and Theoretical Modeling, and Searching for the Best Crystals: Integration of Synchrotron-Based Crystal Quality Measurements and Structure Determination (PCG-STES-MM, Expeditions 4, 5)

Geoffrey Chang, Scripps Research Institute, La Jolla, Calif., Crystallization of the Integral Membrane Protein Using Microgravity (PCG-STES-IMP, Expedition 5)

Barbara Golden, Purdue University, West Lafayette, Ind., Engineering a Ribozyme for Diffraction Properties (PCG-STES-RDP, Expedition 5)

Ronald Kaplan, The Chicago Medical School, Chicago, Ill., Crystallization of the Mitochondrial Metabolite Transport Proteins (PCG-STES-MMTP, Expedition 5)

Bill Thomas, Universities Space Research Association, Huntsville, Ala., Crystal Growth Model System for Material Science (PCG-STES-MS, Expedition 6)

Gerald Bunick, Oak Ridge National Laboratory, Oak Ridge, Tenn., Regulation of Gene Expression (PCG-STES-RGE, Expedition 6)

Associate Investigators: Wayne Shultz, Hauptman-Woodward Institute, Buffalo, N.Y.; Debashis Ghosh, Hauptman-Woodward Institute, Buffalo, N.Y.; D. A. A. Myles, European Molecular Biology Laboratory, Grenoble, France; Naomi Chayen, Imperial College, London; Jean-Paul Declercq, University of Louvain, Louvain-la-Neuve, Belgium

Research Area Protein Crystal Growth

The Protein Crystal Growth-Single Locker Thermal Enclosure System (PCG-STES) is a suite of protein crystal growth investigations that was performed in the station's U.S. Destiny laboratory. Multiple independent and collaborating principal investigators contributed samples and evaluated the technology for crystal growth in space. In general these studies sought to grow crystals of target proteins that would be of superior quality to similar crystals that were grown on the ground. The sedimentation and convection forces that cause many Earth-grown crystals to be irregular in shape and small in size are absent in microgravity.

SCIENCE AND APPLICATIONS (Carter)
These investigations focused on the hardware that provided a suitable environment for crystal growth in microgravity. Samples were housed in the PCG-STES and within two different types of crystallization hardware: the Protein Crystallization Apparatus for Microgravity (PCAM) or the Diffusion-Controlled Crystallization Apparatus for Microgravity (DCAM).

PCAMs consist of nine trays, each containing seven vapor-equilibration wells. The nine trays are sealed inside a cylinder. Crystals are formed by the "sitting drop" method of vapor diffusion. Each sample well holds a drop of protein solution and precipitant (salts or organic solvents, which draw water away from the protein solution) that is mixed together. A surrounding moat holds a reservoir, filled with an absorbent fluid, that draws moisture away from the mixed solution. Crystals form as the moisture is absorbed. A rubber seal pressed into the lip of the reservoir keeps crystals from forming on Earth or from bouncing out of their wells during transport. Each cylinder holds 63 experiments for a total of 378 experiments inside the Single-locker Thermal Enclosure System (STES). PCAM was used for all samples during Expeditions 2, 4, and 5.

DCAMs, which are slightly smaller than a 35mm film canister, each contained two cylindrical chambers that are connected by a tunnel. One chamber holds the precipitant solution and the other contains the protein sample. A thin semipermeable membrane covers the protein sample that allowed the precipitant to pass through at a controlled rate. The rate of diffusion was controlled by a porous plug that separates the two chambers. This is referred to as the liquid-liquid diffusion method. Eighty-one DCAMs, which were housed inside the STES, were used for all samples during Expedition 6.

The STES provides a controlled temperature environment between 1°C (32°F) and 40°C (104°F) in which to grow large, high-quality crystals. Its Thermal Control System (TCS) regulates the temperature inside the payload chamber. A fan pulls cabin air through an intake on the front panel, causing the air to flow across the heat exchanger fans and then out the rear left side of the unit. Pushbuttons and a liquid crystal display (LCD) on the front panel allow the station crew to command the unit. STES can also be commanded from the ground.

These investigations also focused on key proteins of the circulatory system, such as human serum albumin, human antithrombin III, and human peroxiredoxin 5 (PRDX5). In addition, S-layer protein from *Bacillus sphaericus*, cytochrome p450, and c-phycocyanin from *Synechococcus elongatus* were flown as part of associate investigations during Expedition 5.

Vapor Equilibration Kinetic Studies (Achari)
In a technical investigation that was associated with the use of PCG-STES, crystallization conditions were varied experimentally, including days to activate the crystallization, droplet volumes, and different precipitants. By characterizing the time that crystallization took to reach equilibrium in microgravity, the goal was to help identify optimum crystallization conditions.

Improved Diffraction Quality of Crystals (Kundrot)
These experiments focused on the growth of better-quality crystals for X-ray diffractions analysis. During ISS Expedition 2, two types of samples were selected. The first was a complex of basic fibroblast growth factor (bFGF) (a fibroblast is a cell from which connective tissue develops) and 19t2mod, a 42-nucleotide DNA that inhibits bFGF activity (inhibition of bFGF activity is a type of anti-cancer therapy). The second sample type was the plant protein thaumatin. Thaumatin, which is a protein from the katemfe fruit of West Africa, is an extremely potent sweetener. The objective of this sample type was to determine whether some of the chemical conditions that do not produce crystals on Earth would produce crystals in microgravity. In addition to the samples above, two further proteins, rDerf2 and glucocerebrosidase, were flown as part of an associate investigation during Expeditions 4 and 5. A deficiency of β-glucocerebrosidase leads to Gaucher's disease, which is a rare chronic disorder of cerebroside metabolism that is characterized by enlargement of the spleen, skin pigmentation, and bone lesions.

Superoxide Dismutase (Borgstahl)
During Expedition 4, this investigation focused on the crystallization of human Replication Protein A (RPA) and manganese superoxide dismutase (MnSOD), which are to be used in ground-based X-ray diffraction studies to understand the atomic structure. Human RPA is a single-stranded DNA binding protein that is used in DNA metabolism (replication, transcription, recombination, and repair). SODs are antioxidant enzymes that protect living cells against oxide radicals that are associated with cell damage.

Integral Membrane Proteins (Chang)
This Expedition 5 investigation focused on the crystallization of two transporter proteins, Escherichia coli (*E. coli*) MsbA and EmrE. *E. coli* MsbA is a protein transporter that is responsible for transporting lipopolysaccharides and phospholipids from the inner membrane of the bacteria cell to the outer membrane, increasing cell wall strength. It is theorized that antibiotics are transported out of the bacteria cells using this type of transporter protein and that knowledge of the structure of this protein could help develop a new class of drugs to supplement antibiotics. *E. coli* EmrE is of interest because its production is associated with antibiotic resistance.

RIBOZYME FOR DIFFRACTION PROPERTIES (Golden)

This Expedition 5 investigation focused on the crystallization of a molecular engineered RNA enzyme. RNA enzyme, which is also called a ribozyme, is an RNA molecule that is responsible for catalyzing its own cleavage or the cleavage of other RNA strands. For this investigation, a ribozyme was engineered to be used to examine the Group I introns (a segment of RNA that is not coded for a gene) through X-ray crystallography to get a detailed view of the ribozyme active site at the atomic level.

MITOCHONDRIAL METABOLITE TRANSPORT PROTEINS (Kaplan)

This Expedition 5 investigation focused on the crystallization of the mitochondrial citrate transporter protein (CTP). Mitochondria are round or rod-shaped organelles that are located in most cells and that produce enzymes for the metabolic conversion of food to energy (citric acid cycle). CTP is an important part in cellular metabolism.

CRYSTAL GROWTH MODEL SYSTEM (Thomas)

This Expedition 6 investigation used Ferritin and Apoferritin as a crystal growth model system to look at fundamental protein biochemistry.

REGULATION OF GENE EXPRESSION (Bunick)

This Expedition 6 investigation focused on the crystallization of two different proteins: nucleosome core particle and D-xylose ketol-isomerase (xylose isomerase). The nucleosome core particle is a building block of chromatin, which is found in the nucleus and is responsible for gene expression and housing DNA. This investigation examined the structure-function relationship in chromatin. Xylose isomerase is an enzyme that is used in the food industry to convert glucose to fructose.

RESULTS

PCG-STES is a suite of nine experiments with additional shared samples for associated investigators. Samples were taken to and from station five times for crystallization during Expeditions 2, 4, 5, and 6. The logistical considerations of space flight affected some of the results, as flight delays compromised some samples and a jarring drop of the hardware shortly after return on 11A/STS-113 probably destroyed any larger crystals that had formed during that set of runs. PCG-STES samples in DCAM were on orbit prior to the Space Shuttle *Columbia* accident, and then spent an unprecedented 981 days (Nov 2002–Aug 2005) on ISS before being returned on the next space shuttle flight.

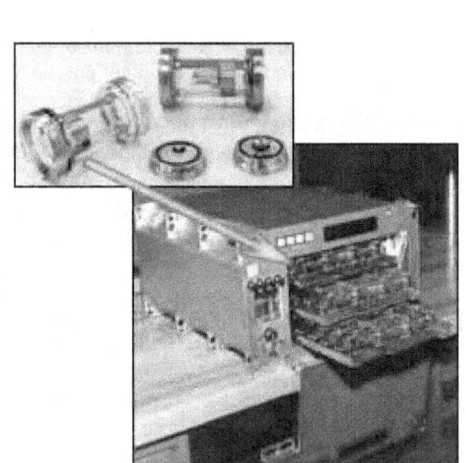

DCAM trays shown in the STES.

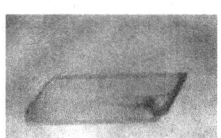

An MnSOD crystal that was grown in microgravity. The pink color is a result of oxidized manganese in the active site.

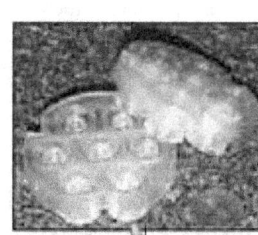

(a) PCAM trays with seven experimental cells.
(b) Nine trays are housed in one cylinder.
(c) Six cylinders fit into a thermal carrier and are housed in an ExPRESS rack on board ISS.

Not surprisingly, given the wide array of materials and objectives, some samples did produce large crystals while other samples produced crystals no better than those produced on Earth. Yet other samples failed to crystallize at all.

Crystals of MnSOD, which were produced during Expedition 4, exhibited an 80-fold volume increase when compared to the crystals that were produced on Earth. The crystals that were produced in orbit ranged from small, needle-like crystals to large three-dimensional crystals. These crystals were used for Synchrotron X-ray analysis, the use of a high-energy, adjustable particle beam used for crystal diffraction. Through this analysis it was determined that the diffraction resolution and quality of data for the crystals that were produced in microgravity were increased when compared to the diffraction resolution of the crystals grown on Earth (Vahedi-Faridi et al. 2003).

High-resolution structural data were also obtained from human albumin and human antithrombin III crystals, and publications of new structural information is anticipated. Analyses of the samples retuned in Aug 2005 is ongoing.

PUBLICATION(S)
Vahedi-Faridi A, Porta J, Borgstahl G. Improved three-dimensional growth of manganese superoxide dismutase crystals on the International Space Station. *Acta Crystallographica*. Section D, Biological Crystallography. 59:385–388, 2003.

ISS005E21531 — Astronaut Peggy A. Whitson, Expedition 5 science officer, works the PCG-STES hardware on board ISS.

Toward Understanding Pore Formation and Mobility During Controlled Directional Solidification in a Microgravity Environment (PFMI)

Principal Investigator(s): Richard Grugel, NASA Marshall Space Flight Center, Huntsville, Ala.
Expeditions 5, 7, 8

Research Area Materials Science

On Earth, bubbles that form in molten materials rise to the surface and release trapped gas prior to solidification. In microgravity where there is no buoyancy or convection, bubbles can become trapped inside the material, leaving pores as the material solidifies. These pores can greatly reduce the finished material's strength and structural integrity, making it a less desirable product. One of a couple of experiments investigating melting and solidification of materials, the Pore Formation in Microgravity (PFMI) experiment, was designed to learn how bubbles form and move during phase change (from liquid to solid) inside molten material. The PFMI experiment used succinonitrile (SCN), which is a clear organic compound that is a transparent metal analog material, and SCN-water (1%) mixtures to observe bubble formation and bubble movement. The experiment was designed to methodically investigate pore formation and growth using SCN that was loaded with an excess amount of dissolved nitrogen gas, and examine the role of hermocapillary forces in transporting the bubbles away from the solidification interface.

Experiments were conducted inside the MSG, which is a sealed and ventilated work volume in the U.S. *Destiny* laboratory. Samples were melted inside a thermal chamber with temperature-controlled hot zones and one thermoelectric cold zone. Flow visualization technology was used in support of the experiment to observe bubble movement.

Results

The PFMI experiment used glass tubes (1 cm inner diameter and 30 cm in length) filled with SCN and water in concentrations ranging from pure SCN to 1% SCN-water mixture. Of the 24 experiment runs, 21 were successful. The data from this experiment were provided by downlinked images during real-time operations on ISS. In addition, images were recorded using a videotape recorder (VTR) inside the MSG.

Grugel et al. (2004) observed bubble migration up the temperature gradient due to thermocapillary forces and reported that thermocapillary forces do play a role in bubble removal during solidification, thereby providing a potential mechanism for avoiding porosity in space processing.

ISS008E06301 — C. Michael Foale, Expedition 8 mission commander and ISS science officer, installs equipment in the MSG for the PFMI experiment.

Strutzenberg et al. examined the morphological evolution of the solidification of the PFMI samples (0.25% water to SCN mixture and 0.50% water to SCN mixture). Direct comparison between the ground-based thin (two-dimensional) samples and the flight bulk (three-dimensional) samples showed significant differences in the interface morphology. The flight samples achieved planar growth— an emergence of dendrites (crystallizes in the shape of a tree or branch)— in less time than the ground-based samples. When comparing the planar interface recoil, the flight sample was steeper than the ground-based sample. Additionally, the dendrite spacings in the flight bulk samples were closer together than the ground-based thin samples. This highlights the researchers' premise that the use of two-dimensional (thin) samples in one-g to obtain quantitative data for comparison with theoretical models has significant shortcomings.

Strutzenberg et al. concluded that the thin samples are not adequate to provide data on the initial planar front dynamics, which is the dynamical condition for the planar interface instability, and the steady-state primary dendrite spacing. Solidification of bulk samples in a microgravity environment and in the lab setting is necessary for a suitable comparison. The flow visualization images that were obtained for the PFMI experiment allowed Grugel et al. (2005) to study bubble formation in SCN. The bulk solidification samples, which were filled with SCN, were melted and re-solidified to observe the bubbles that formed. During controlled re-solidification, aligned tubes of gas were seen to be growing perpendicular to the planar solid/liquid interface, inferring that the nitrogen that had previously dissolved into the liquid SCN was now coming out at the solid/liquid interface and forming the little-studied liquid=solid+gas eutectic-type reaction. The flight sample results could not be duplicated in the ground-based samples.

Cox et al. (2006) are now attempting to better replicate the space phenomena in a ground-based lab by using small-diameter channels to minimize bulk convection and buoyant bubble rise effects.

The experiment team expects that the results will be directly applicable to understanding solidification for materials processing by providing insights into fundamental behavior of bubbles.

PUBLICATION(S)

Grugel RN, Anilkumar AV, Lee CP. Direct observation of pore formation and bubble mobility during controlled melting and re-solidification in microgravity, solidification processes and microstructures. A Symposium in Honor of Wilfried Kurz. The Metallurgical Society, Warrendale, Penn. 111–116, 2004.

Grugel RN, Anilkumar AV. Bubble Formation and Transport during Microgravity Materials Processing: Model Experiments on the Space Station. Proceedings of the 42nd AIAA Aerospace Sciences Meeting and Exhibit, Reno, Nev. AIAA 2004-0627, Jan 5–8, 2004.

Grugel RN, Anilkumar AV, Cox MC. Morphological Evolution of Directional Solidification Interface in Microgravity: An Analysis of Model Experiments Performed on the International Space Station. Proceedings of the 43rd AIAA Aerospace Sciences Meeting and Exhibit, Reno, Nev. AIAA 2005-917, Jan 10–13, 2005.

Strutzenberg LL, Grugel RN, Trivedi R. Observation of an Aligned Gas-Solid Eutectic during Controlled Directional Solidification aboard the International Space Station—Comparison with Ground-based Studies. Proceedings of the 43rd AIAA Aerospace Sciences Meeting and Exhibit, Reno, Nev. AIAA 2005-919, Jan 10–13, 2005.

Cox MC, Anilkuman AV, Grugel RN, Hofmeister WH. Isolated Wormhole Growth and Evolution during Directional Solidification in Small Diameter Cylindrical Channels: Preliminary Experiments, Proceedings of the 44th AIAA Aerospace Sciences Meeting and Exhibit, Reno, Nev. AIAA 2006-1140, Jan 9–12, 2006.

Grugel RN, Luz P, Smith G, Spivey R, Jeter L, Gillies D, Hua F, Anilkumar AV. Materials research conducted aboard the International Space Station: Facilities overview, operational procedures, and experimental outcomes. *Acta Astronautica*. 2008; 62;491–498. (*Also presented at the 57th International Astronautical Congress IAC-06-A2.2.10*)

Image of the solid liquid interface of the PFMI experiment. Bubbles detach from the interface and move up the temperature gradient (right to left), courtesy of Dr. Richard Grugel, NASA Marshall Space Flight Center.

SOLIDIFICATION USING BAFFLE IN SEALED AMPOULES (SUBSA)

Principal Investigator(s): Alexander Ostrogorsky, Rensselaer Polytechnic Institute (RPI), Troy, N.Y.
Expedition 5

Research Area Materials Science

Material melt-growth experiments have been difficult to run in the space environment because there is just enough residual micro-acceleration (g-jitter) to produce natural convection that interferes with the structure and purity of the material. This convection is responsible for the lack of reliable and reproducible solidification data and, thus, for gaps in the solidification theory. The Solidification Using Baffle in Sealed Ampoules (SUBSA) experiment tested an automatically moving baffle (driven by melt expansion during freezing) that was designed to reduce thermal convection inside an ampoule to determine whether the baffle significantly reduces convection. Ground studies showed that the baffle reduces the movement of the material during its liquid phase, making the process easier to analyze and allowing more homogenous crystals to form.

The key goal of SUBSA was to clarify the origin of the melt convection in space and to reduce the magnitude to the point that it does not interfere with the transport phenomena.

RESULTS

The baffle proved successful. Eight single crystals of indium antimonide (InSb), which were doped with tellurium and zinc, were directionally solidified in microgravity. The molten semiconductor material solidified as expected, without separating from the ampoule walls or releasing the undesirable bubbles that have been reported in several previous microgravity investigations. Semiconductor crystals with reproducible, nearly identical composition were obtained for the first time in space.

PUBLICATION(S)

Churilov A, Ostrogorsky A. Solidification of Te and Zn Doped InSb in Space. Proceedings of the 42nd AIAA Aerospace Sciences Meeting and Exhibit, Reno, Nev. AIAA 2004-1388, Jan 5–8, 2004.

Churilov AG, Ostrogorsky A. Model of Tellurium- and Zinc-Doped Indium Antimonide Solidification in Space. *Journal of Thermophysics and Heat Transfer*. 0887-8722, 19(4):5420–541, 2005.

Ostrogorsky A. Solidification Using a Baffle in Sealed Ampoules (SUBSA). Proceedings of the 41st AIAA Aerospace Sciences Meeting and Exhibit, Reno, Nev. AIAA 2003-1309, Jan 6–9, 2003.

Spivey RA, Gilley S, Ostrogorsky A, Grugel R, Smith G, Luz P. SUBSA and PFMI Transparent Furnace Systems Currently in Use in the International Space Station Microgravity Science Glovebox. 41st Aerospace Sciences Meeting and Exhibit, Reno, Nev. AIAA 2003-1362, Jan 6–9, 2003.

ISS005E07172 — Astronaut Peggy Whitson, NASA science officer, in shown in front of the MSG as she works on SUBSA during Expedition 5.

ZEOLITE CRYSTAL GROWTH (ZCG)

Principal Investigator(s):   Albert Sacco, Jr., Northeastern University, Boston, Mass.
Expeditions 4–6

Research Area Materials Science

Zeolites are hydrated aluminosilicate minerals, with a micro-porous crystalline framework that includes a network of interconnected tunnels, creating a structure similar to a honeycomb. A sort of mineral sponge, which is their microporous structure, enables zeolites to very specifically absorb and release chemical species, including various cations and gas molecules such as H_2O, CO_2, N_2, Freon, and more.

Zeolites and zeotypes (similar microporous silicates) are widely used for many critical industrial processes, such as commercial ion exchangers, adsorbents, and catalysts. For example, virtually all of the world's gasoline is produced or upgraded using zeolites. In a very different application, zeotype *Engelhard titanosilicate* structure (ETS) titanosilicates contain natural quantum wires, which are of great potential in electronic and optical applications and in photochemistry, in their structures. However, scientists have insufficient understanding of how zeolite and zeotype materials grow, and how the desirable aspects of their porous structure (size, morphology, purity) can be controlled during crystal growth for commercial applications.

The Zeolite Crystal Growth (ZCG) experiment examined the growth of zeolite crystals in microgravity, where the role of convection and sedimentation is minimized, providing a growth environment that would produce fewer crystal defects.

The ZCG furnace, which was derived from earlier shuttle models, can grow zeolites, zeotype titanosilicate materials, ferroelectrics, and silver halides—all materials that are of commercial interest. The unit consists of a cylinder-shaped furnace, the Improved Zeolite Electronic Control System (IZECS), which includes a touchpad and data display as well as autoclaves. Two precursor growth solutions are placed into each autoclave, which mix during their stay in the furnace.

Zeolite Beta was grown from precursor solutions of sodium aluminate and colloidal silica that had been heated to 403 K (130°C) on station. The samples were characterized by X-ray diffraction to determine the crystal structure. The performance of zeolite Beta as a Lewis acid catalyst was evaluated using a standard set of chemical reduction reactions known as the Meerwein-Pohhdorf-Verley (MPV) reactions.

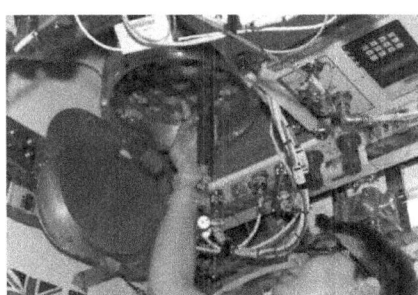

ISS005E19055 — Peggy Whitson, Expedition 5 flight engineer, places a cartridge into the ZCG experiment in the U.S. *Destiny* laboratory.

RESULTS

ZCG produced zeolite crystals with a high degree of crystalline perfection in microgravity. During ISS Expedition 6, 19 zeolite samples were mixed and incubated. These samples were returned to Earth at the conclusion of Expedition 6 and sent back to the principal investigator for analysis.

Results from the samples that were mixed on ISS suggest that the Lewis acid catalytic sites are altered in microgravity, as indicated by lower catalytic activity in the MPV probe reaction compared to Earth-grown zeolite. This further suggests that the control of fluid dynamics during crystallization may be important in making better industrial catalysts. Although space-grown zeolites had the same particle morphology and identical surface framework as zeolites that were grown on Earth, the average zeolite size of the space-grown crystals was 10% larger than crystals that were grown on Earth (Akata et al. 2004).

Larger zeolite crystals allow researchers to better define the structure and understand how they work, with a goal of producing improved crystals on Earth. Improved zeolites may have applications in storing hydrogen fuel, reduction of hazardous byproducts from chemical processing, and more efficient techniques for petroleum processing.

PUBLICATION(S)
Akata B, Yilmaz B, Jirapnogphan SS, Warzywoda J, Sacco Jr. A. Characterization of zeolite Beta grown in microgravity. *Microporous and Mesoporous Materials*. 71:1–9, 2004.

BIOLOGICAL SCIENCES IN MICROGRAVITY

The ISS laboratories enable scientific experiments in the biological sciences that explore the complex responses of living organisms to the microgravity environment. The lab facilities support the exploration of biological systems ranging from microorganisms and cellular biology to integrated functions of multicellular plants and animals. Several of the biological sciences experiments have facilitated new technology developments that allow growth and maintenance of living cells, tissues, and organisms.

Animal Biology

Animal studies in microgravity can be beneficial for framing research that can be applied to understanding the human system in space. While many animal studies are performed in analog environments on the Earth, the ISS provides facilities that allow for research on animals in a microgravity environment.

CELLULAR BIOLOGY AND BIOTECHNOLOGY

In a microgravity environment, physical controls on the directionality and geometry of cell and tissue growth can be dramatically different to those on Earth. Various experiments have used the culture of cells, tissues, embryos, and small organisms on orbit as a tool to increase our understanding of biological processes in microgravity.

MICROBIOLOGY

New developments in microbiology have led to an increasingly exciting set of experiments that tests how microorganisms change in space, including analysis of changes in the gene expression of different pathogens. These tests are directly relevant to the health of astronauts during their missions, and hold great potential for understanding the mechanisms that are employed by disease-causing microorganisms,

PLANT BIOLOGY IN MICROGRAVITY

Studies of plant physiology in microgravity provide insight into the basic biology of plants, and into how plants might be used as part of future Exploration missions. Successfully growing plants in microgravity presents challenges—from predictable distribution of water and nutrients to the reliability of biomass production.

Biological Experiments Performed on the International Space Station, Grouped by Discipline

Animal Biology
FIT (Fungal Pathogenesis, Tumorigenesis, and Effects of Host Immunity in Space)
Cellular Biology and Biotechnology
ADF-Otolith (Avian Development Facility – Development and Function of the Avian Otolith System in Normal Altered Gravity Environments)
ADF-Skeletal (Avian Development Facility – Skeletal Development in Embryonic Quail)
CBOSS-01-02-Renal (Cellular Biotechnology Operations Support Systems: Human Renal Cortical Cell Differentiation and Hormone Production)
CBOSS-01-Colon (Cellular Biotechnology Operations Support Systems: Use of NASA Bioreactor to Study Cell Cycle Regulation: Mechanisms of Colon Carcinoma Metastasis in Microgravity)
CBOSS-01-Ovarian (Cellular Biotechnology Operations Support Systems: Evaluation of Ovarian Tumor Cell Growth and Gene Expression)
CBOSS-01-PC12 (Cellular Biotechnology Operations Support Systems: PC12 Phenochromocytoma Cells – A Proven Model System for Optimizing 3-D Cell Culture Biotechnology in Space)
CBOSS-02-Ertythropioetin (Cellular Biotechnology Operations Support Systems: Production of Recombinant Human Erythropoietin by Mammalian Cells)
CBOSS-02-HLT (Cellular Biotechnology Operations Support Systems: The Effect of Microgravity on the Immune Function of Human Lymphoid Tissue)
CBOSS-FDI (Cellular Biotechnology Operations Support Systems: Fluid Dynamics Investigation)
CGBA-APS (Commercial Generic Bioprocessing Apparatus – Antibiotic Production in Space)
CGBA-KCGE (Commercial Generic Bioprocessing Apparatus – Kidney Cell Gene Expression)
CGBA-SM (Commercial Generic Bioprocessing Apparatus – Synaptogenesis in Microgravity)
MEPS (Microencapsulation Electrostatic Processing System)
StelSys (StelSys Liver Cell Function Research)
Microbiology
MDRV (Microbial Drug Resistance Virulence) ((inc 16))
Microbe (Effect of Spaceflight on Microbial Gene Expression and Virulence)
NLP-Vaccine-1A (National Lab Pathfinder – Vaccine – 1A) ((inc 16))
POEMS (passive Observatories for Experimental Microbial Systems)
SPEGIS (*Streptococcus pneumoniae* Expression in Genes in Space)
Yeast-Gap (Yeast-Group Activation Packs)
Plant Biology in Microgravity
AdvAsc (Advanced Astroculture™)
BPS (Biomass Production System) Technology Validation Test (TVT)
Gravi (Threshold Acceleration for Gravisensing)
Multigen (Molecular and Plant Physiological Analyses of the Microgravity Effects on Multigeneration Studies of *Arabidopsis thaliana*)
ORZS (The Optimization of Root Zone Substrates for Reduced Gravity Experiments Program)
PESTO (Photosynthesis Experiment and System Testing and Operation)
PGBA (Plant Generic Bioprocessing Apparatus)
Tropi (Analysis of a Novel Sensory Mechanism in Root Phototropism)

Avian Development Facility (ADF), Two Investigations

Principal Investigator(s): J. David Dickman, Washington University, St. Louis, Mo., Development and Function of the Avian Otolith System in Normal and Altered Gravity Environments (ADF-Otolith)

Stephen B. Doty, Hospital for Special Surgery, New York, N.Y., Avian Development Facility-Skeletal Development in Embryonic Quail (ADF-Skeletal)

Expedition 4

Research Area Cellular Biology and Biotechnology

The avian hatching habitat with the exterior panels removed.

The avian development biology experiment, which is a tool for the study of embryogenesis in space, provided the support hardware that is needed for researchers to better understand and mitigate or nullify the forces of altered gravity on embryo development. Avian eggs are ideal for studying embryo development since they are self-contained and self-sustaining, and can be nurtured without a maternal host. The Avian Development Facility (ADF) provided incubation of avian eggs under controlled conditions (humidity, temperature, and gas environment) on orbit, and the fixation of the eggs for study while minimizing the effects of launch and landing. Up to 36 eggs in centrifuge carousels could be exposed to simulated gravity of zero-g to one-g in 0.1-g increments.

During its flight on space shuttle mission STS-108 to the ISS, the ADF housed two investigations: the Development and Function of the Avian Otolith System in Normal and Altered Gravity Environments (ADF-Otolith) and the Skeletal Development in Embryonic Quail on the ISS (ADF-Skeletal) investigations.

ADF-Otolith (Dickman)

The otolith system (small bones of the inner ear) in all vertebrates functions to detect head position and movement relative to gravity depending on neuromotor responses. The avian otolith system offers an excellent model with which to study the effects of gravity upon development due to the short maturation period following fertilization and due to the extensive knowledge of otolith system structure and function in birds.

One of two ADF carousels, each featuring 18 sample containers. During space missions, one carousel rotates at 77.3 revolutions per minute (RPM) to simulate a one-g gravity field. The other carousel remains motionless to provide a microgravity environment for the specimens inside. In addition to eggs, the ADF can carry fish, plants, insects, or cells in its sample containers.

Otoliths are part of the vestibular system (balance system) in vertebrates, and are an essential component in the production of movement-related responses that are critical for daily function and survival. During space flight, vestibular disturbances are frequently reported by astronauts, with approximately 80% to 90% of current astronauts experiencing some symptoms of space motion sickness (disorientation and nausea) during the first 48 to 72 hrs of weightlessness.

Many scientists have suggested that lack of gravity as a constant stimulus during space flight produces significant changes in vestibular system function. Preliminary studies indicated that the structure and function of the vestibular system is affected by exposure to microgravity. For example, changes in size of otoconia in the receptor-afferent morphology, hair cell conductance, vestibular afferent responsiveness, vestibular central pathways, and vestibular-related neuromotor responses have all been observed in both adult and developing animals that are exposed for brief periods to either microgravity or hypergravity.

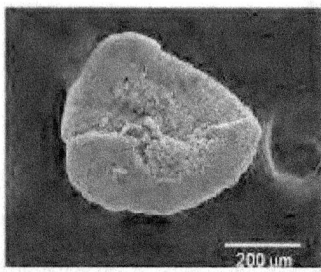 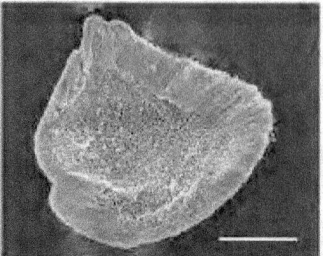 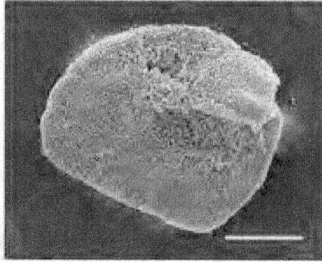

Scanning electron micrographs of fused otoconial stones from embryonic day 12 quail embryos. Three saccule stones are shown, one each from an embryo that was raised from fertilization in zero-g (left), one-g flight (middle), and two-g laboratory centrifugation (right).

RESULTS
The inner ear bones in the embryos that developed in microgravity appeared to be larger than those found in the controls that remained on Earth. There are some indications that the fan-shaped arrangement of receptor cells may also be altered under the influence of microgravity. Conclusive data from this investigation are pending further analysis.

ADF-SKELETAL (Doty)
The objective of this experiment was to define the effects of space flight on embryonic skeletal development. This investigation was a stepping-stone in determining the effect of microgravity on the molecular and cellular biology of bone formation and loss. Many of the biological processes that are observed in bone formation during embryogenesis (development and growth of an embryo) also occur in the adult skeleton during fracture repair. Furthermore, previous space flight studies identified bone demineralization and bone density loss in embryo quails, which was similar to what is observed in adult humans with osteoporosis. However, up to a certain developmental age embryonic quail bones will rapidly recover from this condition when re-exposed to gravity; whereas humans suffering from bone loss can take years to recover. Therefore, the data gained from this study should provide a foundation of future studies on bone demineralization and density loss and insight into the mechanisms that are involved in the full re-mineralization of bones.

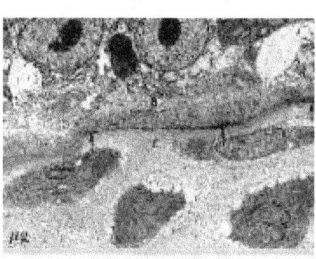
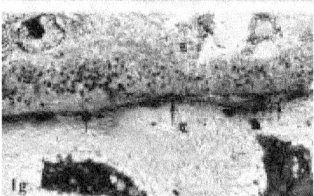

Electron micrographs of quail limb bones formed under the influence of microgravity show decreased mineralization compared to bones formed in normal gravity. The bone that developed during space flight (top) shows less mineral compared to the control sample (bottom); the control sample clearly shows mineral deposits (dark spots) that are absent in the space flight sample.

RESULTS
No space flight effects were observed for osteocalcin levels in the day 12 embryos, based on bone matrix stating. Since osteocalcin reflects the degree of bone mineralization, this would suggest that mineralization is not affected in an older embryo. However, direct mineralization quantitative studies have not been reported for day 7 and day 12 embryos, which should provide definitive evidence for whether osteocalcin-associated processes are affected.

The second finding was that the space flight embryos that are on the spinning carousel or stationary carousel had a reduced level of collagen-synthesizing activity as compared to the ground control specimens, although the sample size was small. If this trend is validated, it would suggest that space flight has a component that can affect collagen synthesis that is not correctable by an applied one-g force. These insights might be important for the development of appropriate countermeasures for space travel. Final analyses and publication of results are pending.

PUBLICATION(S)
There are no publications at this time.

ADVANCED ASTROCULTURE (AdvAsc)

Principal Investigator(s): Weijia Zhou, Wisconsin Center for Space Automation and Robotics, University of Wisconsin at Madison, Wis.
Expeditions 2, 4, 5

Research Area Plant Biology in Microgravity

AdvAsc was a commercially sponsored payload that provided precise control of environmental parameters for plant growth, including temperature, relative humidity, light, fluid nutrient delivery, and CO_2 and ethylene concentrations. AdvAsc hardware was used in a series of tests over three Expeditions (2, 4, and 5). First, AdvAsc demonstrated the first "seed-to-seed" experiment in space, growing *Arabidopsis thaliana* through a complete life cycle. (*Arabidopsis thaliana* (thale cress) is a model system in plant biology studies that has a short life cycle, a completely sequenced genome, and a history of space experiments.) Next, 35% of the space-grown seeds and 65% of wild *Arabidopsis* seeds were grown. Finally, soybean plants were also grown through an entire life cycle.

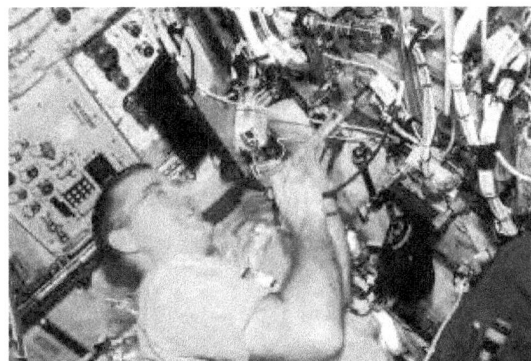

ISS002E6300 — James S. Voss, Expedition 2 flight engineer, works with the AdvAsc condensate fluid syringe at the AdvAsc growth chamber in the U.S. *Destiny* laboratory.

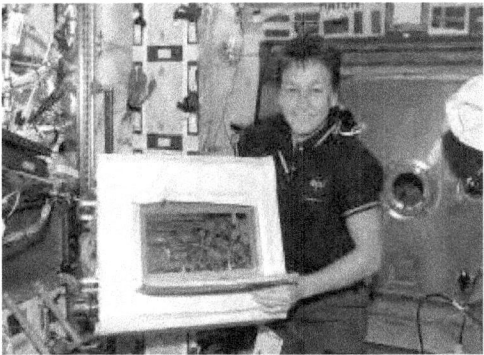

ISS005E07209 — Astronaut Peggy A. Whitson, Expedition 5 NASA ISS science officer, displays a first crop of soybeans growing inside the AdvAsc. Photos of the growing plant have been a useful tool for the ground-based science team, which uses them to determine optimal time for cross-pollination and harvesting.

RESULTS

Arabidopsis thaliana was successfully grown from seed-to-seed on ISS. During a 2-month growth period, the plants progressed from seed hydration to germination, vegetative, and reproductive stages, producing mature seeds. Ninety percent of the seeds germinated in space, although only 70% of the plants grew to maturity.

Some of the seeds that were harvested from the plants that were grown in microgravity were planted in a ground study. These seeds produced typical plants without any visible abnormalities (Link et al. 2003). During a second AdvAsc run, second-generation seeds were produced and tissues were harvested and preserved for RNA and complementary deoxyribonucleic acid (cDNA) analysis. Detailed results of the germination and harvesting of space-grown seeds in the AdvAsc growth chamber in the U.S Laboratory *Destiny* have not been released.

In the third AdvAsc run, which took place over approximately 95 days on ISS, soybeans were grown from seed-to-seed for the first time in space. Biomass production in the space seeds was approximately 4% larger than ground controls. Flight and grounds controls produced nearly identical numbers of seeds, but the space seeds were larger on average. Scientists found that the seeds that were produced in space were healthy, the germination rates were comparable to those on Earth, and no major morphological differences were evident. Phytochemical analysis of commercially important components such as oils, amino acids, proteins, carbohydrates, and phytoestrogens have not yet been released.

Publication(s)

Link BM, Durst SJ, Zhou W, Stankovic B. Seed-to-seed growth of *Arabidopsis thaliana* on the International Space Station. *Advances in Space Research*. 31(10):2237–2243, 2003.

Zhou W, Durst SJ, DeMars M, Stankovic B, Link BM, Tellez G, Meyers RA, Sandstrom PW, Abba JR. Performance of the Advanced Astroculture™ plant growth unit during ISS-6A/7A mission. SAE Technical Paper Series. Paper no. 02ICES-267, 2002.

Zhou W. Advanced Astroculture™ Plant Growth Unit: Capabilities and Performances. 35th International Conference on Environmental Systems, Rome, Italy. Jul 11–14, 2005.

Dried *Arabidopsis thaliana* plants, from ISS Expedition 4, upon return to Earth.

BIOMASS PRODUCTION SYSTEM (BPS) TECHNOLOGY VALIDATION TEST (TVT)

Principal Investigator(s):   Robert C. Morrow, Orbital Technologies Corporation Space Center, Madison, Wis.
Expedition 4

Research Area Plant Biology in Microgravity

The BPS was developed as a precursor to systems that are capable of maintaining plant growth in microgravity for more than 90 days (e.g., planetary missions). It is comprised of multiple plant growth chambers that are independently controlled. The BPS objective was to validate plant growth system hardware functionality and performance, plant productivity and health, information acquisition, and experiment operations and support in microgravity. BPS demonstrated the ability to accurately measure the variables that are associated with plant photosynthesis, respiration, and transpiration, and to control the growth variables and measure the responses so that the effects of microgravity on plant growth and function can be isolated. The BPS housed two experiments: the Technology Validation Test (TVT) experiment and the Photosynthesis Experiment and System Testing and Operation (PESTO) experiment.

ISS004E11721 — View of Brassica plants from PGC 2 being harvested as part of the technical validation test of the BPS that was conducted during ISS Expedition 4.

Brassica rapa (field mustard) and *Triticum aestivum* (super dwarf wheat) were the test species for the BPS. The BPS plant growth chambers (PGCs) contained plants that were started on the ground and had already developed their photosynthetic apparatus, such as stoma, guard cells, and other structures that are found in the leaves. Samples taken from the plants were compared to data taken from previous ground-based experiments that were conducted using BPS. Over the course of the 73-day test, additional sets of plants were germinated and grown in microgravity conditions. In-flight progress of plant growth was monitored through image collection; harvested plants were frozen or fixed for later analysis on the ground.

BPS tested the hypothesis that environmental control subsystems would provide a stress-free growing environment in microgravity. These technology validation studies provide a foundation on which to base the design of future plant growth units for station or future Exploration missions. These results can lead to the development of regenerative life support systems on future missions to the moon or Mars. While creating useful technology and science, BPS allowed students in grades Kindergarten through 12 to work as co-investigators on real space research. This research, which is known as "Farming in Space," examined the basic principles and concepts that are related to plant biology, agricultural production, ecology, and the space environment. Activities that are associated with this research encouraged curiosity in the sciences while teaching good scientific methodology.

RESULTS

Multiple criteria were used to evaluate BPS technology; nearly all of the performance requirements (plant health, temperature and humidity control, atmospheric composition control, nutrient and water delivery, lighting, data acquisition of CO_2 levels, water use, biofouling) were met successfully (Iverson et al. 2003). While researchers noted a few performance parameters required additional work (e.g., elevated temperatures of the root zones), these indicators were identified and documented, and will be built in to the design modification. Overall the BPS hardware performed as expected, and may provide a viable use in the development for regenerative life support systems for future spacecraft development.

Thirty-two germinating *Brassica rapa* plants were launched inside the BPS for the TVT of the hardware. The *Brassica rapa* plants were grown over two growth cycles on ISS. *Brassica rapa* tissue from BPS was analyzed for general morphology, seed anatomy and storage reserves, foliar carbohydrates, and chlorophyll and root zone hypoxia analysis. Some of the wheat plantings were evaluated for growth, germination, weight, chlorophyll concentration, and root appearance (Morrow et al. 2004). By the end of the 73-day experiment, BPS TVT produced a

Video screen shot of *Brassica rapa*, 36 days after planting on ISS during Expedition 4.

total of eight harvests, seven primings, and a plant tissue archive of more than 300 plants.

Gross measure of growth, leaf chlorophyll, starch, and soluble carbohydrates confirmed comparable performance by the plants on the station with ground controls. Of particular interest were the differences between the immature seedlings. Immature seeds from station had higher concentrations of chlorophyll, starch, and soluble carbohydrates than the ground controls. Seed protein was significantly lower in the ISS material. Also, microscopy of immature seeds fixed on ISS showed embryos to be at a range of developmental stages, while ground control embryos had all reached the same stage of development. These differences could be attributable to differences in water delivery or reduced gas exchange due to lack of convection. These results suggest that the microgravity environment may affect flavor and nutritional quality on potential space produce (Musgrave et al. 2005).

An ancillary study tested for bacterial and fungal communities in the BPS chambers and root modules; these cultures were compared to ground control bacterial and fungal growth. Analysis indicated that more species of both bacteria and fungus were identified in the flight samples than in the ground samples. The populations were common airborne species found on Earth. The significance of the difference is uncertain (Frazier et al. 2003).

PUBLICATION(S)
Musgrave ME, Kuang A, Tuominem LK, Levine LH, Morrow RC. Seed Storage Reserves and Glucosinolates in *Brassica rapa L*. Grown on the International Space Station. *Journal of the American Society for Horticultural Science*. 130(6): 848–856, 2005.

Morrow RC, Iverson JT, Richter RC, Stadler JJ. Biomass Production System (BPS) Technology Validation Test Results. *Transactions Journal of Aerospace* 2004 1:1061–1070.

Frazier CM, Simpson JB, Roberts MS, Stutte GW, Fields NW, Melendez-Andrade J, Morrow RC. Bacterial and fungal communities in BPS chambers and root modules. *SAE Technical Paper Series*. 2003 Paper No. 03ICES-147.

Iverson JT, Crabb TM, Morrow RC, Lee MC. Biomass Production System Hardware Performance. *SAE Technical Paper Series*. 2003 Paper No. 03ICES-67.

CELLULAR BIOTECHNOLOGY OPERATIONS SUPPORT SYSTEM (CBOSS), SEVEN INVESTIGATIONS

Payload Developer: Johnson Space Center, Houston
Expeditions 3, 4 for cell cultures, Expeditions 8, 10, 12, ongoing for Fluid Dynamics Investigation

Principal Investigator(s): Timothy G. Hammond, Tulane University Medical Center, New Orleans, La., Human Renal Cortical Cell Differentiation and Hormone Production (CBOSS 01 02-RENAL, Expeditions 3, 4)

J. Milburn Jessup, Georgetown University Medical Center, Washington, D.C., Use of NASA Bioreactor to Study Cell Cycle Regulation Mechanisms of Colon Carcinoma Metastasis In Microgravity (CBOSS-01-Colon, Expedition 3)

Jeanne L. Becker, University of South Florida, Tampa, Fla., Evaluation of Ovarian Tumor Cell Growth and Gene Expression (CBOSS-01-Ovarian, Expedition 3)

Peter Lelkes, Drexel University, Philadelphia, Penn., PC12 Pheochromocytoma Cells: A Proven Model System for Optimizing Three Dimensional Cell Culture Biotechnology in Space (CBOSS-01-PC12, Expedition 3)

Arthur J. Sytkowski, Beth Israel Deaconess Medical Center, Boston, Mass., Production of Recombinant Human Erythropoietin by Mammalian Cells (CBOSS-02-Erythropoietin, Expedition 4)

Joshua Zimmerberg, National Institutes of Health, Bethesda, Md., Simulated Microgravity Applications Towards the Study Of HIV: The Effect of Microgravity on the Immune Function of Human Lymphoid Tissue (CBOSS-02-HLT, Expedition 4)

J. Milburn Jessup, Georgetown University Medical Center, Washington, D.C., Joshua Zimmerberg, National Institutes of Health, Bethesda, Md., Cellular Biotechnology Operations Support Systems: Fluid Dynamics Investigation (CBOSS-FDI, Expeditions 8, 10, 12, ongoing)

Research Area Cellular Biology and Biotechnology

The purpose of the CBOSS study was to support biotechnological research on board ISS by providing a stable environment in which to grow cells. The system was a multi-component cell incubator that was intended to grow three-dimensional clusters of cells in microgravity. A self-contained apparatus, CBOSS was designed to allow multiple experiments to be performed, thereby enabling scientists to study various types of cells operating simultaneously.

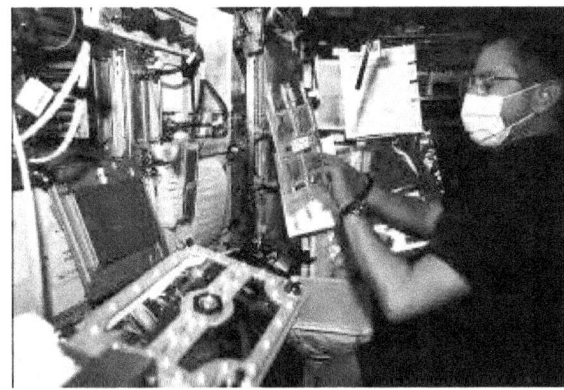

ISS003328030 — Frank L. Culbertson, Jr., Expedition 3 mission commander, works at the biotechnology specimen temperature controller (BSTC) for the CBOSS in the U.S. Destiny laboratory on ISS.

In the human body, cells normally grow within a scaffolding of protein and carbohydrate fibers that creates a three-dimensional structure. But outside the body, cells tend to grow in flat sheets and are incapable of duplicating the structure that they normally hold, which can make them behave differently in the laboratory than they would in the body. Past research has shown that cells that are grown in a microgravity environment arrange themselves into three-dimensional shapes that more closely duplicate how they would behave in the body. Cell culture in microgravity thus becomes a tool for studying cells in a state that is closer to that which occurs normally in the body.

HUMAN RENAL CORTICAL CELLS (Hammond)

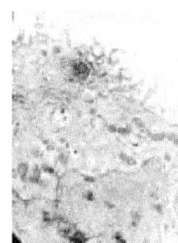

To better understand the mechanisms that cause several kidney disorders, human renal cortical epithelial cell lines were grown on station. This experiment used kidney cells to study the mechanism by which the kidney reabsorbs proteins that are filtered from the blood. The goal of this ISS experiment was to again create three-dimensional growth of normal human renal cells, and to assess the production of erythropoietin and vitamin D_3 while assessing the model for production of commercial applications.

Microscopic image of human renal cortical epithelial cells.

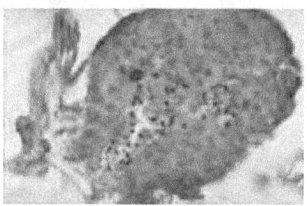

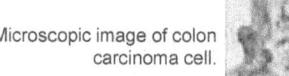
Microscopic image of colon carcinoma cell.

COLON CARCINOMA CELLS (Jessup)

Human colorectal carcinoma cells were grown to test the hypothesis that three-dimensional growth in microgravity facilitates the reprogramming of signal transduction pathways and gene expression as cells differentiate into the two major colonic cell lineages. This differentiation is important because it may inhibit cancer growth and, if applied to developing cancers, may block the emergence of new cancer formation. The unique environment of microgravity can provide insight into the growth, maturation, and death of this type of cancer cells.

OVARIAN TUMOR CELLS (Becker)

The goal of this study was to characterize the complex three-dimensional development of the human Müllerian ovarian (LN1) tumor cell line to characterize morphological changes that occur during three-dimensional growth. This study also sought to determine accompanying alterations in cell cycle kinetics, cell cycle proteins, and cellular oncoproteins. Cells were preserved in RNAlater, a fixative that allows cells to remain stable at refrigerator temperatures (4°C (39°F)) for up to 30 days. Following return to Earth after 3 months on ISS, the cells were analyzed for antigenic stability after removal of RNA using the RNAqueous kit. Knowledge that was gained from this experiment could help define mechanisms in tumor cell development that can be targeted for treatment of patients with ovarian cancer.

PC12 PHEOCHROMOCYTOMA CELLS (Lelkes)

Neuroendocrine cells (PC12) are cells that receive electric signals from the nervous system and chemical signals secreted from glands. As they differentiate, the cells are known to produce catecholamines, which are key to normal function and pain suppression. Evidence of differentiation is seen in ground-based (simulated) microgravity rotating wall systems. In this experiment, the ability and extent of the differentiation in actual microgravity was assessed by the growth and subculture of these cells to provide a greater understanding of neural regeneration and pain suppression.

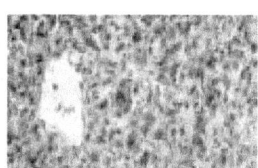
Stained microscopic image of neuroendocrine cells.

ERYTHROLEUKEMIA CELLS (Sytkowski)

Reduced immune system function and anemia that are related to decreased red blood cell production are two problems that face astronauts after extended durations in space. This experiment was designed to study cells in space to gain insight into the way microgravity affects blood cell formation. EMS-3 cells are Rauscher murine erythroleukemia cells, derived originally from mice infected with the virus that causes erythroleukemia. The EMS-3 cell line serves as an important model system for studying the cellular and molecular aspects of erythropoiesis (red blood cell formation), including the mechanism of action of erythropoietin in vitro under controlled conditions. EMS-3 cells were selected for culture in CBOSS to advance our knowledge of the effects of microgravity on the hematopoietic system and to suggest possible in-flight countermeasures and treatments for ground-based disease states.

HUMAN LYMPHOID TISSUE (Zimmerberg)

Impaired immune function has been observed in astronauts during flight, and these observations are bolstered by evidence of lymphocyte dysfunction in the rotating wall vessel (RWV) culture system (the ground-based analog or model for cell culture in microgravity conditions). Using human lymphoid tissue cells that have been isolated from human tonsils and derived from five donors for the experiment, the goal of this study was to determine whether microgravity was detrimental to the immune responses of human lymphoid tissue cell suspensions.

FLUID DYNAMICS INVESTIGATION (Jessup and Zimmerberg)

When cells arrive for culture on ISS, they are thawed, injected into a static tissue culture module (TCM) with media (nutrition), and then additional media is added at different times while waste liquid is removed. When these cells are injected or additional media is added to the TCM, it is important that the entire contents of the TCM be uniformly distributed. If cells in one corner of the TCM are not receiving nutrients they will die, causing a leaching of waste products that can be toxic to other cells. There is also the potential for bubble formation in the semi-permeable TCM, which could be deleterious to cells in culture, so procedures are also being developed for their removal. CBOSS-FDI [Fluid Dynamics Investigation] involves a series of experiments that is aimed at optimizing CBOSS fluid-mixing and bubble-removal operations while contributing to the characterization of the CBOSS stationary bioreactor vessel (the TCM) in terms of fluid dynamics in microgravity. These experiments will validate the most efficient fluid-mixing and bubble-removal techniques on orbit; these techniques are essential to conducting cellular research in microgravity, and will enhance the probability of success for future investigations.

RESULTS

The CBOSS hardware supported six cell culture investigations with different detailed scientific objectives. There were problems in the growth and preservation of all of the cell lines grown on Expeditions 3 and 4. The PC12 and erythroleukemia cells did not survive well in long-term culture, so no scientific results are expected from these experiments. It was found that there was more bubble formation than expected that may lead to cell death at the air-liquid interface. Although not well documented in this experiment, it was noted that poor mixing of cells/tissues and medium occurred in the other CBOSS payloads as well. Both the poor mixing and greater-than-expected bubble formation were important lessons learned that led to the addition of the CBOSS-FDI to study mixing and bubble formation in microgravity on later Expeditions.

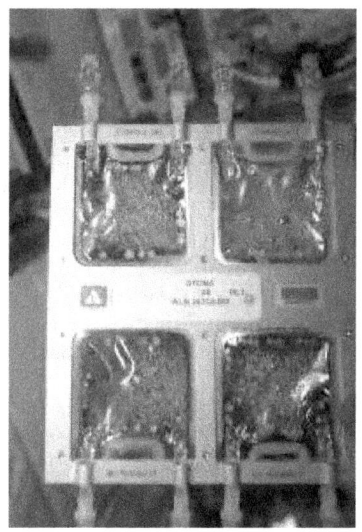

ISS004E5210 — Image of the Quad Tissue Culture Module Assembly (QTCMA)-5 on ISS Expedition 4 after activation of the cells. A syringe was used to inject cells into the pink nutrient growth media. When the samples completed their growth cycle, the crew transferred the cells from the incubator to the biotechnology refrigerator, where they were stored until they were examined at a ground-based laboratory.

Renal cortical cells returned were treated with an RNA stabilizing agent (RNAlater-Ambion) that enabled analyses of both RNA and immunoreactive proteins. The space and ground control cell cultures exhibited similar immunoreactivity profiles for the antibodies that were tested. These data provide evidence that the techniques used can be generalized to other cell lines, and that RNAlater will provide long-term storage of proteins at 4°C (39°F) for long-duration investigations (Hammond et al. 2006).

Analyses of the returned colon carcinoma cells revealed that the cells had died on orbit. However, ground-based research led to an appreciation of a novel mechanism by which microgravity may kill cells as well as of the role of tumor marker carcinoembryonic antigen (CEA) on preventing cell death. It has been shown that CEA interacts with death receptors on the cell membrane to reduce cell death. Since CEA is important to many of the cancers that afflict men and women in the United States, this is a critical finding that was in large part initiated by studies of growth in simulated microgravity. These results are not yet published, but were presented by Jessup at the Keystone Symposium on "Stem Cells, Senescence and Apoptosis" (Singapore, Oct 25–30, 2005).

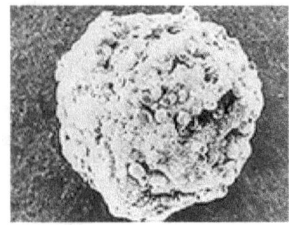

Scanning electron micrograph of a human Müllerian ovarian cell cancer nurtured in microgravity conditions. The three-dimensional structure shown is much closer in true size and form and natural tumor cells found in cancer patients.

The LN1 ovarian cell cultures on board station did not survive in long-term culture. However, the cells that ere grown on ISS were found to have produced reduced amounts of cytokines (small secreted proteins that mediate and regulate immunity, inflammation, and hematopoiesis) compared to the ground controls. The proteins were recovered after the RNA had been removed from the cells via filtration. The novel proteins, vimentin and epithelial membrane antigen (EMA) proteins, were extracted from filtrate of the RNA extraction. Vimentin is the main intermediate filament protein in embryonic cells. It plays an important role in the differential diagnosis of undifferentiated neoplasms (abnormal tissue growths). EMA, which belongs to a family of proteins known as human milk fat globulemembrane proteins, is considered a broad-spectrum antibody that is reactive against many types of adenocarcinomas. The data that were obtained from protein extraction indicate the presence of the antigenic proteins, vimentin and EMA, in RNA-stabilized LN1 cells following long-duration storage at 4°C (39°F). The vimentin and EMA proteins showed similar profiles at different times between the flight and ground samples. These data provide confirmation that the techniques that were used can be generalized to other cell lines, and that RNAlater will provide long-term storage of proteins at 4°C (39°F) for long-duration investigations (Hammond et al. 2005).

The human lymphoid tissue cultures were activated on board ISS but did not survive in longer-term culture. Early preliminary results, which were in agreement with RWV ground studies (microgravity simulation), indicated that the human tonsil cell suspensions show impaired immune responses in microgravity and that the extent of impairment depended on the activation state of the cells. Cells in all conditions showed metabolic activity, indicating that they were alive. Cells that were activated in microgravity did not demonstrate any increases in antibody or cytokine production; however, if the cells were activated prior to exposure to microgravity, they did demonstrate such responses. These results indicated that microgravity suppresses humoral immune responses in a not dissimilar fashion to that of Human Immunodeficiency Virus (HIV) on Earth, and that this phenomenon may reflect the immune dysfunction that was observed in astronauts during space flights (Fitzgerald et al. 2006).

For CBOSS-FDI, a series of procedures was performed on Expeditions 8, 10, and 12 to optimize particle mixing and bubble removal. A mixing protocol for particles has been found that appears to be effective and time-efficient, and crew feedback has been very valuable in these studies. Two bubble-removal methods were tested. Future experiments will help determine their effectiveness, and a protocol for bubble removal can be created for future tissue culture investigations. This investigation is critical for optimizing cell culture in space and ensuring the success of future investigations.

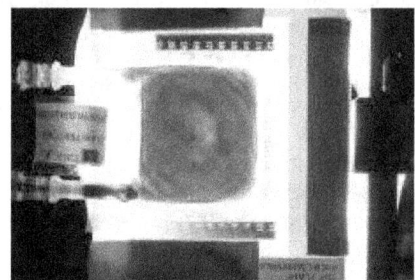

ISS008E09941 — A CBOSS-FDI tissue culture module that was photographed during Expedition 8.

Publication(s)

Hammond DK, Elliott TF, Holubec K, Baker TL, Allen PL, Hammond TG, Love JE. Proteomic Retrieval from Nucleic Acid Depleted Space-Flown Human Cells. *Gravitational and Space Biology.* 19(2), Jun 2006.

Hammond DK, Becker J, Elliot TF, Holubec K, Baker TL, Love JE. Antigenic Protein in Microgravity-Grown Human Mixed Müllerian Ovarian Tumor (LN1) Cells Preserved in a RNA Stabilizing Agent. *Gravitational and Space Biology.* 18(2):99–100, 2005.

COMMERCIAL GENERIC BIOPROCESSING APPARATUS (CGBA), THREE INVESTIGATIONS

Payload Developer: BioServe Space Technologies, Boulder, Colo., Antibiotic Production in Space Expeditions 0 (prior to human occupation of ISS), 2, 4

Principal Investigator(s): Timothy G. Hammond, Tulane University Medical Center, New Orleans, La., Kidney Cell Gene Expression (CGBA-KCGE, Expedition 0 [prior to permanent human occupation of ISS])

Haig Keshishian, Yale University, New Haven, Conn., Synaptogenesis In Microgravity (CGBA-SM, Expedition 0 [prior to permanent human occupation of ISS])

Louis Stodlieck and David Klaus, BioServe Space Technologies, Boulder, Colo., Antibiotic Production in Space (CGBA-APS, Expeditions 2, 4)

Research Area Cellular Biology and Biotechnology

The CGBA provided automated processing for biological experiments. It can contain up to eight containers that house the experiments, and each container is programmable and temperature-controlled between 4°C (39°F) and 37°C (99°F). The CGBA was equipped with data, video, and telemetry electronics to allow telescience remote operation. Three experiments, which are discussed below, were conducted in the CGBA.

ISS004E11048 — The CGBA isothermal containment module (ICM) v.3, installed in ExPRESS rack just above astronaut Dan Bursch's extended left arm; photograph taken during Expedition 4.

KIDNEY CELL GENE EXPRESSION (Hammond)
The primary objective of CGBA-KCGE [Kidney Cell Gene Expression] was to assess how microgravity alters the genes that control protein production in kidney cells. The investigator hoped to be able to manipulate the kidney cells to produce specific tissues that can be used in models when developing medicines or in humans. The kidney cell samples were drawn into the test tubes containing a preservative approximately 2 hours after reaching orbit. Once the samples were drawn, a messenger RNA (mRNA) preservative was added to the cell cultures for postflight analysis.

SYNAPTOGENESIS IN MICROGRAVITY (Keshishian)
CGBA-Synaptogenesis in Microgravity (SM) used the CGBA hardware to examine how microgravity affects the neuronal development of fruit flies, *Drosophila melanogaster*. This investigation used *D. melanogaster* embryos and larvae to observe how nerves that control movement navigate through an embryonic central nervous system (CNS) and attach to muscle fibers. Investigators observed how the synapses, which are the

The adult *Drosophila melanogaster* is a species of fruit fly that was used in the CGBA-SM investigation.

junction between two nerve cells where signals are transferred from one nerve to another, developed both during and after the embryonic stage.

ANTIBIOTIC PRODUCTION IN SPACE (Stodieck and Klaus)

The objective of this experiment was to determine whether secondary metabolite production in microbes was impacted by long-duration space flight. Previous research, which was conducted during short-duration space shuttle flights, identified significant potential for antibiotic production by microorganisms in orbit. The CGBA-Antibiotic Production in Space (APS) experiment was the first ISS investigation to test whether long-duration exposure to microgravity stimulated antibiotic production in microorganisms. CGBA-APS spent a total of 72 days in orbit on ISS. The experiment used *Streptomyces plicatus* to produce the antibiotic compound actinomycin D. Actinomycin D is an anti-tumor antibiotic that is used to treat tumors of the bone, urogential tract, skeletal muscle, kidney, and testis.

CBGA ICM with eight containers. The ICM provides both the computer and the thermal control for the samples.

RESULTS

CGBA-KCGE: Preliminary results indicate that an average of 60% of the kidney cell samples from CGBA-KCGE were drawn into the Vacutainers. Although the sample size was smaller, the samples were sufficient for postflight analysis. For the synaptogenesis experiment with fruit flies, preliminary results indicated that although the CGBA hardware operated successfully, there were unexpected temperature drifts above the planned temperature in two of the seven containers. While ground tests were completed for comparison to the in-flight samples, final data analysis has not been released.

Postflight images show MOBIAS [Multiple Orbital Bioreactor with Instrumentation and Automated Sampling, a fermentation, cell culture, and tissue engineering apparatus] tray with waste bag and samples visible (dark substance indicates actinomycin D).

CGBA-APS: This experiment originally flew on Expedition 2 but was unable to function due to technical issues. Its re-flight took place during Expedition 4 where the hardware performed as planned. Samples of antibiotic were taken at 4-day intervals. A total of 48 samples of *Streptomyces plicatus* were used to produce the antibiotic compound actinomycin D for a span of 72 days on orbit. The initial production of actinomycin D from on-orbit samples was higher than those produced during the ground tests. This was true for samples that were taken on day 8 (15.6 % increase) and day 12 (28.5% increase) of the investigation. Beginning at day 16, the ground experiment produced more antibiotic than the on-orbit experiment. This trend continued for the remainder of the experiment. The causes for the higher yield during the first 12 days of the experiment are still unknown. One theory is that there is a shorter lag phase, which allowed ISS samples to reach the growth and production phases sooner than the ground samples (Benoit et al. 2005).

Identifying the mechanism that caused increased production of antibiotics while in microgravity and applying them to production on Earth could be advantageous to the pharmaceutical industry. A method for transferring the microgravity research results to Earth-based production has not yet been identified.

PUBLICATION(S)

Klaus D, Benoit M, Bonomo J, Bollich J, Freeman J, Stodieck L, McCllure G, Lam KS. Antibiotic Production in Space Using an Automated Fed-Bioreactor System. AIAA International Space Station Utilization – 2001, Cape Canaveral, Fla., Oct 15–18, 2001.

Benoit MR, Li W, Stodieck LS, Lam KS, Winther CL, Roane TM, Klaus DM. Microbial antibiotic production aboard the International Space Station. *Applied Microbiology Biotechnology*. 2006 70 (4) 403–411.

FUNGAL PATHOGENESIS, TUMORIGENESIS, AND EFFECTS OF HOST IMMUNITY IN SPACE (FIT)

Principal Investigator(s): Sharmila Bhattacharya, Ph.D., NASA Ames Research Center, Moffett Field, Calif.
Expedition 13

Research Area Animal Biology

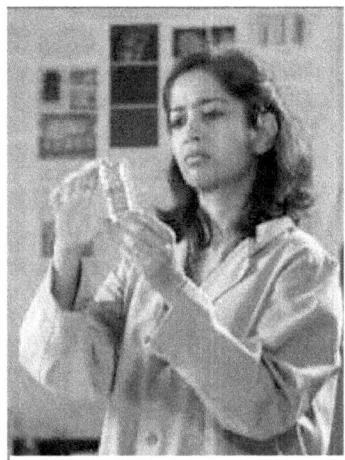

Dr. Sharmila Bhattacharya, checks the health of the fly culture in readiness for the shuttle flight experiment. Image courtesy of NASA Ames Research Center

The Fungal Pathogenesis, Tumorigenesis, and Effects of Host Immunity in Space (FIT) investigation took place on board the space shuttle during the STS-121/ULF1.1 mission. FIT addressed a series of human health risks that is associated with space flight using *Drosophila melanogaster* (fruit flies) and the fungus *Beauveria bassiana (B. bassiana* fungus occurs in soils and can behave as a parasite on a number of insect species). Specifically, this research examined tumor progression and the compounding effect of radiation, and the progression of an immune response in the host in response to a pathogen in space.

D. melanogaster were used to study the progression of oncogenic (cancerous) and benign melanotic tumors in sensitized mutant lines that show an increase in the incidence of tumor formation. Postflight samples were analyzed for changes in blood cell, hematopoietic organ (lymph gland), and fat body (liver) morphologies. The effect of radiation exposure was factored into this study.

Additionally, fungal pathogens (*B. bassiana)* were exposed to radiation and the space environment; space-flown samples were used for postflight infection of *D. melanogaster* hosts.

RESULTS
The flies were returned to Earth on STS-121/ULF1.1; analysis of the flies is ongoing, and results are expected in the near future.

PUBLICATION(S)
There are no publications at this time.

Threshold Acceleration for Gravisensing (Gravi)

Principal Investigator(s): Dominique Driss-Ecole, Ph.D., University Pierre-et-Marie Curie, Paris, France
Expedition: 14, 19, 20

Research Area Plant Biology in Microgravity

On Earth, plants respond to gravity so that stems grow up and roots grow down; lack of gravity causes plant growth problems in space. The goal of the Threshold Acceleration for Gravisensing (Gravi) study is to investigate the three parameters of the gravitropic response (a tendency to grow toward or away from gravity) for better understanding of the perception of the gravistimulus (reorientation within the gravity field) during initial plant growth. The three parameters are:
- presentation dose (minimum quantity of stimulation to provoke a significant curvature)
- gravisensing threshold acceleration (minimum acceleration that can be perceived)
- minimum deviation from the gravity vector, which leads to a reorientation of the root

Gravi uses lentil seedling roots that have been subjected to centrifugal acceleration levels from 10^{-2}g to 10^{-3}g in microgravity to determine the threshold of acceleration to which the roots respond.

Gravi was divided into two phases to be operated on board the ISS. Gravi-1, which was completed during Expedition 14, was the observation-only phase of the investigation. The development of roots by the lentil seeds during acceleration was videotaped by the European Modular Cultivation System (EMCS) for a designated period of time. The tapes were returned to Earth for analysis. Gravi-2, which is the second phase of the investigation, is scheduled for future operation on ISS and will allow for automated fixation of the lentil roots after acceleration is complete. The videotapes and the preserved roots of Gravi-2 will be returned to Earth for analysis.

Results

The Gravi-1 experiment was launched to the ISS in Dec 2006 and returned in Jan 2007. The goal was to understand the mechanism of gravisensing in roots by determining the threshold acceleration and the presentation of the seedling.

In the first run, lentil seedlings were hydrated and grown in the microgravity environment for 15 hr, then were subjected to small centrifugal accelerations for 13 hr and 40 min (.29 to .99 × 10^{-22}g). In the second run, seedlings grew for 30.5 hr or 21.5 hr, and were then placed in an accelerating environment for 9 hr (1.2 to 2.0 × 10^{-2} g). The orientation and shape of the roots were captured in time series photographs for analysis. The video tapes from Gravi-1 are currently undergoing analysis (position of root tip and root curvature) by the investigator team.

The figure below provides graphic results from the Gravi-1 phase of the experiment. The top row of photographs documents seedlings that did not experience acceleration. They show the initial root growth away from the seed, then later growth to straighten out the root. The second row shows the effects of accelerations administered by the centrifuge. The root curves back in the environment of acceleration. When the position and curvature of the roots vs. acceleration was analyzed, the threshold acceleration was determined to be 3.8 × 10^{-4} g. The microgravity-grown plants had greater sensitivity to accelerations than plants grown in one g.

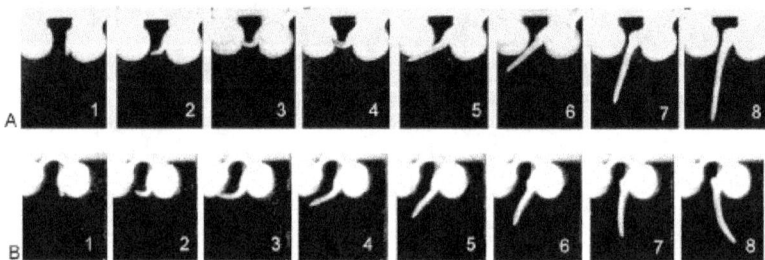

Time lapse photography of lentil seedling root growth in microgravity without (A) and with (B) being subjected to centrifugal accelerations.

PUBLICATIONS(S)
Driss-Ecole D, Legue V, Carnero-Diaz E, Perbal G. Gravisensitivity and automorphogenesis of lentil seedling roots grown on board the International Space Station. *Physiologia plantarum*. 2008; Epub.

MICROENCAPSULATION OF ANTI-TUMOR DRUGS (MEPS)

Principal Investigator(s):   Dennis Morrison, NASA Johnson Space Center, Houston
Expedition 5

Research Area Cellular Biology and Biotechnology

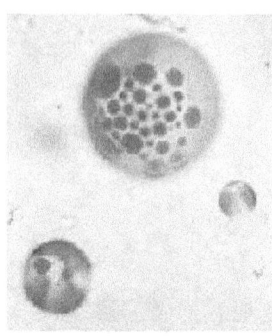

Micro-balloons containing anti-tumor drugs and small amounts of radio-contrast oil were created during MEPS operations. This oil, which is traceable by radiograph, allows doctors to follow the microcapsules as they travel to the tumor. The permeable outer skin releases the drug slowly, giving the microcapsule plenty of time to reach its destination. This slow release prevents artery damage as the drug travels to its destination.

MEPS is an automated system that is used to produce liquid-filled micro-balloons. It works through the use of microcapsules—unique capsules resembling miniature liquid-filled balloons the size of blood cells—that deliver Food and Drug Administration (FDA)-approved anti-cancer drugs by injection into the bloodstream. The microgravity environment on ISS was vital to the development of these capsules, enabling the balloon structures to form and encase immiscible liquids containing the pharmaceutical enables the pharmaceutical and its outer membrane to form spontaneously.

MEPS was designed with flexibility in mind. The system can process a wide range of experiments. For example, it can handle volumetric proportions of up to six chemical constituents; it can transfer liquids back and forth, at variable rates, between its six reservoirs and two main chambers; it can apply different electrical fields to the enclosed experiments; and it can be programmed to use filters or membranes of different porosity between chambers. Electrical fields charge the surface of the microcapsules, making it less recognizable as a foreign invader to the immune system.

The use of microcapsules will benefit the treatment of several diseases. For example, diabetes patients can use implanted microcapsules as treatment instead of daily insulin shots. Another Earth application of microcapsules is the delivery of drugs directly to cancer cells. This method can be used as a substitution for traditional anti-cancer treatments, such as chemotherapy, that involve large quantities of drugs that affect the entire body. The microcapsules contain a smaller dose of medication that directly targets tumors; smaller doses and targeted drug delivery help reduce the unwanted side effects that are currently produced by chemotherapy.

RESULTS

MEPS experiments were conducted during Expedition 5. Eight samples were processed using various methods to mix dissimilar liquids to form micro-balloons/microcapsules. The recovered micro-balloons were analyzed for size and drug content. Additionally, studies included the effects of temperature and internal pressure on the size of the micro-balloons. Ground-based medical investigations revealed that when using these microcapsules, the growth of human prostate and lung tumors can be inhibited with only a few local injections. When anti-cancer microcapsules are injected following cryosurgery, the combined treatment can completely destroy 1- to 2-cm-size tumors in just 3 weeks.

In addition to the successful demonstration of production of microcapsules and their utilization for drug delivery, MEPS and follow-on technologies that were developed have been awarded patents.

PUBLICATION(S)

Le Pivert P, Haddad RS, Aller A, Titus K, Doulat J, Renard M, Morrison DR. Ultrasound Guided, Combined Cryoablation and Microencapsulated 5-Fluorouracil, Inhibits Growth of Human Prostate Tumors in Xenogenic Mouse Model Assessed by Fluorescence Imaging. *Technology in Cancer Research and Treatment*. 3(2):135–42, 2004.

Morrison DR, Haddad RS, Ficht A. Microencapsulation of Drugs: New cancer therapies and improved drug delivery derived from microgravity research. Proceedings of the 40th Space Congress, Cape Canaveral, Fla. Apr 2003.

EFFECT OF SPACE FLIGHT ON MICROBIAL GENE EXPRESSION AND VIRULENCE (Microbe)

Principal Investigator(s): Cheryl A. Nickerson, Ph.D., Arizona State University, Tempe, Ariz.
Expedition 13

Research Area Microbiology

A human presence in space, whether permanent or temporary, is accompanied by the presence of microbes. The extent of changes to microorganisms in response to space flight conditions is not completely understood. Because the length of human space missions is increasing, orbiting humans are increasing at risk of infectious disease events occurring in flight. Previous studies have indicated that space flight weakens the immune system in both humans and animals. As astronauts and cosmonauts live for longer periods in a closed environment and use recycled water and air, there is an increase in the potential for negative impacts of microbial contamination upon the health, safety, and performance of crewmembers. Therefore, understanding how the space environment affects microorganisms and their disease-causing potential is critically important for space flight missions and requires further study.

The Microbe experiment employed three model microbes—*Salmonella typhimurium*, *Pseudomonas aeruginosa*, and *Candida albicans*—to examine the global effects of space flight on microbial gene expression and virulence attributes. These represent different types of bacteria and yeast. *Salmonella* is the most common agent for gastroenteritis in humans. Sanitation procedures are used to eliminate *Salmonella* from food sent to orbit, but if some of the *Salmonella* were missed, the impact on crew health could be significant. *Pseudomonas* has been detected as a contaminant in the water system of spacecraft, and was once a cause of crewmember infection during the Apollo era. *Candida*, which is yeast that s present as part of the natural human flora, has the potential for harmful overgrowth if microbial communities were to change over time in space.

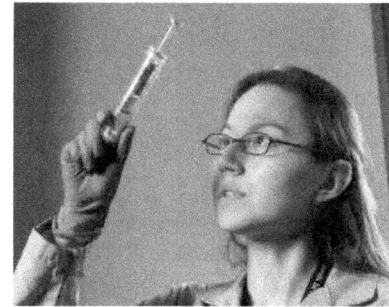

Researcher Hami Teal, Ph.D. examines hardware for the Microbe experiment. Image courtesy of NASA Ames Research Center.

The experiment was flown inside self-contained culture chambers that can be activated manually by a crewmember turning a hand crank to release growth media into the cell chamber. After 24 hours of growth at ambient temperatures, the growth was stopped by a crewmember turning the hand crank once more. Upon landing, one-third of the samples were recovered as soon as possible; the live cells will be used immediately for the virulence studies while the remaining stabilized samples were frozen at –80°C (–112°F). Ground analysis focused on identifying differences in growth rates and patterns, changes in gene expression, and changes in the virulence of the microbes in space compared to Earth.

By understanding the changes that microorganisms undergo in the space environment, these studies may lead to the development of vaccines and other novel countermeasures for the treatment and prevention of infectious diseases occurring during space flight and on Earth.

RESULTS

The Microbe experiment was performed in Sep 2006 during the STS-115/12A mission to the ISS; it tested three microbial pathogens; *Salmonella typhimurium*, *Pseudomonas aeruginosa*, and *Candida albicans*. Initial data from *S. typhimurium* showed that a total of 167 genes were expressed differently in flight when compared with ground controls. The data indicate that bacteria respond to the microgravity environment with widespread alterations of gene expression (process by which DNA is made into a protein), alterations in microbial morphology (shape and form of microbes), and increased virulence (disease-causing potential).

The changes in gene expression in *S. typhimurium* indicated that the Hfq protein plays an important role in the response of this organism to the space flight environment. Hfq is a protein that binds to mRNA, which creates the blueprint for proteins, to regulate gene expression. Sixty-four genes that are involved in the expression of Hfq were altered in flight; this is 32% of the total genes that were identified and that were expressed differently in flight. In addition, by using a ground-based model of space flight conditions on Earth, it was possible to reproduce the Hfq regulation of some of the *Salmonella* responses that were observed in flight. Collectively, these data indicate that Hfq is involved in globally regulating the *S. typhimurium* space flight response.

Scanning electron microscope (SEM) analysis of *S. typhimurium* was performed and showed no apparent difference in the size and shape of individual cells in the ground samples that were compared to the space samples. However, the space samples did demonstrate clear differences in microbial morphology, as exhibited by the cellular aggrega-

tion and clumping that was associated with the formation of the extracellular matrix that provides structural support to increase survival. Because extracellular matrix formation can help to increase the survival of bacteria under various conditions, this phenotype indicates a change in bacterial responses that are potentially related to the increased virulence of the flight-grown *S. typhimurium*.

S. typhimurium from flight and ground cultures were harvested and immediately used to inoculate rodents on the same day as the STS-115/12A landing. Rodents that were infected with bacteria from the space cultures displayed a decreased time to death and an increased percent mortality at each infection dosage compared with those that were infected with ground controls. These data indicate increased virulence for space flight *S. typhimurium* samples. The results of this work were recently published in the Proceedings of the National Academy of Sciences (Wilson et al. 2007). Data for *P. aeruginosa* and *C. albicans* samples are currently being analyzed.

S115e07274 — Astronaut Heidemarie M. Stefanyshyn-Piper, mission specialist, is shown holding the microbe group activation pack containing eight fluid-processing apparatuses in the middeck of the Space Shuttle *Atlantis* during Expedition 13 and STS-115 joint operations.

PUBLICATIONS(S)

Wilson JW, Ott CM, Hoener zu Bentrup K, Ramamurthy R, Quick L, Porwollik S, Cheng P, McClelland M, Tsaprailise G, Radabaugh T, Hunt A, Fernandez D, Richter E, Shah M, Kilcoyne M, Joshi L, Nelman-Gonzalez M, Hing S, Parra M, Dumars P, Norwood K, Bober R, Devich J, Ruggles A, Goulart C, Rupert M, Stodieck L, Stafford P, Catella L, Schurr MJ, Buchanan K, Morici L, McCracken J, Allen P, Baker-Coleman C, Hammond T, Vogel J, Nelson R, Pierson DL, Stefanyshyn-Piper HM, Nickerson CA. Space flight alters bacterial gene expression and virulence and reveals a role for global regulator Hfq. Proceedings of the National Academy of Sciences of the United States of America. 2007; 104(41):16299–16304.

Molecular and Plant Physiological Analyses of the Microgravity Effects on Multigeneration Studies of *Arabidopsis thaliana* (Multigen)

Principal Investigator(s): Tor-Henning Iversen, Ph.D., Norwegian University of Science and Technology, Trondheim, Norway

Expeditions 15, 16

Research Area Plant Biology in Microgravity

Multigen tested a novel method of immobilization of biological samples (e.g., plant seeds). A similar immobilization system was tested on the ground with positive results at the Plant BioCentre, NTNU, Trondheim, Norway. The Multigen experiment proposed to observe how this system will work in microgravity conditions.

The experiment used a water-soluble polyvinyl alcohol (PVA) membrane to fix plant seeds to a surface, taking care that the fixation method does not impact the science (e.g., growth pattern and biocompatibility) and that it is compatible with the experimental setup (auto-immunization and hardware in general). Experiment protocols include observation of the behavior of the membrane as it is dissolved, looking for any movement of the seeds. Multigen tested and compared PVA membranes of varying thicknesses, and also tested the set-up with different types of seeds, although the main work is done on *Arabidopsis thaliana*.

Housed in the EMCS, Multigen grows *Arabidopsis thaliana* to determine the effects of microgravity on plants. This image was taken a few days after germination of the seeds during Expedition 15. Multigen is a cooperative investigation with ESA. Image provided by ESA.

Results

Multigen samples were returned to Earth for analysis by the investigator team in Apr 2008. Final results of the investigation are pending data analysis of the returned samples.

Publication(s)

There are no publications at this time.

Housed in the EMCS, Multigen grows *Arabidopsis thaliana* to determine the effects of microgravity on plants. This image was captured on GMT 299 and shows the flower stalk (stem) with two to three flower bulbs on top and a few stem leaves below. Image provided by ESA.

THE OPTIMIZATION OF ROOT ZONE SUBSTRATES (ORZS) FOR REDUCED-GRAVITY EXPERIMENTS PROGRAM

Principal Investigator(s): Gail Bingham, Ph.D., Utah State University, Space Dynamics Laboratory, North Logan, Utah
Expeditions 14–16, 18

Research Area Plant Biology in Microgravity

Many long-term space flight life-support scenarios assume the use of plants to provide food supplies for crewmembers, as well as to recycle waste products. To date, *Brassica rapa* (field mustard plant), *Triticum aestivum* (super dwarf wheat), and four species of salad plants have grown in microgravity throughout their useful life cycles, with *Triticum* and *Brassica* producing viable seeds in nearly normal amounts and quality. These successes came at the end of nearly a decade of repeated efforts using the same equipment to arrive at optimal settings for substrate moisture and oxygen.

Optimization of Root Zone Substrates (ORZS) for Reduced-Gravity Experiments is managed at the Space Dynamics Laboratory (SDL) as part of the laboratory's Space Plant Technology Program, and was developed to provide direct measurements and models for plant rooting media that will be used in future advanced life support (ALS) plant growth experiments. The goal of this program is to develop and optimize hardware and procedures to allow optimal plant growth to occur in microgravity for future space exploration beyond LEO.

ISS016E027955 — Astronaut Peggy Whitson, Expedition 16 commander, checks the progress of plants growing in the Russian Lada greenhouse in the *Zvezda* service module of the ISS.

The key to this effort is validating wet substrate oxygen diffusion calculations in microgravity. While the measurements appear simple and well studied in agricultural soils on Earth, collecting repeatable results at high-water contents in the coarse-textured growth media that are required for microgravity requires a modified approach, due to the dominance of capillary rise in microgravity and the greater root zone density of plants that are grown in space. Collecting accurate data under both one-g and microgravity conditions and correctly interpreting these data at reasonable cost requires careful management and organization, and expert technical microgravity experience. Only a few microgravity substrate water-management experiments have been conducted. ORZS will be the first experiment to directly measure oxygenation parameters in wet substrates.

ORZS data collection will use the following two existing flight programs to meet experiment requirements:
- Institute of Biomedical Problems' (IBMP's) (Moscow, Russia) "Growth and Development of Higher Plants through Multiple Generations" program authority, using the "Substrate Technology Development Section"
- The Lada Space Growth Chamber, which was developed jointly by SDL and IBMP

SDL has a cooperative agreement with IBMP to conduct a joint experimental effort to satisfy the ORZS scientific and modeling requirements.

RESULTS
Initial samples from the ORZS investigation have been returned to Earth and are currently undergoing analysis by the investigator team. Data from the flight system of gas diffusion cells with different substrates will be compared with results from similar ground control gas diffusion cells to describe gas diffusion in microgravity (Jones, et al. 2005).

PUBLICATION(S)
Jones SB, Heinse R, Bingham GE, Or D. Modeling and Design of Optimal Growth Media from Plant-Based Gas and Liquid Fluxes, 2005, SAE International 2005-01-2949.

PHOTOSYNTHESIS EXPERIMENT AND SYSTEM TESTING AND OPERATION (PESTO)
Principal Investigator(s): Gary Stutte, Dynamac Corporation, NASA Kennedy Space Center, Fla.
Expedition 4

Research Area Plant Biology in Microgravity

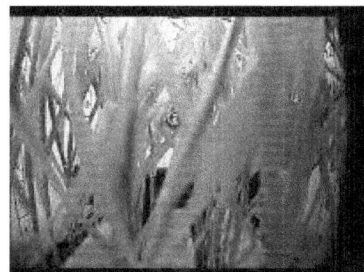

Video screen shot of *Triticum aestivum* 20 days after planting inside the BPS on station during Expedition 4.

The PESTO investigation, which was conducted in concert with technology verification for the BPS hardware, measured the canopy photosynthesis (the production of oxygen and carbohydrates from CO_2 and water in the environment) of *Triticum aestivum* (super dwarf wheat). The wheat was grown under high light and controlled CO_2 conditions in microgravity. The investigation also measured the metabolic effects on the photosynthetic apparatus to quantify the effects on metabolism and to model the impact of microgravity on biological approaches to atmospheric regeneration.

To test the hypothesis that the carbon exchange rates would be the same in microgravity as on Earth, investigators measured and characterized the CO_2 and light response curves for a wheat photosynthetic canopy that was grown in microgravity. Data came from various sources: gas samples that were taken from the closed atmosphere inside the chambers; liquid samples that were taken through ports in the chambers; plant tissue samples that were extracted by the crew during different points in the growth cycle; and, finally, plant tissue that was extracted from the live plants that were returned to Earth inside the BPS. The investigators analyzed the plant tissue postflight for primary photosynthesis parameters, such as electron transport, carbohydrate partitioning, and photosystem (the biochemical pathway for photosynthesis). Measurements were taken over a range of relative humidity conditions to discover whether atmospheric vapor pressure deficits affect gas exchange in microgravity.

RESULTS

During ISS Expedition 4, PESTO grew 32 plants for 73 days inside the plant growth chambers of BPS. Following return to Earth, these plants were compared to ground controls that were grown in BPS plant growth chambers on Earth.

ISS004E10138 — Close-up view of apogee wheat plants with a scale as backdrop to exhibit the development of the plants grown as part of the PESTO experiment on ISS Expedition 4.

The PESTO investigation supported research that resulted in a more complete picture of microgravity influences on photosynthesis, including chloroplast development, changes in gene expression, and resulting gas exchange functions, partitioning, and carbohydrate metabolism. CO_2 and light response curves allowed researchers to establish whether canopy photosynthetic responses were affected by space conditions. This is noteworthy since plants can be used to regenerate the atmosphere in space conditions through removal of CO_2 and production of oxygen. In addition, the tests that evaluated movement of water via transpiration are important since they are indicative of the stomatal responses that regulate photosynthesis. Further, the impact of microgravity on transpiration was significant since plants can be used to purify water under space flight conditions. These studies involving gas exchange at elevated CO_2 concentrations increased our understanding of the biological impacts of increasing levels of atmospheric CO_2 on Earth-based ecosystems. Furthermore, an understanding of plant responses under a range of CO_2 and light conditions has potential benefits to commercially controlled environment agriculture industries. Additional studies examined physiological changes in chloroplasts and evidence for changes in gene expression.

The growth and development of the dwarf wheat plants on the space station was similar to the growth and development of plants on Earth. Analysis of the plants indicated that the microgravity-grown plants were 10%

taller than plants grown on Earth, although the growth rate of dwarf wheat leaves was very similar to the plants that were grown on Earth. The near-real-time video data that were provided by BPS allowed for validation of the growth data in microgravity when compared to the controls. Design applications can be made to the BPS to allow for successful plant production on ISS and future long-duration missions to the moon and Mars (Stutte et al. 2003).

To effectively farm in space, multiple redundant plant growth chambers will be needed to acquire the maximum yield of food, oxygen, and water. PESTO evaluated the transpiration (water) and photosynthesis (oxygen) processes of the dwarf wheat plant in microgravity and found that microgravity did not affect either the transpiration or the photosynthesis processes of the plants (Monje et al. 2005).

When environmental controls such as temperature, relative humidity, CO_2, and water are effectively maintained, microgravity does not affect canopy growth of dwarf wheat plants. Slight differences in photosystem I (photosynthesis in which light of up to 700 nm is absorbed and reduced to create energy) and photosystem II (photosynthesis in which light of up to 680 nm is absorbed and its energy is used to split water molecules, giving rise to oxygen) were noted and are being evaluated further (Stutte et al. 2005).

Additional studies on leaf samples examined microgravity effects on leaf morphology, cell structure, cell function, and gene expression (Stutte et al. 2006). Some changes in morphology were seen in leaf cells and cell walls—leaves had thinner cross sections, chloroplasts were more ovoid, and cells were packed more densely. However, no changes were observed in gene expression or cell function, and levels of starch, soluble sugars, and lignins in the flight samples were statistically indistinguishable from the ground control samples. These results suggest that the physiological changes that were observed in the leaf cells did not affect the metabolic functions.

When conducting biological studies, it is important to maintain the integrity of the samples. The standard method to preserve samples is quick freezing at low temperatures (–80°C (–112°F) and below), but strict temperature control of samples on station is not always uniform or possible. Therefore a preservative is needed that will maintain the integrity of biological samples before cooling. RNAlater was used to preserve some of the PESTO samples on station. The viability of the samples that were preserved with RNAlater was greater than that of the samples that were preserved using formalin. To carry out long-term studies aboard ISS, a fixative such as RNAlater is needed to maintain the integrity of samples at the varying temperatures that are experienced in microgravity (Paul et al. 2005).

PUBLICATION(S)
Monje O, Stutte GW, Goins GD, Porterfield DM, Bingham GE. Farming in Space: Environmental and Biochemical Concerns. *Advances in Space Research*. 31:151–167, 2003.

Monje O, Stutte G, Chapman D. Microgravity does not alter plant stand gas exchange of wheat at moderate light levels and saturating CO_2 concentration. *Planta*. Online, Jun 2005.

Paul A, Levine HG, McLamb W, Norwood KL, Reed D, Stutte GW, Wells HW, Ferl RJ. Plant molecular biology in the space station era: Utilization of KSC fixation tubes with RNAlater. *Acta Astronautica*. 56:623–628, 2005.

Stutte GW, Monje O, Anderson S. Wheat (Triticum Aesativum L. cv. USU Apogee) Growth Onboard the International Space Station (ISS): Germination and Early Development. *Proceedings of the Plant Growth Regulation Society of America*. 30:66–71, 2003.

Stutte GW, Monje O, Goins GD, Tripathy BC. Microgravity effects on thylakoid, single leaf, and whole canopy photosynthesis on dwarf wheat. *Planta*. 1–11, 2005.

Stutte GW, Monje O, Hatfield RD, Paul A-L, Ferl RJ, Simone CG. Microgravity effects on leaf morphology, cell structure, carbon metabolism and mRNA expression of dwarf wheat. *Planta*. 224:1038–1049, 2006.

PLANT GENERIC BIOPROCESSING APPARATUS (PGBA)
Principal Investigator(s):   Gerard Heÿenga, NASA Ames Research Center, Moffett Field, Calif.
Expedition 5

Research Area Plant Biology in Microgravity

The PGBA is used to grow and monitor plants in microgravity experiments. PGBA is a self-contained plant growth chamber that provides preset or remotely controlled temperature, humidity, nutrient delivery, and light. The PGBA venting system also supplies the plants with ambient air and controls ethylene buildup.

The objective during Expedition 5 was to grow two crops of *Arabidopsis thaliana* (thale cress). The first crop was to be harvested when it reached maturity and placed into cold storage. The second crop was to be started at the harvest of the first crop and returned to Earth while it was still growing. Scientific objectives were focused on understanding lignin production.

Arabidopsis thaliana (Brassica family) plants grown under controlled conditions in a plant cultivation module in the Bioserve Laboratories.

RESULTS
The returned plant material did not develop in a normal manner, and the primary scientific objectives were not met. The study did, however, help to identify the need for greater regulation of air quality within a PGC to ensure uniform plant growth. Although no results will be published from this ISS activity, lessons learned from this study are being applied to the development of subsequent plant growth investigations and improved space flight plant chamber design (Heÿenga et al. 2005).

PUBLICATION(S)
Heÿenga G, Stodieck L, Hoehn A, Kliss M, Blackford C. Approaches in the Design of a Space Plant Cultivation Facility for *Arabidopsis thaliana*. 34th International Conference on Environmental Systems (ICES), Jul 2004, Colorado Springs, Colo., SAE-paper no. 2004-01-2459, 2004. (*New designs influenced by lessons learned*)

Heÿenga G, Kliss M, Blackford C. The Performance of a Miniature Plant Cultivation System Designed for Space Flight Application. 35th International Conference on Environmental Systems, Rome, Italy, SAE 2005-01-2844, 2005. (*New designs influenced by lessons learned*)

PASSIVE OBSERVATORIES FOR EXPERIMENTAL MICROBIAL SYSTEMS (POEMS)

Principal Investigator(s):   Michael Roberts, Ph.D., Dynamic Corporation, Cape Canaveral, Fla.
Expeditions: 13, 14

Research Area Microbiology

The Passive Observatories for Experimental Microbial Systems (POEMS) investigation was designed to demonstrate a passive system for microbial cultivation in the microgravity environment. The secondary objective of POEMS was to observe the generation and maintenance of genetic variation within microbial populations in microgravity.

POEMS supported experiments that determined the growth, ecology, and performance of diverse assemblages of microorganisms in the microgravity environment. These studies of microorganisms are required to maintain human health and bioregenerative function in support of NASA Exploration systems requiring ALS. For this experiment, replicate cultures were inoculated on the ground and launched on board the space shuttle. Half of the cultures were returned at the close of the shuttle mission; the remaining cultures were transferred to the ISS where they were preserved (frozen) at successive time-points over the course of 6 months. The cultures were returned to Earth and compared to ground controls to determine how microgravity affected the physiology and genetic composition of the microorganisms.

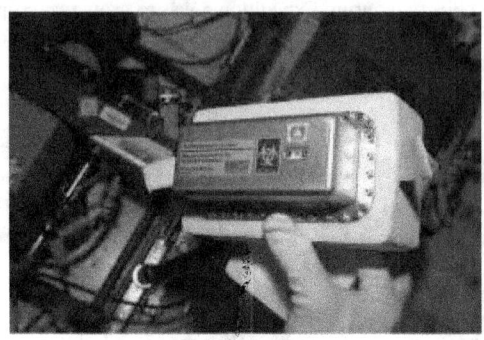

ISS014E05120 — Astronaut Michael E. Lopez-Alegria, Expedition 14 commander and ISS science officer, prepares to work with a POEMS sample container in the *Destiny* laboratory module.

Bacillus subtilis was introduced by a small liquid inoculum into the void volume of an Opti-cell during preflight integration, thereby providing a uniform inoculation front to the semi-solid media. Bacterial growth proceeded in a planar fashion down the long axis of the culture chamber, thereby providing a physically traceable chronology of bacterial growth during flight. As bacteria propagated through the media, they encountered genomic DNA from related *Bacillus* species that have been marked with an antibiotic resistance gene. Some cells have the capability to bind, transport, and recombine the DNA via transformation. By assimilating the resistance marker, these transformants naturally obtain a competitive advantage in the selective medium. The rate and extent of genetic transformation during growth in flight is compared to ground controls. Genetic and physiological screening of isolates identifies changes during flight in vegetative cells. If cells form spores during flight due to nutrient exhaustion or other environmental stresses, the spores are isolated for analysis. Post-flight analyses seek to understand the effects of space flight on DNA repair mechanisms, the ability to assimilate exogenous DNA, and the ecology of a diversifying assemblage of bacteria cells in the space flight environment.

RESULTS

Ten POEMS canisters were launched on STS-121 in a middeck locker (ambient environment). As a passive, sortie experiment, five canisters were returned on the same shuttle flight and represent a data point for approximately 11 to 12 days of exposure to space flight. The remaining canisters were transferred to an ambient location on ISS. Of these, three canisters were successively transferred into the Minus Eighty Degree Laboratory Freezer for ISS (MELFI) freezer and maintained at –68°C (–90°F) or colder. Two of the frozen canisters returned home on STS-115, and the third canister, as well as one of the ambient canisters, returned on STS-116.

POEMS performed as expected, and data (temperature, thermal offsets and humidity) have been recovered from data-loggers. Bacterial cells and genetic transformants were recovered in all returned canisters. Fixed gases and volatile organic compound analyses of canister headspace are under way.

ISS014E05124 — Expedition 14 commander and NASA science officer, Astronaut Michael E. Lopez-Alegria, is shown inserting a sample container for the POEMS payload in the MELFI in the *Destiny* laboratory of the ISS. MELFI is a low-temperature freezer facility with nominal operating temperatures of –80°C (–112³F), –26°C (–15°F), and 4°C (39°F) that will preserve experiment materials over long periods.

The investigation team continues to isolate and characterize rifampicin-resistant bacteria and to enumerate auxotroph mutants (lysine methionine tryptophan) that ere recovered from sortie mission canisters (five canisters) and three canisters from the ISS mission that were returned on STS-115. Viable cell and spore counts are completed; total direct counts and genetic characterization of transformants is ongoing. Transformation efficiency was lower than expected in both flight and ground experiments, but rates were sufficient to yield >102 rpoB transformants. Data analysis will continue for 6 months after recovery of all POEMS canisters from ISS.

Trends in early analyses suggest that microbe population densities and transformation rates may be slightly elevated in flight samples compared to ground controls, but the preliminary conclusion is that the effects of the space environment on the rate of horizontal gene transfer are not statistically significant for *Bacillus subtilis*. Full data analysis is pending.

PUBLICATIONS(S)
Roberts MS, Reed DW, Rodriguez JI. 2005. Passive Observatories for Experimental Microbial Systems (POEMS): Microbes Return to Flight. 35th Annual International Conference of Environmental Systems (ICES). SAE 05ICES-214.

STREPTOCOCCUS PNEUMONIAE EXPRESSION OF GENES IN SPACE (SPEGIS)

Principal Investigator(s):   David W. Niesel, Ph.D., University of Texas Medical Branch at Galveston, Texas
Expedition 15

Research Area Microbiology

The *Streptococcus pneumoniae* Expression of Genes in Space (SPEGIS) experiment investigated *S. pneumoniae* gene expression and protein production in the space environment. *S. pneumoniae* is an important human pathogen (disease causing) and the leading cause of bacterial pneumonia (inflammation of the lungs with congestion), meningitis (inflammation of the membrane covering the brain and spinal cord), and otitis media (ear infection). Importantly, this pathogen has been isolated previously from the crew preflight, and related bacteria are found in the spacecraft environment. Experiments were performed to identify and characterize *S. pneumoniae* genes and proteins, which are differentially expressed in response to the space environment, and compare microgravity-induced genes and proteins to those expressed during postflight rodent infection.

S. pneumoniae is a respiratory microbe that is normally found in the upper respiratory tract of approximately 40% of the healthy human population. The identification of specific *S. pneumoniae* virulence factors and cellular and molecular processes may aid scientists in furthering the understanding of how this bacteria causes infection. These data may aid in the design and development of new antimicrobial drugs. This experimental approach will result in new information about a significant human pathogen, add to our knowledge about the *S. pneumoniae* pathogenic mechanism, and provide basic information on the bacterial model system of gene and protein expression in the space environment.

RESULTS
SPEGIS is a recently completed investigation that is undergoing data analysis; final results are pending.

PUBLICATION(S)
There are no publications at this time.

StelSys Liver Cell Function Research (StelSys)

Principal Investigator(s): Albert Li, StelSys LLC, Baltimore, Md.
Expedition 5

Research Area Cellular Biology and Biotechnology

The liver filters potentially harmful substances from the blood and breaks these substances down into water-soluble forms that can be washed from the body. The liver is, therefore, a difficult organ to treat because medications can be broken down and removed before they have an opportunity to provide effective treatment. The purpose of this experiment was to allow the investigator to observe how human liver cells react to the presence of drugs in microgravity, and to compare these results to a control experiment that was conducted on the ground.

Human liver cells were launched inside a caddy that was held at freezing temperatures within a Dewar. The experiments were conducted in the CBOSS, including the biotechnology specimen temperature controller (BSTC), the gas supply module, and syringes. Individual cell cultures were grown in the temperature-controlled environment of the BSTC. When the experiments were complete, they were stored in the ARCTIC freezer until the end of the Expedition.

Results

The samples that were returned from space were analyzed by specialized mass spectrometry equipment to determine the amount of drug metabolites that were formed by the liver cells from the drug substances added. Overall this analysis showed that the rate of metabolism by the liver cells in space was lower than that of the liver cells that were maintained under similar conditions on Earth. This was true for all of the drug substances tested as well as for cells from three different liver donors. These results indicate that microgravity may well retard the rate of drug metabolism in the human liver, although the mechanism for this effect is yet unknown.

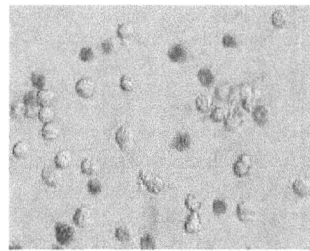

Human liver cells at the completion of a ground-control experiment. A stain has been applied to indicate the live vs. dead (blue-colored) cells.

Returned samples were also analyzed by gene array to determine whether genetic expression differed for cells in microgravity. Differences were found, including 9,200 of 13,000 genes that had at least two-fold greater expression in space as compared to Earth and 9,800 genes that had decreased expression in space. This large body of data is being analyzed for clues as to how liver cell function changes in specific ways in the microgravity environment of space.

Publication(s)

There are no publications at this time.

ANALYSIS OF A NOVEL SENSORY MECHANISM IN ROOT PHOTOTROPISM (Tropi)

Principal Investigator(s): John Kiss, Ph.D., Miami University, Oxford, Ohio
Expedition 14

Research Area Plant Biology in Microgravity

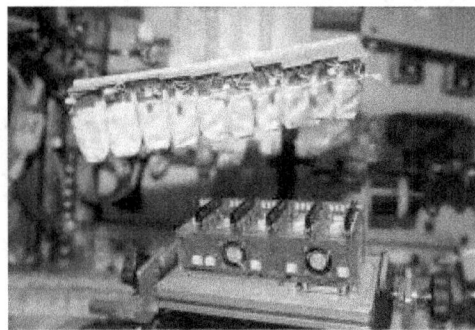

ISS014E10652 — View of the Tropi seedling cassette for the EMCS experiment container (EC) in the *Destiny* laboratory module during Expedition 14.

Tropi consists of dry *Arabidopsis thaliana* (thale cress) seeds that are stored in small seed cassettes. *A. thaliana* is a rapidly growing, flowering plant in the mustard family. The seed cassettes were contained inside the EMCS. The seeds were installed into the EMCS dry and at ambient temperature until hydrated by an automated system of the EMCS. At specified times during the experiment, the plants were stimulated by different light spectrums and different gravity gradients. The only operations that were required by the crew were to replace video tapes and harvest the plants when they were grown. Once the plants were harvested, they were stored in the MELFI until their return to Earth (STS-116 and STS-117). Part of the experiment relies on a biochemical analysis of the returned plants, but the investigators also gathered data from videotaped images of the plants, watching their roots as they developed in the EMCS.

RESULTS

Preliminary analysis of the Tropi samples that were returned on STS-116 in Dec 2006 resulted in little yield of total RNA, which is responsible for the transfer of information from DNA. However, samples that were returned on STS-117 in Jun 2007 resulted in good-quality RNA. For the samples that were returned on STS-117, a 3-minute requirement was established: the seedlings in the Tropi cassettes would be transferred from MELFI to the NASA Johnson Space Center cold bag within 3 minutes on the ISS. This new requirement resulted in the success of the cold transfer procedures and a successful yield of RNA.

PUBLICATION(S)

There are no publications at this time.

YEAST-GROUP ACTIVATION PACKS (Yeast-GAP)

Principal Investigator(s): Cheryl A. Nickerson, Arizona State University, Tempe, Ariz.
Expedition 8, 13

Research Area Microbiology

This experiment was designed to study how individual genes respond to microgravity conditions. To achieve this, scientists studied yeast cells—eukaryotic cells, or cells that contain a distinct nucleus bound by a cell membrane. Mammalian cells have a similar eukaryotic structure, and the results of this experiment could aid in understanding more complex mammalian cell response to microgravity. Yeast cells are far simpler than mammalian cells because they have a well-characterized, much-smaller genome. This makes it easier for scientists to study how microgravity alters the makeup of the cells and their potential function.

Yeast is an ideal candidate for such a study because it is hardy enough to resist the rigors of flight, requires no refrigeration, and poses little risk to ISS crewmembers. The experiment used genetically engineered cells of brewer's yeast (*Saccharomyces cerevisiae*) and a special cell growth chamber called a GAP, which was developed by BioServe Space Technologies. The goal is to identify the precise genes of yeast that are affected by growth in microgravity to understand differences in the growth of yeast cells in space and on Earth.

RESULTS
Samples were returned on space shuttle flight STS-114/LF1 in Aug 2005. Samples also flew as a sortie mission on STS-115 and returned in Sep 2006. Further analysis is ongoing.

PUBLICATION(S)
There are no publications at this time.

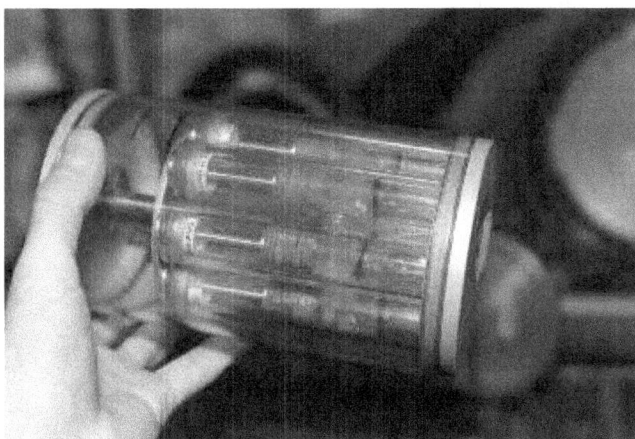

ISS008E14397 — ISS science officer Mike Foale is holding the GAP for the Yeast-GAP experiment during Expedition 8.

Human Research and Countermeasure Development for Exploration

ISS is being used to study the risks to human health that are inherent in space exploration. Many research investigations address the mechanisms of the risks—the relationship to the microgravity and radiation environments—and other aspects of living in space, including nutrition, sleep, and interpersonal relationships. Other experiments are used to develop and test countermeasures to reduce these risks. Results from this body of research are critical enablers for missions to the lunar surface and future Mars exploration missions.

The overarching strategy that guides the ISS-based Human Research experiments focuses on filling in specific knowledge gaps and testing proposed countermeasures to determine their effectiveness and evaluate their operational feasibility for mitigating known human health problems in space.

Over the first 15 Expeditions, 32 research investigations were initiated in the discipline of Human Research and Countermeasure Development for Exploration. These studies are complemented by ground-based studies. Most investigations require large numbers of subjects; many of the early experiments are ongoing. Vetted results will continue to grow over the coming years.

These experiments fall into sub-disciplines that address specific element of human health in space. The table provides a breakdown of the Human Research and Countermeasure Development experiments by sub-discipline.

Physiological Studies: Bone and Muscle

Astronauts experience significant physiological effects on bone and muscle after spending months in a microgravity environment. To date, the primary countermeasure to loss of bone and muscle while on space station has been exercise. Exercise equipment that was deployed on ISS is more mature and the prescribed exercise regimens are more rigorous than for any previous U.S. space missions. Early station studies have evaluated the effects of long-duration space flight on bone and muscle in the context of these exercise protocols. The transit to Mars from Earth will be equivalent in duration to the typical 6-month space station Expedition. This allows us to draw insight into the health status that can be expected when a crew lands on Mars.

Physiological Studies: Cardiovascular and Respiratory Systems

When astronauts move between different gravity environments, a number of acute physiological responses in their cardiovascular and respiratory systems can affect their overall health and performance. Studies of the process of adapting to changes in gravity are important for mission success for future exploration missions as astronauts may transition from Earth to interplanetary transit, to the moon or Mars, and back. Other physiological changes in microgravity—from changes in immune function, pharmacology, and clinical diagnostic measures—are also key areas of study.

Integrated Physiological Studies: Immune, Neurological, and Vestibular Systems

The complex, integrated changes that are observed in astronaut physiology result in other changes in astronaut health, including diminished immune systems, new visual-neural connections, and more. Several investigations are designed to document the more complex cause-effect relationships in overall crew health during long-duration tours on the ISS.

Behavior and Performance Studies

Long-duration space missions, by their very nature, exert many pressures on groups and individuals that could compromise mission success. Studies of the behavior of individuals and teams under the conditions of space flight are important for determining the best team composition and interaction models for future Exploration missions.

Radiation Studies

Cosmic radiation is one of the most critical risks to human health in space flight. Once we venture beyond the protective shielding of the Earth's atmosphere, we are exposed to a wider spectrum of radiation that does not normally threaten us. Exposure to radiation found in LEO and beyond can cause cataracts and cancer, damage the reproductive organs and nervous system, and cause genetic damage.

Human Research and Countermeasure Development Experiments by Sub-discipline

Physiological Studies: Bone and Muscle
Biopsy (Effect of Prolonged Space Flight on Human Skeletal Muscle)
CTBM (Commercial Biomedical Testing Module: Effects of Osteoprotegerin on Bone Maintenance in Microgravity)
CBTM-2 (Commercial Biomedical Test Module-2)
Foot (Foot Reaction Forces During Space Flight)
H-Reflex (Effects of Altered Gravity on Spinal Cord Excitability)
HPA (Hand Posture Analyzer)
Renal Stone (Renal Stone Risk During Space Flight: Assessment and Countermeasure Validation)
Subregional Bone (Subregional Assessment of Bone Loss in the Axial Skeleton in Long-term Space Flight)
Physiological Studies: Cardiovascular and Respiratory Systems
CCISS (Cardiovascular and Cerebrovascular Control on Return from ISS)
Midodrine-SDBI (Test of Midodrine as a Countermeasure Against Post-flight Orthostatic Hypotension – Short-Duration Biological Investigation)
PuFF (The Effects of EVA and Long-Term Exposure to Microgravity on Pulmonary Function)
Xenon1 (Effect of Microgravity on the Peripheral Subcutaneous Veno-Arteriolar Reflex in Humans)
Immune System and Integrated Studies
ADUM (Advanced Diagnostic Ultrasound in Microgravity)
CCM-Immune Response (Cell Culture Module – Immune Response of Human Monocytes in Microgravity)
CCM-Wound Repair)Cell Culture Module – Effect of Microgravity on Wound Repair: In Vitro Model of New Blood Vessel Development)
Epstein-Barr (Space Flight-Induced Reactivation of Latent Epstein-Barr Virus)
Integrated Immune (Validation of Procedures for Monitoring Crewmember Immune Function)
Integrated Immune-SDBI (Validation of Procedures for Monitoring Crewmember Immune Function – Short-Duration Biological Investigation)
Latent Virus (Incidence of Latent Virus Shedding During Space Flight)
Nutrition (Nutritional Status Assessment)
Repository (National Aeronautics and Space Administration Biological Specimen Repository)
SWAB (Surface, Water and Air Biocharacterization – A Comprehensive Characterization of Microorganisms and Allergens in Spacecraft Environment)
Neurological and Vestibular Systems
ELITE-S2 (ELaboratore Immagini TElevisive – Space 2)
Mobility (Promoting Sensorimotor Response Generalizability: A Countermeasure to Mitigate Locomotor Dysfunction After Long-Duration space Flight)
PMZ (Bioavailability and Performance Effects of Promethazine During Space Flight)
Behavior and Performance Studies
Interactions (Crewmember and Crew-Ground Interaction During International Space Station Missions)
Journals (Behavioral Issues Associated with Isolation and Confinement: Review and Analysis of ISS Crew Journals)
Sleep-Long (Sleep-Wake Actigraphy and Light Exposure During Spaceflight-Long)
Sleep-Short (Sleep-Wake Actigraphy and Light Exposure During Spaceflight-Short
Radiation Studies
ALTEA (Anomalous Long-Term Effects in Astronauts' Central Nervous System)
BBND (Bonner Ball Neutron Detector)
Chromosome (Chromosomal Aberrations in Blood Lymphocytes of Astronauts)
DOSMAP (Dosimetric Mapping)
EVARM (A Study of Radiation Doses Experienced by Astronauts in EVA)
Stability (Stability of Pharmacotherapeutic and Nutritional Compounds)
Torso (Organ Dose Measurement Using the Phantom Torso)

ADVANCED DIAGNOSTIC ULTRASOUND IN MICROGRAVITY (ADUM)

Principal Investigator(s): Scott A. Dulchavsky, Henry Ford Health System, Detroit, Mich.
Expeditions 8–11

Research Area Immune System and Integrated Studies

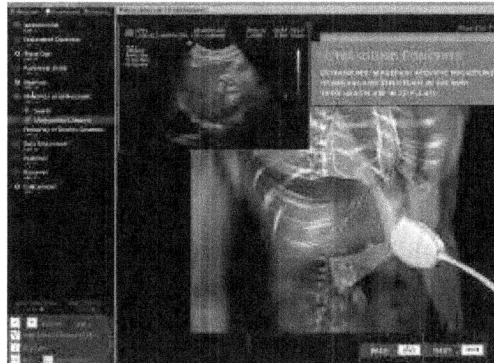

Screen shot of the ADUM on-board proficiency enhancer that is used to teach crewmembers how to conduct an ultrasound scan of a new organ or system.

Advanced Diagnostic Ultrasound in Microgravity (ADUM) tests the accuracy of using ultrasound technology in the novel clinical situation of space flight. This investigation includes assessing health problems in the eyes and bones, as well as sinus infections and abdominal injuries. ADUM further tests the feasibility of using an in-flight ultrasound to monitor bone density during long-duration space flights. Another objective of the experiment is determining how well nonmedical crewmembers can learn to use an ultrasound device with CD-ROM training manuals and remote guidance from Earth. The intent of the ADUM investigation is to develop methods by which an individual who is untrained in medicine can use an ultrasound machine with remote diagnostician assistance to evaluate a vast array of medical problems.

Expedition crews used the ISS HRF ultrasound machine and four scan sets: the cardio/thoracic scan, which focuses on the heart but also can scan the lungs; the abdominal/retroperitoneal scan, which focuses on the organs of the abdomen, including the liver, spleen, kidneys, and bladder; the dental scan, which can image the mouth, teeth, gums, facial bones and sinuses, and eyes; and the bone scan, which images bones and characterizes bone loss during flight. In addition to the ultrasound machine and probes, another key component of ADUM on station is the on-board proficiency enhancer (OPE)—a software application that is used to train crewmembers on the methods that are employed for each scan.

RESULTS

The ISS crews, which began their work with ADUM on Expedition 8 and completed it during Expedition 11, have demonstrated that minimal training along with audio guidance from a certified sonographer can produce ultrasound imagery of diagnostic quality. The ISS crewmembers, acting as operators and subjects, have completed comprehensive scans of the cardiothoracic and abdominal organs as well as limited scans of the dental, sinus, and eye structures. They also have completed multiple musculoskeletal exams, including a detailed exam of the shoulder muscles. To date, analyses of ultrasound video downlinked to ground teams at the NASA Johnson Space Center have yielded excellent results that are beginning to appear in the scientific literature.

Ultrasound technology is now deployed in many trauma centers around the world as a first-line diagnostic procedure; results have been accurate even when performed by non-radiologists. The use of ultrasound technology as a diagnostic tool on the ISS required an OPE program, visual cue cards, procedures, and direction from ground-based trained radiological personnel. The Expedition 8 crew was able to capture high-fidelity images of the thoracic, cardiac, and vascular systems with minimally trained nonmedical personnel. This investigation has laid the groundwork for using ultrasound as a diagnostic tool in microgravity and remote locations on Earth when a physician is not readily available. A scientific paper discussing these results was submitted by the crewmembers directly from orbit (Foale et al. 2005).

Ultrasound images of the shoulder during Expedition 9 showed that ultrasound that was performed by crewmembers obtained diagnostic-quality imagery for evaluation of shoulder integrity. An application of this technology would be if a crewmember were to injure his/her shoulder during a strenuous EVA (or spacewalk), these techniques would allow evaluation and diagnosis of possible injuries (Fincke et al. 2005).

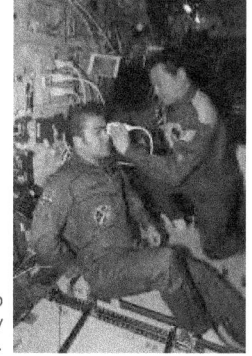

ISS010E18770 — ISS commander and science officer Leroy Chiao performs an ADUM scan on the eye of flight engineer Salizhan Sharipov during ISS Expedition 10.

Following a traumatic event to the head or face, eye examination is a very important component of the physical examination. This examination may be difficult due to significant orbital or facial swelling. The Expedition 10 crew used ultrasound technology to examine the eye through a closed eyelid. This examination could determine a number of problems with the eye that are signs of other, more significant trauma to the head (Chiao et al. 2005).

In addition to the importance of establishing ultrasound techniques for examination and diagnosis on ISS, this study is establishing ultrasound as a key tool for clinical medicine on future vehicles, the moon, and, eventually, Mars. The success of ADUM may also lead to additional applications of ultrasound on Earth. The remote guidance paradigm can be adapted on Earth for patients in rural/remote areas, disaster relief, and the military. Using existing communication systems, a person (e.g., nurse, physician's assistant, military medic) who is minimally trained in ultrasound could perform an ultrasound exam on a patient with guidance from an expert at a medical facility hundreds or thousands of miles away. This would expand the tools for the rural medical community, provide the ability to triage a mass casualty, and help in the decisions to conduct medical transport of patients. This capability was demonstrated with a National Hockey League team and at the Olympic Training Facilities. Trainers were provided training that was similar to that received by ISS astronauts. Images of various locations (groin, knee, elbow, etc.) from athletes were obtained and transmitted for diagnostic interpretation by remotely located experts (Kwon et al. 2007).

PUBLICATION(S)

Chiao L, Sharipov S, Sargsyan AE, Melton S, Hamilton DR, McFarlin K, Dulchavsky SA. Ocular examination for trauma; clinical ultrasound aboard the International Space Station. *Journal of Trauma*. 58(3):885-889, 2005.

Fincke EM, Padalka G, Lee D, van Holsbeeck M, Sargsyan AE, Hamilton DR, Martin D, Melton SL, McFarlin K, Dulchavsky SA. Evaluation of Shoulder Integrity in Space: First Report of Musculoskeletal US on the International Space Station. *Radiology*. 234(2):319–322, 2005.

Foale CM, Kaleri AY, Sargsyan AE, Hamilton DR, Melton S, Margin D, Dulchavsky SA. Diagnostic instrumentation aboard ISS: just in time training for non-physician crewmembers. *Aviation Space and Environmental Medicine*. 76:594–598, 2005.

Sargsyan AE, Hamilton DR, Jones JA, Melton S, Whitson PA, Kirkpatrick AW, Martin D, Dulchavsky SA. FAST at MACH 20: Clinical Ultrasound Aboard the International Space Station. *The Journal of Trauma, Injury, Infection, and Critical Care*. 2004; 58(1):35–39.

Kwon D, Bouffard JA, van Holsbeeck M, Sargsyan AE, Hamilton DR, Melton SL, Dulchavsky SA. Battling fire and ice: Remote guidance ultrasound to diagnose injury on the International Space Station and the ice rink. *American Journal of Surgery*. 2007; 193(3):417–20.

Anomalous Long-term Effects in Astronauts' Central Nervous System (ALTEA)

Principal Investigator(s):   Livio Narici, Ph.D., University of Rome Tor Vergata and INFN, Rome, Italy
Expeditions: 13–15

Research Area Radiation Studies

Long-duration space flights result in increased cosmic radiation exposure to astronauts. The Anomalous Long-term Effects in Astronauts' Central Nervous System (ALTEA) hardware is designed to measure particle radiation in the space environment, and determine how this radiation impacts the CNS of the crew. The experiment is comprised of a helmet-shaped device holding six silicon particle detectors that are designed to measure cosmic radiation passing through the brain. The detectors measure the trajectory, energy, and species of individual ionizing particles. At the same time, an electroencephalograph (EEG) will measure the brain activity of the crewmember to determine whether radiation strikes cause changes in the electrophysiology of the brain in real time.

ISS014E16208—An astronaut wears the ALTEA experiment helmet while conducting the experiment in the *Destiny* Laboratory module.

A common effect of radiation exposure that is reported by astronauts is the perception of light flashes. The actual mechanism of these light flashes is not understood. Earlier studies on the *Mir* space station suggest that both heavy nuclei and protons trigger abnormal CNS responses (Casolino et al. 2003). A visual stimulator tests the astronaut's overall visual system, including dark adaptation stimuli to monitor visual status. While not manned, the ALTEA hardware provides a continuous measure of the cosmic radiation in the station's U.S. laboratory *Destiny*. The neurophysiological effects of cosmic radiation in long-term space travel have never been explored with the depth of the ALTEA experiment. Data that are collected will help to quantify risks to astronauts on future long-duration space missions and propose optimized countermeasures.

Results

Preliminary data that were compiled from three Expeditions, are currently in being fully analyzed. A long run of dosimetry data was collected during Expeditions 13 and 14; including definable particle and solar events. During Expedition 13, the visual stimulator malfunctioned, resulting in a loss of data from this instrument. During Expedition 14, variable results were obtained from different crewmembers. Overall, the crew reported a lower frequency of light flash events than expected; this was anecdotally attributed to a lack of dark adaptation. The astronauts did perceive a higher rate of light flashes in their sleeping quarters.

Publication(s)

There are no publications at this time.

BONNER BALL NEUTRON DETECTOR (BBND)

Principal Investigator(s): Tateo Goka, Japan Aerospace Exploration Agency, Tokyo, Japan
Expeditions 2, 3

Research Area Radiation Studies

The Bonner Ball Neutron Detector (BBND), which was developed by the Japan Aerospace Exploration Agency (JAXA), was used inside the ISS to measure the neutron energy spectrum. It consisted of several neutron moderators enabling the device to discriminate neutron energies up to 15 mega electron volts (MeV). This BBND characterized the neutron radiation on ISS during Expeditions 2 and 3.

Due to their neutral charge, neutrons are not affected by magnetic forces, which allows the particles to penetrate deeper into materials. Potentially, the particles may deposit a larger fraction of their energy into human tissue than charged-particle radiations. Neutrons in the space environment are present in galactic background and solar events (primary particles), neutrons are also produced by the interaction of particles in the materials that are used to construct spacecraft (secondary particles). High-energy secondary neutrons are produced by the interactions of high-energy charged particles with spacecraft materials and planetary surfaces.

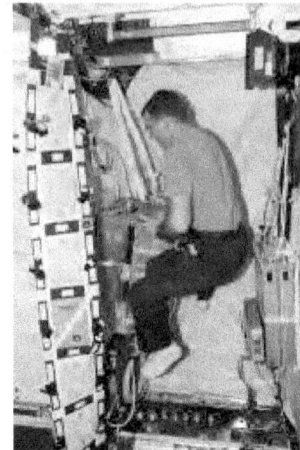

ISS002E5714 — Astronaut James S. Voss, Expedition 2 flight engineer, sets up the BBND inside the U.S. *Destiny* laboratory. He holds the control unit (nearest the wall) and the detector unit (DU) (in his hands).

BBND experiment hardware consisted of two assemblies: the BBND control unit, which stored radiation measurements and controlled data quality; and the BBND DU, which measured neutron radiation via a series of six stainless-steel spherical shells. These six spheres were divided as follows: four spheres were thermal neutron detectors that were covered in polyethylene of different thickness; one detector was covered in gadolinium; and one detector was uncovered. The gadolinium-covered sphere acted as a control; neutrons are unable to penetrate the dense gadolinium. The data that were collected by the uncovered sphere were used to determine the difference between pulses that were created by neutrons and those that were created by protons. Data collected from the polyethylene-covered spheres show that the amount of hydrogen that was surrounding the detector affects the amount of radiation penetration.

RESULTS

BBND characterized the neutron radiation on ISS during Expeditions 2 and 3 and determined that galactic cosmic rays were the major cause of secondary neutrons measured inside ISS. The neutron energy spectrum was measured from Mar 23, 2001 through Nov 14, 2001 in the U.S. Laboratory module of the ISS. The time frame enabled neutron measurements to be made during a time of increased solar activity (solar maximum) and allowed results to be derived from observation of a solar flare on Nov 4, 2001.

BBND results showed that the overall neutron environment at the ISS orbital altitude was influenced by highly energetic galactic cosmic rays, except in the South Atlantic Anomaly (SAA) region where protons that are trapped in the Earth's magnetic field cause a more severe neutron environment. However, in the data that were obtained by BBND, the number of particles measured per second per square centimeter (cm^2) per MeV is consistently lower than that of precursor investigations. The average dose-equivalent rate that was observed throughout the investigation was 3.9 micro Sieverts/hour (Sv/hr) or about 10 times the rate of radiological exposure to the average U.S. citizen. In general, radiation damage to the human body is indicated by the amount of energy that is deposited in living tissue modified by the type of radiation that is causing the damage; this is measured in units of Sieverts. The background radiation dose that is received by the average person in the U.S. is approximately 3.5 milliSv/yr. Conversely, an exposure of 1 Sv/hour can result in radiation poisoning, and a dose of 5 Sv/hr will result in death in 50% of exposed individuals. The average dose-equivalent rate that was observed through the BBND investigation is 3.9 micro Sv/hr, or about 10 times the average U.S. surface rate. The highest rate, which was 96 microSv/hr. was observed in the SAA region.

The Nov 4, 2001 solar flare and its associated geomagnetic activity caused the most severe radiation environment that was experienced inside the ISS during the BBND experiment. The increase of neutron dose-equivalent due to those events was evaluated to be 0.19 mSv, which is less than 1% of the measured neutron dose-equivalent that was measured over the entire 8-month period.

Although this experiment did not characterize the neutron radiation environment outside of Earth's magnetic field, the BBND sampling equipment provided results without the return of equipment to Earth and proved that similar measurement systems could be used on missions to the moon and Mars to monitor real-time radiation risks (Expedition 2 and 3 One Year Postflight Report).

PUBLICATION(S)
Koshiishi H, Matsumoto H, Koga K, Goka T. Evaluation of Low-Energy Neutron Environment inside the International Space Station. Technical Report of Institute of Electronics, Information, and Communications Engineers SANE2003-79:11–14, 2003. [*in Japanese*]

Effect of Prolonged Space Flight on Human Skeletal Muscle (Biopsy)

Principal Investigator(s): Robert Fitts, Marquette University, Milwaukee, Wis.
Expeditions 5–7, 9–11

Research Area Physiological Studies: Bone and Muscle

ISS004E6331 – Expedition 4 commander Yury Onufrienko exercises on a treadmill in the *Zvezda* service module.

It is well established that space flight can result in loss of skeletal muscle mass and strength know as atrophy. This condition continues throughout a crewmember's mission, even if crewmembers adhere to a strict exercise regime. What researchers do not understand, however, are the effects that prolonged stays in microgravity have on skeletal muscles. Biopsy evaluated changes in calf muscle function over long-duration space flights (30 to 180 days).

For the Biopsy investigation, a specially designed torque velocity dynamometer was used to measure muscle strength before and after flight. Biopsies were also taken from the calf muscles (soleus and gastrocnemius) of the participants. These tissue samples allowed for determination of the cell size and structural properties of individual fast and slow muscle fibers. Chemical analysis of the biopsies determined the muscle fiber structural changes involving myosin, which is a protein "molecular motor" that drives muscle contractions and cell divisions, enzymes, and substrates. Electron microscopy was used to determine the relationship between thick and thin filament, the amount of myofilament loss, and changes in membrane-associated protein complexes that are found in skeletal muscle fibers and connective tissue and that help the muscle resist stretch-induced damage

Results

Preliminary results were presented at the 2004 American Physiological Society Intersociety Meeting: Integrative Biology of Exercise in three abstracts (see Publications on next page). Summarizing data that were collected from the first five subjects, microgravity produced a 47% decrease in the peak power of postflight muscle fiber samples compared to preflight muscle fiber samples. This decrease was due to the combined effects of reduced fiber size and a decline in the size of the myofibrils that make up the fiber.

Further examination of the data that were collected from the crew indicated that astronauts who performed high treadmill exercise (greater than 200 minutes/week) vs. low treadmill exercise (less than 100 minutes/week) exhibited a smaller decrease in peak power. Astronauts who performed high treadmill exercise showed a 13% decrease compared to a 51% decrease in peak power of astronauts who performed low treadmill exercise. Sample analysis of the muscle fibers indicated that the ratio of myosin and actin proteins in the muscle fibers was not affected by long-duration space flight. Although exercise slowed the onset of atrophy and loss of strength in muscle fibers, a significant amount of muscle volume and strength loss still occurred on long-duration missions.

Of the exercise countermeasures that are currently being employed, treadmill exercise appeared the most effective in protecting the calf muscles from loss of strength and atrophy. Final publication is pending an analysis of data from all of the subjects.

PUBLICATION(S)

Fitts RH, Romatowski JG, Lim W-M, Gallagher P, Trappe S, Costill D, Riley DA. Microgravity, exercise countermeasures, and human single muscle fiber function (Abstract). Paper 21.16, American Physiological Society Intersociety Meeting: Integrative Biology of Exercise. *The Physiologist.* 47(4):320, 2004.

Gallagher P, Trappe S, Costill D, Riley DA, LeBlanc A, Evans H, Peters JR, Fitts RH. Human muscle volume and performance: the effect of 6-mo of microgravity (Abstract). Paper 21.19, American Physiological Society Intersociety Meeting: Integrative Biology of Exercise. *The Physiologist.* 47(4):321, 2004.

Riley DA, Bain JL, Gallagher P, Trappe S, Costill D, Fitts RH. The effect of 6-mo microgravity on human skeletal muscle structure (Abstract). Paper 21.17, American Physiological Society Intersociety Meeting: Integrative Biology of Exercise. *The Physiologist.* 47(4):320, 2004.

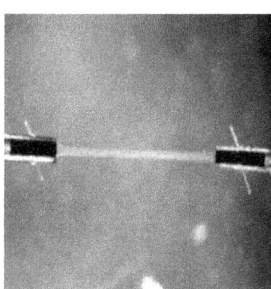

A single muscle fiber. Each muscle is composed of thousands of these fibers. Samples of muscles fibers will be extracted and tested as part of the Biopsy experiment. Image courtesy of NASA JSC.

Commercial Biomedical Testing Module (CBTM): Effects of Osteoprotegerin (OPG) on Bone Maintenance in Microgravity

Principal Investigator(s):   Ted Bateman, Clemson University, Clemson, S.C.
Expedition 4

Research Area Physiological Studies: Bone and Muscle

Osteoporosis is a debilitating disease that causes a reduction in bone mineral density that leads to weakened bones. One of the physiological changes that is experienced by crews during space flight is the accelerated loss of bone mass due to the lack of gravitational loading on the skeleton—a loss that is similar to the osteoporosis that is experienced by the elderly population on Earth. Osteoprotegerin (OPG), which is a bone metabolism regulator, was evaluated by the FDA as a new treatment for osteoporosis.

The Commercial Biomedical Testing Module (CBTM) examined the effects of OPG on bone maintenance in microgravity using aged mice (older than 9 months) as test subjects. Bone changes that were observed in older mice closely reflect the bone changes that are observed in older humans. The mice were housed in three animal enclosure modules (AEMs) that provide the animal subjects with everything necessary to maintain health. Half of the mice were treated with OPG (a novel protein that regulates bone resorption) and half were treated with a placebo.

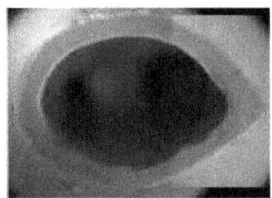

Fluorescent image of femur diaphysis from space flight placebo-treated mouse, indicating greatly decreased bone formation (calcein label indicates where bone was forming at the time of launch, which allows quantification of bone formation rates during flight).

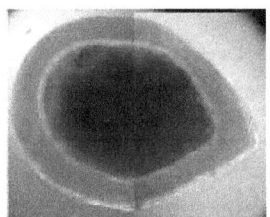

Fluorescent image of femur diaphysis from ground control placebo-treated mouse, indicating greatly decreased bone formation (calcein label indicates where bone was forming at the time of launch, which allows quantification of bone formation rates during flight).

Results

During ISS Expedition 4, 24 female mice were flown to ISS on shuttle flight STS-108 in three AEMs. The AEMs remained on STS-108 throughout the 12-day mission.

Mice that were exposed to microgravity exhibited a 15% to 20% decline in femur elastic strength and a 40% to 60% decrease in bone formation when compared to the controls. The femur elastic strength decline was caused by three mechanisms: reduced bone formation, increased bone resorption, and inhibition of mineralization. Mice that were treated with OPG and before being exposed to microgravity exhibited no discernable decline in femur elastic strength, and bone resorption was significantly increased (Bateman 2004).

Mechanical testing data were complimented by serum, mRNA, and histological analyses that indicated a decline in bone formation and an increase in bone resorption in addition to an inhibition of mineralization. OPG mitigated the decline in mechanical strength by preventing increase in resorption and maintaining mineralization. In addition to this detailed analysis of skeletal properties, a secondary analysis of calf muscles from placebo-treated specimens was performed to collect baseline data to validate space-flown mice as an appropriate model for sarcopenia (age-related muscle loss). Space flight caused a 15% to 30% decline in muscle fiber diameter size compared to appropriate ground controls (Harrison et al. 2003).

Data obtained from the mice following return to Earth also indicated some alternations in immune functions. Analysis of the spleenocytes (immune cells produced by the spleen) indicated an increase in B-cell (a white blood cell that matures in the bone marrow and, when stimulated by an antigen, differentiates into plasma cells) production compared to T-cells (white blood cells that complete maturation in the thymus and have various roles in the immune system). A slightly lower white blood-cell count in the flight animals compared to the controls was not statistically significant. The spleen mass was 18% to 28% lower in flight mice compared to controls. Results also indicate that flight mice weighed 10% to 12% less than ground controls (Pecaut et al. 2003).

The ability to survive a major physical trauma in microgravity may be compromised due to an altered immune system. Platelets (a constituent of blood that promotes clotting at the site of injury) are the primary cells involved in the wound healing process. The animals that were studied had significantly higher platelet levels but low volume compared to the controls. This indicates that the lack of platelets in the wound healing process is not a problem, but that platelets formed in microgravity have a decreased functionality in the wound-healing process. Data indicate that a short stay in microgravity can induce significant changes in immune defense mechanisms, hematopoiesis (blood cell formation), and other aspects of health (Gridley et al. 2003).

PUBLICATION(S)

Bateman TA, Morony S, Ferguson VL, Simske SJ, Lacey DL, Warmington KS, Geng Z, Tan HL, Shalhoub V, Dunstan CR, Kostenuik PJ. Molecular therapies for disuse osteoporosis. *Gravitational and Space Biology Bulletin.* 17:83–89, 2004.

Gridley DS, Nelson GA, Peters LL, Kostenuik PJ, Bateman TA, Morony S, Stodieck LS, Lacey DL, Simske SJ, Pecaut MJ. Genetic models in applied physiology: selected contribution: effects of spaceflight on immunity in the C57BL/6 mouse. II. Activation, cytokines, erythrocytes, and platelets. *Journal of Applied Physiology.* 94(5):2095–2103, 2003.

Harrison BC, Allen DL, Girten B, Stodieck LS, Kostenuik PJ, Bateman TA, Morony S, Lacey DL, Leinwand LA. Skeletal muscle adaptations to microgravity exposure in the mouse. *Journal of Applied Physiology.* 95(6):2462–2470, 2003.

Pecaut MJ, Nelson GA, Peters LL, Kostenuik PJ, Bateman TA, Morony S, Stodieck LS, Lacey DL, Simske SJ, Gridley DS. Genetic models in applied physiology: selected contribution: effects of spaceflight on immunity in the C57BL/6 mouse. I. Immune population distributions. *Journal of Applied Physiology.* 94(5):2085–094, 2003.

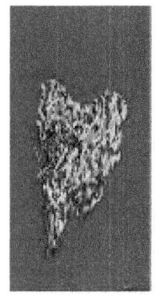
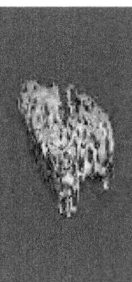

Image on the left shows a microCT image of trabecular bone from proximal tibia from a space flight mouse compared to a ground control mouse on the right.

COMMERCIAL BIOMEDICAL TEST MODULE-2 (CBTM-2)

Principal Investigator(s): H.Q. Han, M.D., Ph.D., Amgen Research, Thousand Oaks, Calif.
David Lacey, M.D., Amgen Research, Thousand Oaks, Calif.
Expedition 15

Research Area Physiological Studies: Bone and Muscle

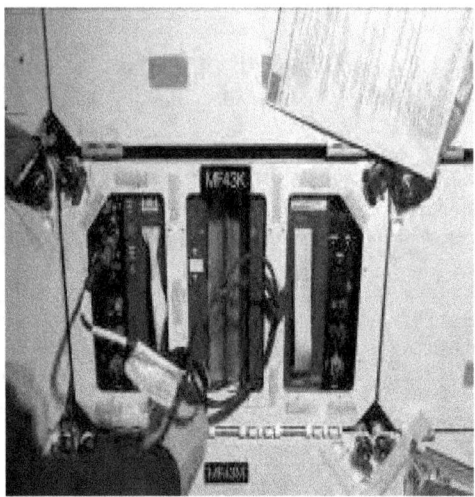

Currently, no pharmaceutical countermeasures for muscle loss that are designed for use in crewmembers during space flight have been approved. To counter muscle atrophy on orbit, crewmembers engage in daily exercise. Exercise routines last 1.5 to 3 hours per day, which can causes a reduction in crewmember productivity. These regimes also do not completely alleviate the muscle loss that occurs as a result of extended stays in space. Developing and approving a pharmaceutical countermeasure to this condition could be instrumental to long-duration human exploration missions.

CBTM-2 examined the effectiveness of an experimental therapeutic in preventing muscle loss in mice that were exposed to microgravity. This was the first time that an experimental therapeutic for muscle loss was investigated in space; an important and significant step in developing a more effective countermeasure to microgravity induced muscle changes.

S118E09308—The CBTM-2 hardware that is seen in this image flew on board STS-118/13A.1 in Aug 2007. CBTM-2 will test the effectiveness of an experimental therapeutic as a possible countermeasure for muscle atrophy.

Physical deficits from muscle atrophy also affect millions of Americans and result in significant health care expenditures. This experimental therapeutic has the potential to benefit NASA crewmembers, and could be a precursor to the development of potential interventions for muscle wasting that is related to a range of diseases, including cancer, kidney failure, and age-related frailty.

CBTM-2 used AEMs. Three AEMs were flown, each containing eight mice for a total of 12 treated and 12 control mice. Three AEMs, which were in the same configuration as on station, were operated on the ground as ground controls and ran in parallel with the space flight modules.

This research is also expected to contribute data to the current body of research on microgravity effects on the skeletal, cardiovascular, and immune systems in addition to liver and kidney function as well as other physiological systems through a tissue-sharing program. Every effort will be made to harvest as many different samples and types of tissue from the mice as possible for other mission-specific biomedical research. Positive results from this research may advance our understanding of the mechanistic changes that occur in various physiological systems after exposure to microgravity and support overall efforts to reduce health risks to crewmembers.

RESULTS

To address the impact of space flight on cancer risks, mice were flown on a 13-day space mission. Once returned to Earth, their T-lymphocytes were analyzed and compared with ground control mice. New findings report that the immune system of mice are highly sensitive to space-flight-induced stresses. After examining spleens and thymuses from mice that flew on STS-118 and ground control mice, it was found that phytohemagglutinin-induced splenocyte DNA syntheses were significantly reduced in the flight animals. Certain T- and B-cell counts from the spleens of the flight animals were also low, and the natural killer cells were increased. Analysis of cancer-related genes in the thymus from the flight animals showed that 30 out of 84 genes were expressed differently after flight. The collective results indicate that T-cell distribution and function and gene expression are significantly modified by space flight (Gridley at al. 2008).

PUBLICATIONS(S)
Gridley DS, Slater JM, Luo-Owen X, Rizvi A, Chapes SK, Stodieck LS, Ferguson VL, Pecaut MJ. Spaceflight effects on T lymphocyte distribution, function and gene expression. *J Appl Physiol*. 2008 Nov 6.

Cardiovascular and Cerebrovascular Control on Return from ISS (CCISS)

Principal Investigator(s):   Richard Lee Hughson, Ph.D., University of Waterloo, Waterloo, Ontario, Canada
Expeditions: 15–17, ongoing

Research Area Physiological Studies: Cardiovascular and Respiratory Systems

The ability to maintain arterial blood pressure and brain blood flow upon return to Earth after prolonged space flight is one of the most critical factors for crew health and safety. The inability of a crewmember to maintain blood pressure appears to be related to inadequate increases in peripheral vascular resistance. Cardiovascular and Cerebrovascular Control on Return from ISS (CCISS) will incorporate a new methodology that determines the simultaneous gains of the arterial and cardiopulmonary baroreflexes in the control of peripheral vascular resistance.

A series of six objectives has been identified that allows a more complete understanding of any alteration in cardiovascular or cerebrovascular responses following long-duration space flight. The first two are addressed with very brief in-flight experiments that monitor the heart rate component of the arterial baroreflex and the relationship between heart rate variability and physical activity as indicators of autonomic nervous system control. The remaining objectives are evaluated during a 32-minute test protocol that is conducted preflight and immediately postflight. During this test period, the central vein compliance is monitored. In addition, the arterial and cardiopulmonary baroreflexes are monitored during experiments that use an optimized schedule of lower body negative pressure (LBNP) to manipulate arterial and central venous blood pressures. Within these same experimental sessions, assessments are made of cerebrovascular responsiveness to changes in arterial blood pressure and arterial partial pressure of CO_2. It is anticipated that the ability to regulate blood pressure through baroreflex control of blood vessel constriction may be impaired after space flight. Similarly, it is expected that brain blood flow will be more sensitive to changes in arterial blood CO_2 and, thus, will not be as tightly regulated after space flight.

ISS015E14753 - Expedition 15 Flight Engineer Clay Anderson is seen here working with an Actiwatch reader and computer during hardware setup for the CCISS experiment in the U.S. Laboratory *Destiny*. The continuous blood pressure device (CBPD) is also visible in the background.

Results
Results from CCISS are pending completion of testing on all subjects before conclusive results are prepared.

Publication(s)
There are no publications at this time.

CELL CULTURE MODULE-IMMUNE RESPONSE OF HUMAN MONOCYTES IN MICROGRAVITY (CCM-Immune Response)

Principal Investigator(s):   William Wiesmann, M.D., Hawaii Chitopure, Honolulu, Hawaii
Expedition 15

Research Area Immune System and Integrated Studies

Previous studies indicate that space flight diminishes the body's immune response. Cell Culture Module-Immune Response of Human Monocytes in Microgravity (CCM-Immune Response) examined the human immune response and the effect of newly discovered natural chitosan-based antibacterials to modulate and improve the immune response. By using a monocyte (white blood cell) cell line and examining the gene expression as a result of bacterial-based stimuli in the absence and presence of chitosan-based materials, this experiment examined the role of these materials in modulating the inflammatory responses as well as connected wound-healing activity.

Monocytes were loaded into perfused hollow fiber bioreactors and divided into four groups. The first group received injections of endotoxin, the second group received injections of chitosan-arginine, the third group received injections of chitosan-arginine and endotoxin, and the final group (the control group) received common media injections. Cells were fixed at two time points and their RNA, which is responsible for the transfer of information from DNA, was extracted for examination by gene chip analysis of the human inflammatory and wound healing subset. It is suspected that chitosan derivative injections may demonstrate a beneficial role in mitigating inflammatory responses while stimulating wound healing.

RESULTS

Preliminary results from CCM-Immune Response show that the cells were successfully cultured and returned to Earth. The monocytes without chitosan did not survive the bacterial infection, whereas the monocytes with chitosan were protected and survived. Preliminary analysis shows the potential for a new pharmaceutical to fight large-scale bacterial infections.

PUBLICATION(S)

There are no publications at this time.

Cell Culture Module-Effect of Microgravity on Wound Repair: In Vitro Model of New Blood Vessel Development (CCM-Wound Repair)

Principal Investigator(s): Stuart K. Williams, Ph.D., The University of Arizona, Tucson, Ariz.
James B. Hoying, Ph.D., The University of Arizona, Tucson, Ariz.
Expedition 15

Research Area Immune System and Integrated Studies

Previous microgravity studies indicate that space flight diminishes the wound repair process. Cell Culture Module-Effect of Microgravity on Wound Repair: In Vitro Model of New Blood Vessel Development (CCM-Wound Repair) used primary endothelial cells that were derived from adipose tissue as a wound-repair model to study and potentially mitigate the effects of microgravity on wound repair. When endothelial cells, which are found lining all blood vessels, are in culture, they form cords and tubes that are precursors to new blood vessels. It is believed that the processes of new blood vessel formation and maturation will be impaired when the cells are cultured in microgravity, as compared to Earth-based cultures. It is also expected to see differences in gene expression when the cultures are subjected to genomic analysis.

The endothelial cells were seeded onto custom flatbed bioreactors and perfused. Growth hormone (VEG-F) was injected into half of the samples preflight, and all samples were fixed during space flight. It is suspected that those samples that were exposed to growth hormone will show improved blood vessel formation and maturation compared to those that were not exposed to growth hormone.

Results
Preliminary results from CCM-Wound Repair show that the cells were successfully cultured and returned to Earth. Data analysis is ongoing to determine deviations in cell pathology between the flight and ground cells that were used to characterize the microgravity-induced stresses on the flight samples.

Publication(s)
There are no publications at this time.

CHROMOSOMAL ABERRATIONS IN BLOOD LYMPHOCYTES OF ASTRONAUTS (Chromosome)

Principal Investigator(s): Günter Obe, Markus Horstmann, Christian Johannes, and Wolfgang Goedecke,
University of Essen, Essen, Germany
Expeditions 6–11, 13

Research Area Radiation Studies

Crewmembers are exposed to radiation when they leave the protection of Earth's atmosphere. Ionizing radiation in particular can damage chromosomes, causing mutations such as chromosome aberrations. To assess the genetic impact of this radiation, blood is drawn before and immediately after flight by venous puncture. The blood is then cultured and the lymphocytes are stimulated to undergo mitosis (the process of cell division). In the first mitosis, at about 48 hours of incubation, the process is stopped and the chromosomes are prepared and stained using three different methods of microscopic analysis to assess all types of aberrations induced by ionizing radiations. These methods are:

- Classic Giesma staining, which allows the researcher to investigate changes in the morphology of the chromosomes. Chromosomes have a natural x-shape. Structural changes that are detected using Giesma include dicentric (the two chromatids of each chromosome are attached twice) and ring chromosomes or fragments (chromosome pieces without a centromere).
- Multicolor Fluorescence In-Situ Hybridization (mFISH), which scores reciprocal translocations and insertions (exchange of parts between different chromosomes).
- Multicolor Banding Fluorescence In-Situ Hybridization (mBAND) of the selected chromosome pair 5, which scores for inversions and translocations between homologous chromosomes (exchange or relocation of DNA parts within the same chromosome pair).

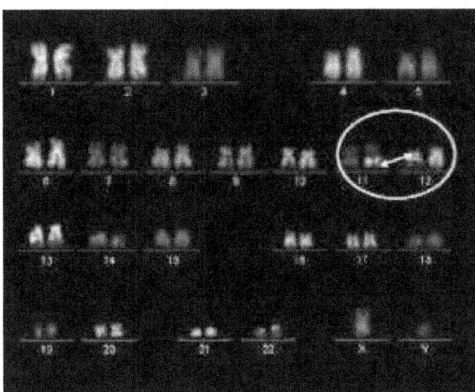

A quantitative comparison between preflight and postflight aberration values provided information about the chromosome-breaking effects of cosmic radiation in blood lymphocytes of space travelers. Information was generated concerning the participation of each chromosome pair in aberration formation as well as the inter- and intrachromosomal distribution of different aberration types. The association of chromosomal aberrations with an enhanced cancer risk stresses the importance of the planned research.

This photograph shows that a reciprocal exchange (translocation between chromosomes 11 and 12) has taken place in the blood lymphocytes of a crewmember after space flight.

RESULTS

In each Expedition where the experiment has been conducted, preflight and postflight blood samples were drawn from each crewmember. To ensure high-quality results, the blood samples arrive at the laboratory within 72 hours after collection. Researchers are currently measuring changes in the genetic material and analyzing their significance and will release preliminary conclusions soon. From this study, scientists may be able to better assess risk factors for genetic damage in space. Understanding and reducing the risk of radiation is important for safe long-duration travel in space, including stays on the moon and journeys to Mars.

PUBLICATION(S)

There are no publications at this time.

DOSIMETRIC MAPPING (DOSMAP)

Principal Investigator(s): Günther Reitz, Deutsches Zentrum fur Luft und Raumfahrt (DLR) Institute for Aerospace Medicine, Cologne, Germany
Expedition 2

Research Area Radiation Studies

Interactions of ionizing radiations with the ISS structure and its contents create a somewhat different radiation field at each location inside the ISS modules. This is in part due to a large contribution from the secondary radiation that is created by particles colliding with the spacecraft materials. Dosimetric Mapping (DOSMAP) was designed to map the total absorbed dose in the U.S. Laboratory module of the ISS. DOSMAP was deployed from Mar through Aug 2001. This time frame included results from the solar particle event (SPE) that occurred on Apr 15, 2001.

To fully characterize the radiation doses that were observed on ISS, the following four detector types were used: thermo-luminescence dosimeter (TLD) chips, CR39 nuclear track detectors with and without converters (NTDPs), a silicon dosimetry telescope (DOSTEL), and four silicon mobile detector units (MDUs). Crewmembers used the MDUs as personal dosimeters. They provided the ability to measure spectral composition with respect to nuclear charge, energy, and rate of energy deposition (linear energy transfer (LET)), as well as to estimate absorbed dose from galactic radiation, radiation belt particles, and the Apr SPE.

ISS002E7814 — James S. Voss, Expedition 2 flight engineer, sets up the HRF DOSMAP power distribution unit in the U.S. *Destiny* laboratory.

RESULTS

In general, radiation damage to the human body is indicated by the amount of energy that is deposited in living tissue, modified by the type of radiation causing the damage; this is measured in units of Sv. The background radiation dose that is received by an average person in the U.S. is approximately 3.5 milliSv/yr. An exposure of 1 Sv/hr can result in radiation poisoning, and a dose of 5 Sv/hr will result in death in 50% of exposed individuals. The average dose that was determined by the data that were collected during Expedition 2 was found to be 532 microSv/day. This is significantly lower than measurements taken on previous space shuttle and *Mir* missions. All dosimeters showed agreement to within 10% of one another, and over 95% of the projected measurements were able to be collected (Reitz 2001, 2005).

Three important conclusions were drawn from this experiment that will help with future monitoring activities:
- The corrections that were needed due to device data storage and readout were negligible to the LET readings and had only a small influence on dose rate estimates.
- About 15% of the tissue-damaging dose (effective dose) is from the short-ranged neutrons and protons that were created within the spacecraft materials.
- About 90% of the crewmember dose is due to particles that deposit less than 150 keV/micron.

PUBLICATION(S)

Reitz G, Beaujean R, Dachev Ts, Deme S, Luszik-Bhadra M, Heinrich W, Olko P. Dosimetric Mapping. Conference and Exhibit on International Space Station Utilization, Cape Canaveral, Fla. AIAA-2001-4903, Oct 15–18, 2001.

Reitz G, Beaujean R, Benton E, Burmeister S, Dachev Ts, Deme S, Luszik-Bhadra M, Olko P. Space radiation measurements on-board ISS – the DOSMAP experiment. *Radiation Protection Dosimetry*. 2005; 116 (1-4):374–379.

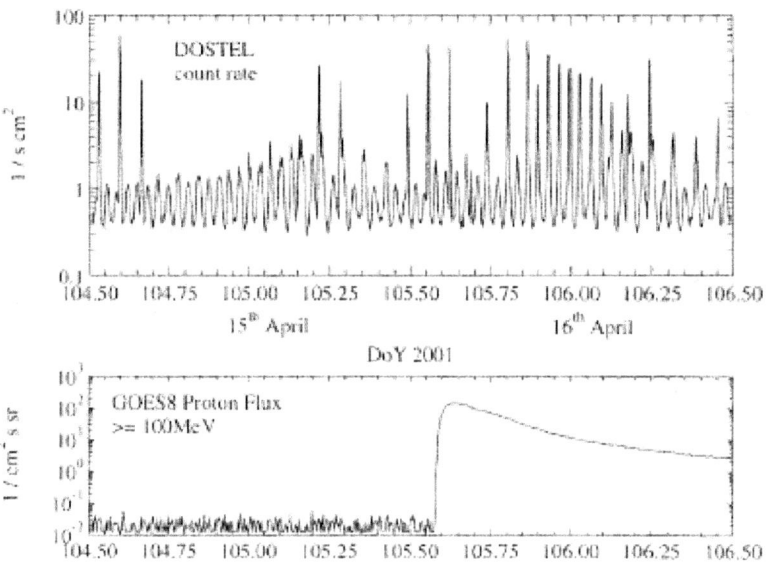

Figure from Reitz et al. (2005) showing the DOSTEL particle count rates Apr 15–16, 2001 (Days 104 and 105) during an SPE. The sinusoidal trend on the upper graph represents count rates on ISS as it transits from high to low latitudes every 45 minutes, and variations due to longitude and the SAA. The lower graph records the proton flux that was measured by the geostationary operational environmental satellite (GOES). The DOSTEL measurements were elevated after the onset of the event (approximately Day 105.75).

Space Flight-Induced Reactivation of Latent Epstein-Barr Virus (Epstein-Barr)

Principal Investigator(s): Raymond Stowe, Microgen Labs, Galveston, Texas
Expeditions 5, 6, 11–17

Research Area Immune System and Integrated Studies

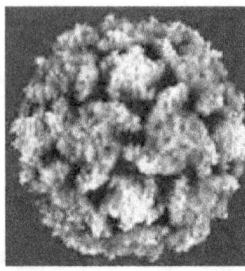

A colored, scanning electron micrograph of an Epstein-Barr particle.

In the United States, approximately 95% of adults have been infected with Epstein-Barr Virus (EBV), which is one of the most common human viruses and a member of the herpes virus family. EBV is an initial infection that establishes a lifelong dormant infection inside the body that can be reactivated by illness or stress. Once active, EBV causes infectious mononucleosis, cancers, and other disorders that are associated with the lymphatic system in people with a compromised immune system.

The decreased cellular immune function that is experienced by astronauts in space flight is likely caused by a combination of the microgravity environment and the stresses that are associated with a mission. With longer-duration missions, it is hypothesized that latent viruses are more likely to be reactivated, placing the crew at risk of developing and spreading infectious illnesses and jeopardizing the mission. Preliminary studies of astronauts have shown increased EBV shedding (the means by which viruses reproduce) in the saliva and increased antibody titers to the virus's proteins.

Epstein-Barr examined levels to which the crews' immune systems were suppressed during space flight and identified conditions under which the virus may reactivate. To conduct Epstein-Barr, investigators collected urine and blood samples preflight and again postflight. The samples were analyzed for the presence of stress hormones and cytokines (messengers of the immune system), EBV replication, and virus-specific T-cell immune function. Epstein-Barr used the levels to which the crews' immune systems are suppressed during space flight to determine the conditions under which EBV may reactivate.

Results

This experiment has recently been completed. Data from subjects will be analyzed collectively before publishing results. Earlier studies during space shuttle missions, which were the predecessors to this one, suggested that virus reactivation results from decreased T-cell function. If Epstein-Barr yields similar results, it will allow for a very specific focus on developing drug therapies that will allow for more rapid treatment for space travelers as well as for humans on Earth.

Publication(s)

There are no publications at this time.

A Study of Radiation Doses Experienced by Astronauts in EVA (EVARM)

Principal Investigator(s): Ian Thomson, Thomson & Nielsen Electronics, Ontario, Canada
Expeditions 4–6

Research Area Radiation Studies

ISS005E22017 — ISS Expedition 5 commander Valery Korzun during an EVA. The EVARM experiment measured the amount of radiation that spacewalkers absorb during EVAs.

A Study of Radiation Doses Experienced by Astronauts in EVA (EVA radiation monitoring (EVARM)) was designed to quantify the radiological dose that is received by astronauts while performing EVAs at the ISS. Extravehicular mobility units (EMUs, or spacesuits), which are worn by spacewalking astronauts, provide less shielding from radiation than the spacecraft. This means that spacewalkers are exposed to higher radiation levels during EVAs than at other times on orbit. When planning EVAs, teams take into account mission parameters, estimated duration, ISS altitude and inclination plus information on space weather conditions (e.g., solar activity, geomagnetic field conditions, proton flux) that are anticipated for that day.

In addition to specific lifetime radiation limits, medical standards specify that radiation doses that are achieved by astronauts should be as low as reasonably achievable (ALARA). To create new and improved shielding for EVAs, researchers must know the type and flux of radiation inside the EMU. EVARM investigated the dose that is received by different parts of the body (skin, eyes, blood-forming organs) during an EVA by measuring dose rate based on the time and position of EVAs as compared to the orbit, altitude, and attitude of the ISS.

As part of EVARM, spacewalkers wore dosimeters that were placed in small pockets along the EMU undergarments. Two dosimeters were placed either inside the thermal comfort undergarment or the liquid-cooled ventilation garment, one dosimeter was placed around the calf to measure absorbed dose to skin, and another dosimeter was worn above the eye inside the communications carrier assembly. EVARM used tiny metal oxide semiconductor field effect transistor (MOSFET) dosimeters, a 0.04-in^2 silicon chip that was placed on a badge made of aluminum. When an MOSFET is exposed to ionizing radiation, a positive charge builds up on the silicon surface, creating a negative shift in threshold voltage. Measurements were taken by comparing the change in threshold voltage with the radiation dose, which was recorded using a photodiode. New dosimeters were worn by the crew during each EVA.

Results

For the EVARM investigation, 10 complete sets of data were collected between Feb and Nov 2002. These badges were compared to radiation monitors already on the ISS as well as to the ESA's Space Environment Information System (SPENVIS).

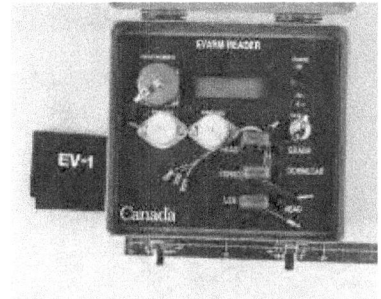

EVARM hardware. The device shown records and processes data sent by radiation detectors in the spacesuits of spacewalkers during EVAs.

The results from EVARM have shown that EVA doses are elevated from those inside the ISS but not significantly. In addition, this time period recorded doses during a time of increased geomagnetic activity (Oct/Nov 2002). It was determined that during this event doses to EVA participants were increased due to elevated levels of electrons in Earth orbit. These electrons are easily shielded by spacecraft materials and, thus, not measured inside the ISS. Fortunately, proper positioning of the spacecraft can dramatically reduce the radiation field that is encountered during EVA missions. A significant finding was that a single detector that was placed at the astronaut's torso was not sufficient to accurately determine organ doses. Results show that the MOSFET detectors are accurate when compared with other monitoring equipment; however, the use of this battery device may present problems in the EVA environment.

PUBLICATION(S)
There are no publications at this time.

Foot/Ground Reaction Forces During Space Flight (Foot)

Principal Investigator(s): Peter R. Cavanagh, The Cleveland Clinic Foundation, Cleveland, Ohio
Expeditions 6, 8, 11, 12

Research Area Physiological Studies: Bone and Muscle

The human body is designed to bear weight. Without the stimulation that is caused by placing weight on lower extremities, whether due to the microgravity environment or lack of use on Earth, bones lose mass and muscles lose strength. The Foot experiment characterized the load that is placed on lower extremities during daily activities on station and examined to what degree mechanical load stimulus, via an in-flight exercise routine, could prevent the muscle atrophy and bone loss that is associated with space flight.

To achieve this objective, Foot had several sensors mounted in a special pair of Lycra exercise pants known as the lower extremity monitoring suit (LEMS). The total force-foot ground interface (TF-FGI) served as an insole that, when placed inside a shoe, measures the amount of force that is placed on the bottom of the foot. Joint excursion sensors (JESs) record joint angles at the ankle, knee, and hip. Electromyography (EMG) electrodes recorded muscle activity, including net neural drive, along the leg (the vastus medialis, rectus femoris, biceps femorics, gastrocnemius, and tibial anterior) and in the right arm (the biceps brachii and triceps brachii). Information was collected by an ambulatory data acquisition system and downloaded into the HRF laptop on board ISS after each session.

ISS006E11018 — Expedition 6 commander Kenneth Bowersox, wearing a body harness, runs on the TVIS as part of the Foot experiment in the Zvezda service module.

Results

Results provided insight into the processes of loss of bone mineral density and muscle mass during long-duration stays on orbit. Knee-joint motion in space was reduced when compared to that on Earth, thus effecting muscle action. In preliminary data analyses of the first subject, significant loss of bone mass was observed. Measurements of forces during exercise suggested that much less force was experienced than would be experienced when exercising on Earth. Detailed data were collected on loads across all exercise hardware settings during Expeditions 11 and 12. Final analysis will help in determining exercise prescriptions for station crewmembers and in the design of future exercise devices for Exploration missions.

Integrating these general results into a contextual overview of bone health on long-duration missions, Cavanaugh and Rice (2007) have compiled a set of summary articles that examines bone health, bone loss, efficacy of exercise and mechanical stimulus, and other factors that are relevant to bone health in space.

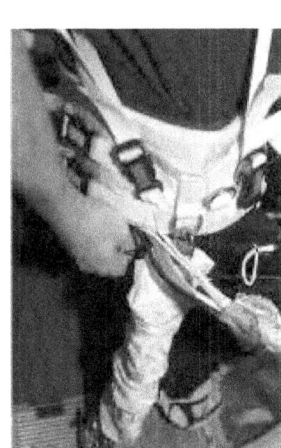

Publication(s)

Cavanagh PR, Maender C, Rice AJ, Gene KO, Ochia RS, Snedeker JG. Lower-Extremity Loading During Exercise on the International Space Station. Transactions of the Annual Meeting of the Orthopaedic Research Society 0395, 2004.

Pierre MC, Genc KO, Litow M, Humphreys B, Rice A, Maender CC, Cavanagh PR. Comparison of Knee Motion on Earth and in Space: An Observational Study. *Journal of NeuroEngineering and Rehabilitation*. 3:8, 2006.

Cavanaugh PR, Rice AJ. Bone Loss During Space Flight: Etiology, Countermeasures, and Implications for Bone Health on Earth. Cleveland Clinic Press, Cleveland, Ohio, 2007, 297 pp.

ISS006E11011 — View of the body harness for the Treadmill Vibration Isolation System (TVIS) and the LEMS for the Foot experiment in the *Zvezda* service module. The apparatus is shown worn by Expedition 6 commander Ken Bowersox.

HAND POSTURE ANALYZER (HPA)

Principal Investigator(s):   Valfredo Zolesi, Kayser Italia SRL, Livorno, Italy
Expeditions 7, 8, 11, 16

Research Area Physiological Studies: Bone and Muscle

The Hand Posture Analyzer (HPA) examined how hand and arm muscles are used differently during grasping and reaching tasks in weightlessness by collecting kinematic and force data on astronaut's upper limbs (hands, wrists and forearms). Three different sets of data were collected: preflight, in-flight and postflight. The measurements, which involved the crewmember manipulating both virtual and concrete objects, is researched to assess the approaching, reaching, and grasping mechanics of the hand and fingers without the effect of gravity.

RESULTS

The HPA was launched to ISS on 12 Progress in Aug 2003. It was performed during Expeditions 7 and 8 on ISS; data from six HPA sessions were collected during Expedition 7 from one crewmember; two preflight collections, two in-flight collections, and two postflight collections. At the end of Expedition 10, ESA astronaut Roberto Vittori performed in-flight data collection with the HPA hardware. These data are being combined with data from the preliminary version of the same hardware (Chiro), which was used on board ISS during an earlier "Marco Polo" mission with astronaut Roberto Vittori in 2002. Together, these experiments assessed the short- and long-term effects of weightlessness on upper limb performance.

PUBLICATION(S)

There are no publications at this time.

ISS008E21605 — NASA science officer Mike Foale was photographed during ISS Expedition 8 as he used hand/wrist position tracking via the hand posture acquisition glove during HPA operations.

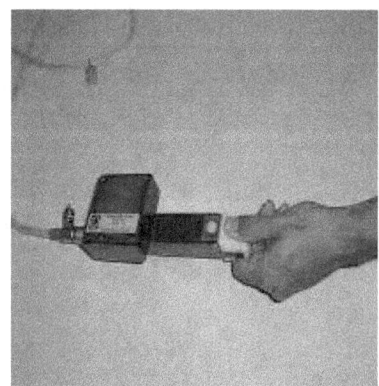

Hand grip dynometer that was used for the muscle fatigue experiment that was part of the HPA investigation on board station.

Effects of Altered Gravity on Spinal Cord Excitability (H-Reflex)

Principal Investigator(s): Douglas Watt, McGill University, Montréal, Canada
Expeditions 2–4

Research Area Physiological Studies: Bone and Muscle

In the weightlessness of LEO, the body loses muscle mass and bone density. The only known countermeasure for this atrophy is exercise. However, as astronauts spend longer durations in space, will exercise continue to be an effective countermeasure? Along with changes in muscle and bone, the neurovestibular system (the complex sensory system that maintains posture, balance, and coordination) adapts to changes in gravity. Researchers hypothesize that, as part of this neurovestibular system adaptation, spinal cord excitability decreases and the spinal cord reacts less to stimuli. If this hypothesis is correct, exercise may become less effective the longer astronauts stay in microgravity, and researchers may have to adjust exercise programs accordingly.

H-Reflex tested this hypothesis by measuring muscle response to mild electrical shocks (40–90 volts). Nerves that are in the leg perceive the electrical shock and send a signal along the spinal cord to the brain. The signal stimulates motor neurons in the brain, which, in turn, send signals that cause leg muscles to contract. The bigger the contraction, the more the neurons are stimulated, indicating the level of spinal cord excitability. Researchers compared measurements that were taken before, during, and after flight to determine whether the spinal cord's ability to respond to stimuli changed over time. The H-Reflex equipment recorded the EMG activity in the muscle—the electrical activity that causes the muscle to move—rather than the movement that follows the electrical activity (as a knee-tap test would), allowing researchers to take more precise measurements.

Results

This study of spinal cord excitability using the Hoffman reflex was completed by a total of eight subjects over ISS Expeditions 2 through 4. H-Reflex measured how excitable the nerve cells were by applying small electrical shocks behind the knee. Each shock produced a reflex response in the calf muscles (the H-reflex response); the data collected indicated that this response decreased significantly while in microgravity. The study found that spinal cord excitability decreased by about 35% in weightlessness, and stayed at this level for the duration of the mission. Although there was notable improvement in the H-reflex response the day after landing, it took about 10 days back on Earth for astronauts to fully recover their muscle strength and spinal cord excitability (Watt and Lefebvre 2001; Watt 2003).

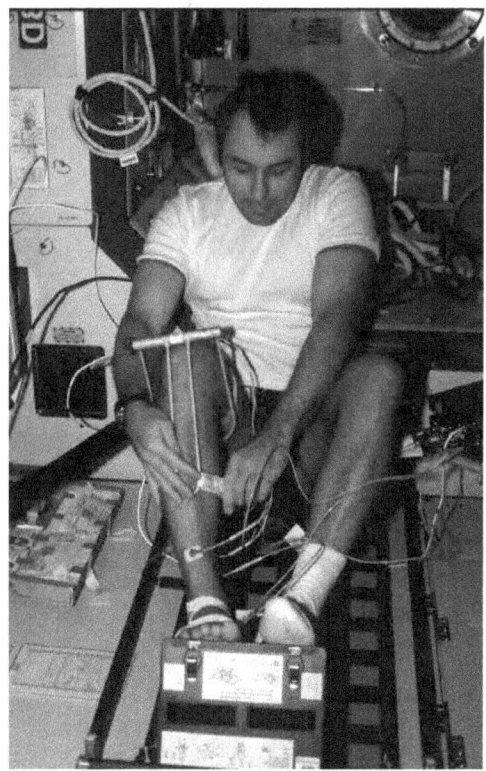

ISS003-330-006 — Cosmonaut Vladimir Dezhurov on ISS Expedition 3 performing H-Reflex activity in the U.S. *Destiny* laboratory.

This difference in excitability means that only a portion of muscle fiber units are contracting in response to signals from the nervous system, and explains functionally why muscle mass declines in weightlessness, even with exercise. Reduced excitability means that there might be limits on the degree to which heart muscle strength, leg muscle tone, and bone density (for which muscle contraction is an important regulating factor) can be maintained through exercise on long-duration missions. Because this decrease in excitability is only observed on orbit and not during bed rest, an analog for weightless space travel, the results highlight the possibility that reduced excitability with corresponding loss of muscle and bone might be partly a nervous system response and not simply due to disuse of the legs.

Based on the results of this study, decreased spinal cord excitability could be an issue for long-duration stays in partial-gravity environments such as are found on the moon and Mars. Future designs of exercise equipment that provide feedback on work actually performed would help crewmembers compensate for decreases in exercise efficiency.

PUBLICATION(S)

Watt DG, Lefebvre L. Effects of altered gravity on spinal cord excitability. First Research on the International Space Station, Conference and Exhibit on International Space Station Utilization, Cape Canaveral, Fla. AIAA 2001-4939, Oct 15–18, 2001.

Watt DG. Effects of altered gravity on spinal cord excitability (final results). Proceedings of the Bioastronautics Investigators' Workshop, Galveston, Texas. Jan 2003.

Watt DG. Effects of prolonged exposure to microgravity on H-reflex loop excitability. Proceedings of the 14th IAA Humans in Space Symposium, Banff, Alberta, Canada. May 2003.

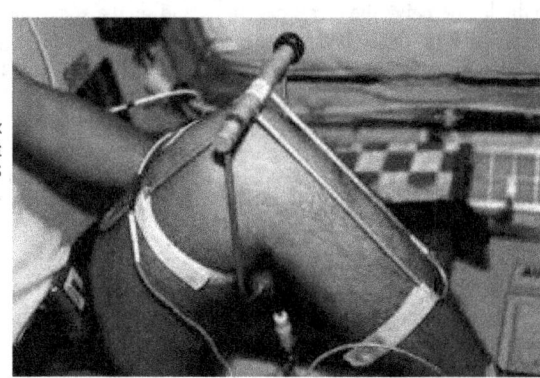

ISS003003011 — Close-up of H-Reflex hardware on the knee of cosmonaut Vladimir Dezhurov during ISS Expedition 3.

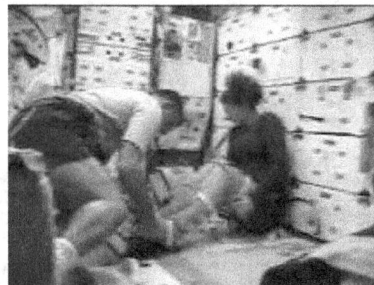

Still frame from a video of Expedition 2 showing astronauts Jim Voss and Susan Helms performing the H-Reflex experiment in the shuttle middeck.

CREW MEMBER AND CREW-GROUND INTERACTIONS DURING INTERNATIONAL SPACE STATION MISSIONS (Interactions)

Principal Investigator(s):   Nick A. Kansas, University of California and Veterans Affairs Medical Center, San Francisco, Calif.
Expeditions 2–5, 7–9

Research Area Behavior and Performance Studies

Isolated in the microgravity and vacuum of near-Earth orbit, the ISS is a potentially risky place in which to work and live. Mission success and crew safety rely on the ability of station crews to communicate and get along with their fellow crewmembers, regardless of their age, gender, nationality, or personal beliefs and preferences. It is also critical that the station crew has good interactions with members of ground operations.

The Interactions study recorded crew and crew-ground activities in an effort to fully understand group dynamics, individual psychological health, and factors that both hinder and help daily life on station. The study consisted primarily of a computerized questionnaire that was filled out weekly by crewmembers in space and by ground personnel at NASA Johnson Space Center, NASA Marshall Space Flight Center, and the Russian Mission Control Center in Moscow. The questionnaire software included a series of questions from three standard mood and interpersonal group climate questionnaires as well as a critical incident log.

JSC2004E25790 — Expedition 9 crewmembers, ISS science officer Michael Fincke (right) and ISS commander Gennady Padalka (left), use video and audio channels to communicate with the Mission Control Center-Houston on Jun 18, 2004. Everyone is celebrating the recent birth of Fincke's daughter. The primary focus of the Interactions investigation is to improve communications between the support personnel and the Expedition crew.

RESULTS

The Interactions experiment observed the day-to-day relations between the ISS crew and the ground support teams in Houston, Huntsville, Alabama, and Moscow, Russia. Data were collected over a period of 4 years during ISS Expeditions 2 through 9. ISS crewmembers and Mission Control personnel responded to questions from three standard mood and interpersonal group climate questionnaires (Profile of Mood States, Group Environment Scale, and Work Environment Scale) and maintained critical incident logs. The questionnaires used well-established psychometric measurements (measures of psychological variables; e.g., intelligence, aptitude, and personality traits). Additionally, crew activities (in particular, Earth photography) were assessed as a mechanism for preserving crew health (Robinson et al. 2006).

Preliminary results were presented in two papers by Kanas et al. (2005) and Ritscher et al. (2005) at the 15th International Academy of Astronautics (IAA) Humans in Space Symposium. Additional results have been presented in subsequent meetings and published forums. Previous studies of crew interactions (e.g., when U.S. crewmembers were added to the Russian space station *Mir* crews) identified important patterns of responses in interactions between and among crews and ground personnel. Not surprisingly, the investigation is also identifying differences in mood and group perceptions between Americans and Russians, as well as between crewmembers and Mission Control personnel.

The early results were replicated after ISS crewmembers were questioned and responses analyzed; no "2nd half" effects were noted (Kanas et al. 2006; 2007). In a separate but related study that was conducted by this research team, ISS crewmembers show evidence of an improvement in mental health as they adjust to the environment (adaptation). The study indicates that crewmembers improve in mood and social climate over the course of their missions. Post-mission surveys of crewmembers are being used to evaluate strategies to enhance their in-flight stress tolerance and postflight adjustment. In addition, other factors, including cultural sophistication and language flexibility, were analyzed among members of both crew and supporting Flight Control teams. While differences in these cultural parameters were noted between U.S. and Russian teams, and the crew and Flight Control teams (crews and Russian controllers exhibited higher scores on cultural sophistication), they did not appear to be related to mood and social climate variables (Ritsher et al. 2006). Finally, flight control teams were interviewed to examine major leadership challenges, impacts to mission management, and solutions to those challenges (Clement et al. 2006). Geographic separation and communication were identified as key challenges requiring continued effort to mitigate. Strong interpersonal relationships, communication, and flexibility are critical for all leaders who are working ISS operations.

ISS007E18044 — Astronaut Edward T. Lu (at musical keyboard), Expedition 7 NASA ISS science officer and flight engineer, and ESA astronaut Pedro Duque of Spain share a light moment during off-shift time in the U.S. *Destiny* laboratory.

Many of the behavioral factors that were studied in this experiment (communication styles, multicultural teams, operational systems) will be important in planning operations systems and relationships between Exploration crews and ground personnel for lunar and Mars missions. Complete and final analysis of the questionnaires is still being conducted.

JSC2003E59333 — Overview of the station flight control room in the NASA Johnson Space Center Mission Control Center. This photograph was taken during rendezvous and docking operations between the Soyuz TMA-3 spacecraft and the ISS.

Publication(s)

Clement J, Ritsher JB. Operating the ISS: Cultural and leadership challenges. 56th International Astronautical Congress. Fukuoka, Japan, IAC-05-A1.5.05. Oct 17–21, 2005.

Kanas N, Ritsher J. Leadership Issues with Multi-cultural Crews on the International Space Station: Lessons learned from Shuttle/Mir. *Acta Astronautica*. 56:932–936, 2005.

Kanas N, Salnitskiy VP, Ritsher JB, Gushin VI, Weiss DS, Saylor S, Marmar C. Human interactions in space: ISS versus Shuttle/Mir. 56th International Astronautical Congress. Fukuoka, Japan, IAC-05-A1.5.02. Oct 17–21, 2005.

Ritsher JB, Kanas N, Gushin VI, Saylor S. Cultural differences in patterns of mood states onboard the International Space Station. 56th International Astronautical Congress. Fukuoka, Japan, IAC-05-A1.5.03. Oct 17–21, 2005.

Ritsher JB, Kanas N, Salnitskiy VP, Gushin VI, Saylor S, Weiss DS, Marmar C. Cultural and Language Backgrounds of International Space Station Program Personnel. Presented at the 57th International Astronautical Congress. Valencia, Spain. Oct 2–6, 2006; IAC-06-A1.1.3.

Kanas NA, Ritsher JB, Saylor SA. Do Psychological Decrements Occur During the 2nd Half of Space Missions? Presented at the 57th International Astronautical Congress. Valencia, Spain. Oct 2–6, 2006; IAC-06-A1.1.02i.

Clement JL, Ritsher JB, Kanas N, Saylor S. Leadership Challenges in ISS Operations: Lessons Learned from Junior and Senior Mission Control Personnel. Presented at the 57th International Astronautical Congress. Valencia, Spain. Oct 2–6, 2006. IAC-06-A1.1.6.

Kanas NA, Salnitskiy VP, Ritsher JB, Gushin VI, Weiss DS, Saylor SA, Kozerenko OP, Marmar CR. Psychosocial interactions during ISS missions. *Acta Astronautica* 2007. 60:329–335.

Kanas, NA, Salnitskiy VP, Boyd JE, Gushin VI, Weiss DS, Saylor SA, Kozerenko OP, Marmar CR. Crewmember and mission control personnel interactions during International Space Station missions. *Aviation Space and Environmental Medicine*. 2007. 78(6):601–607.

ISS008-E-06699 — Astronaut C. Michael Foale (foreground), Expedition 8 mission commander and NASA ISS science officer, and cosmonaut Alexander Y. Kaleri, flight engineer representing Rosaviakosmos, eat a meal in the *Zvezda* service module on the ISS.

Behavioral Issues Associated with Isolation and Confinement: Review and Analysis of ISS Crew Journals (Journals)

Principal Investigator(s):   Jack W. Stuster, Anacapa Sciences, Inc., Santa Barbara, Calif.
Expeditions 8–18, ongoing

Research Area Behavior and Performance Studies

A previous content analysis of journals that were maintained during long-duration expeditions on Earth (e.g., to the Antarctic) provided quantitative data on which to base a rank-ordering of behavioral issues in terms of importance. Journals uses the same content evaluation techniques on journals kept by ISS crewmembers. The objective is to identify equipment, habitat, and procedural features that can help humans when adjusting to isolation and confinement while ensuring that they remain effective and productive during future long-duration space flights.

While on orbit, crewmembers make journal entries at least three times a week in a personal journal. In format, their journal can be either electronic (i.e., using an ISS laptop) or paper. In addition to the journal entries, participating crewmembers also complete a brief electronic questionnaire at the mid-point of their Expeditions.

ISS008E22350 — Astronaut Edward M. (Mike) Fincke, Expedition 9 NASA ISS science officer and flight engineer, works in the U.S. *Destiny* laboratory.

Studies on Earth have shown that analyzing the content of journals and diaries is an effective means of identifying issues that are most important to the person recording his or her thoughts. The method is based on the assumption that the frequency with which an issue is mentioned in a journal reflects the importance of that issue or category to the writer. The tone of each entry (positive, negative, or neutral) and phase of the Expedition are also variables of interest. Study results will lead to recommendations for the design of equipment, facilities, procedures, and training to help sustain behavioral adjustment and performance during long-duration Expeditions on ISS, or to the moon, Mars, and beyond. These studies can also assist on Earth with Antarctic missions, service on submarines, etc.—anywhere humans choose to work in confinement or isolation.

Results
Data collection is ongoing, and the results will be analyzed when all of the journals are available.

Publication(s)
There are no publications at this time.

ISS010E6816 — Expedition 10 crewmembers Leroy Chiao and Salizhan Sharipov participated in the Journals investigation.

INCIDENCE OF LATENT VIRUS SHEDDING DURING SPACE FLIGHT (Latent Virus)
Principal Investigator(s):   Duane L. Pierson, Ph.D., NASA Johnson Space Center, Houston
Expeditions 0, 1, 2, 4, 5, 11, 13–15

Research Area Immune System and Integrated Studies

Latent herpes viruses pose an important infectious disease risk to crewmembers who are involved in space flight and space habitation. The risk certainly increases as mission duration increases. Risks that are associated with most bacterial, fungal, viral, and parasitic pathogens can be reduced by a suitable quarantine period before the flight and by appropriate medical care. However, latent viruses are unaffected by such actions. The observed decrements in the immune response resulting from space flight may allow increased reactivation of the same herpes viruses, and may increase the incidence and duration of viral shedding. Such a result may increase the concentration of herpes viruses in the spacecraft. Particulates (including viruses) do not settle out of the air in the microgravity conditions of space flight. Additional characteristics of space flight, such as living in relatively crowded conditions in a closed environment and using recycled air and water, will increase the potential for transfer of viruses among the crewmembers. This study will help determine the characteristics of viral parameters such as latent virus reactivation, shedding, and crew exchange during space flight, and is an integral part of ongoing efforts to accumulate microbiological data concerning the exposure of astronauts to potentially infectious agents.

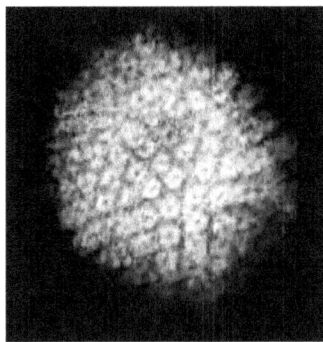

Microscopy image of a herpes virus, one of several latent viruses that will be studied during the Latent Virus investigation. Image courtesy of Linda Stannard of the Department of Medical Microbiology, University of Cape Town, South Africa.

RESULTS
Many of the biological samples that are collected from astronauts immediately before and following space flight have proven valuable for several investigators. Saliva samples that have been collected from crewmembers who are traveling on the shuttle to and from ISS since 2000 have provided preliminary results for the Latent Virus investigation. For this investigation, EBV and Varicella zoster virus (VZV) were studied using the saliva samples. The data that were collected indicate that latent viruses can become infectious under stressful conditions such as space flight.

Thirty-two healthy astronauts have been studied for EBV reactivation on 10 space shuttle missions since 2000. This study revealed that the increased stress of space flight may cause latent virus reactivation in astronauts. The astronauts who were studied served as either commander, pilot, or mission specialist; these are all different positions that carry their own unique stresses. Potential EBV reactivation in astronauts was shown by three measures: EBV presence in saliva, number of copies of viral EBV DNA in saliva, and titer of antibodies to EBV viral antigens. Data revealed that there was no correlation between the shedding frequency of EBV in the saliva and the amount of EBV DNA in the saliva. EBV antibody titers increased before flight and continued to increase 3 days postflight. The amount of EBV DNA increased as the number of days in space increased. The pattern and amount of EBV shedding in the astronauts likely correlated to various events that occurred during space flight. The types, levels, and combination of stresses experienced before, during, and after flight, as well as the different ways in which individuals cope with stress, may result in changes in the EBV shedding frequency (Pierson 2005).

Eight healthy astronauts were studied during three shuttle missions to determine the cause of VZV reactivation in a healthy adult astronaut 2 days before flight. Ten subjects (not astronauts) who remained on Earth were used as controls in this investigation. Polymerase chain reaction (PCR) analysis was performed on DNA samples that were extracted from the saliva of the subjects. Before flight all samples from the experimental subjects were negative for VZV DNA; during flight VZV DNA was detected in 87% of the astronauts; following return to Earth VZV DNA was detected in only 19% of the astronauts who were tested. During this same time frame, no VZV DNA was detected in saliva samples of the control subjects. VZV, like EBV, can reactivate during stressful situations such as space flight (Mehta 2004).

Each virus—EBV and VZV—has its own unique timing when reactivation due to stress. EBV appeared to increase at all phases of space flight (preflight, in-flight, and postflight), while VZV DNA increased as space flight approached and decreased postflight.

PUBLICATION(S)

Mehta SK, Cohrs RJ, Forghani B, Zerbe G, Gilden DH, Pierson DL. Stress-induced Subclinical Reactivation of Varicella Zoster Virus in Astronauts. *Journal of Medical Virology*. 2005; 72:174–179.

Pierson DL, Stowe RP, Phillips TM, Lugg DJ, Mehta SK. Epstein-Barr Virus Shedding by Astronauts During Space Flight. *Brain, Behavior, and Immunity*. 2004; 19:235–242.

TEST OF MIDODRINE AS A COUNTERMEASURE AGAINST POSTFLIGHT ORTHOSTATIC HYPOTENSION-SHORT-DURATION BIOLOGICAL INVESTIGATION (Midodrine-SDBI)

Principal Investigator(s): Steven Platts, Ph.D., NASA Johnson Space Center, Houston
Expeditions 14–17

Research Area Physiological Studies: Cardiovascular and Respiratory Systems

Many astronauts experience postflight orthostatic hypotension, a condition where the blood pressure drops when an individual stands up, resulting in presyncope (lightheadedness) or syncope (fainting). Approximately 20 to 30% of crews on short-duration (fewer than 20 days) missions and 83% of crews on long-duration missions experience some degree of orthostatic intolerance after return to Earth. To date, the countermeasures that have been tested—e.g., fluid loading, the use of LBNP, and the medication Fluronef—have not successfully eliminated postflight orthostatic hypotension.

On Earth, the drug Midodrine has been used extensively to treat low blood pressure. Midodrine acts as a vasopressor (raises blood pressure) by causing constriction (tightening) of the blood vessels, which leads to an increase in blood pressure. Test of Midodrine as a Countermeasure Against Post-flight Orthostatic Hypotension–Short-duration Biological Investigation (Midodrine-SDBI) studies the effectiveness of Midodrine for the treatment of postflight orthostatic hypotension.

RESULTS
Midodrine has been shown to successfully reduce orthostatic hypotension in patients on Earth, as orthostatic hypotension affects people other than astronauts. To date, this investigation is ongoing; future Expeditions will involve testing on more subjects before conclusive results can be determined.

PUBLICATION(S)
There are no publications at this time.

Promoting Sensorimotor Response Generalizability: A Countermeasure to Mitigate Locomotor Dysfunction After Long-duration Space Flight (Mobility)

Principal Investigator(s):   Jacob Bloomberg, NASA Johnson Space Center, Houston
Expeditions 5–12

Research Area Neurovascular and Vestibular Systems

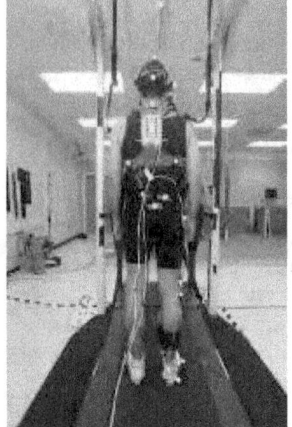

Following space flight, astronauts experience disturbances in balance and walking control during the postflight readaptation period, due in part to changes in the way the CNS processes sensory information as a result of prolonged exposure to microgravity. The goal of this study was to develop an in-flight treadmill training program that facilitates recovery of locomotor function after long-duration space flight.

The proposed training program was based on the concept of adaptive generalization. During this type of training, the subject gains experience producing the appropriate adaptive behavior under a variety of sensory conditions and balance challenges. As a result of this training, the subject learns to solve a class of balance and walking problems rather than producing a single solution to one problem. Therefore, the subject gains the ability to "learn to learn" under a variety of conditions that challenge the balance and walking control systems.

Mobility may help to develop an in-flight countermeasure that is built around ISS treadmill exercise activities. By manipulating the sensory conditions of exercise (e.g., varying visual flow patterns during walking), this training regimen systematically and repeatedly promotes adaptive change in walking performance, improving the ability of the astronaut to adapt to a novel gravity environment. It is anticipated that this training regimen will facilitate neural adaptation to unit (Earth) and partial (Mars) gravity after long-duration space flight.

All participating subjects perform two tests of locomotor performance both preflight and postflight: the Integrated Treadmill Locomotion Test and the Functional Mobility Test.

Results
Data collection on all subjects has been completed and is currently undergoing analysis.

Publication(s)
There are no publications at this time.

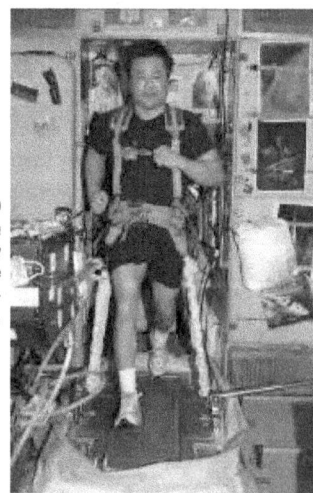

ISS010E24001 —Astronaut Leroy Chiao, Expedition 10 commander and NASA ISS science officer, equipped with a bungee harness, exercises on the TVIS in the *Zvezda* service module of ISS. Crewmembers who complete standard exercise protocols on ISS are the experimental controls for the Mobility experiment.

NUTRITIONAL STATUS ASSESSMENT (Nutrition)

Principal Investigator(s): Scott M. Smith, Ph.D., NASA Johnson Space Center, Houston
Expeditions 14–20, ongoing

Research Area Immune System and Integrated Studies

The NASA Clinical Nutritional Assessment (MR016L) profile, which has been implemented with two *Mir* and all ISS U.S. crewmembers, consists of two preflight and one postflight analysis of nutritional status as well as an in-flight assessment of dietary intake using a Food Frequency Questionnaire. The Clinical Nutrition Assessment (Nutrition) investigation expanded MR016L testing by:
- In-flight blood and urine collection
- Expanding nominal testing to include additional normative markers of nutritional assessment
- Adding a return plus 30-day (R+30) session to allow evaluation of postflight nutrition and implications for rehabilitation

On earlier ISS missions, it was not possible to assess nutritional status during flight because blood and urine could not be collected, stowed frozen, and returned. The altered nutritional status findings for several nutrients postflight are of concern, and require the ability to monitor the status of these nutrients during flight to determine whether there is a specific impetus or time frame for these decrements. In addition to monitoring crew nutritional status during flight, in-flight sample collection would allow for better assessment of countermeasure effectiveness. Nutrition is also designed to expand the current MR016L to include additional normative markers for assessing crew health and countermeasure effectiveness, and extend the current protocol to include an additional postflight blood and urine collection (R+30). Several nutritional assessment parameters are altered at landing, but it is not known whether the changes are still apparent 30 days postflight.

Additional markers of bone metabolism (helical peptide, OPG, receptor activator of NF-B ligand (RANKL), insulin-like growth factor 1 (IGF-1)) are measured to better monitor bone health and countermeasure efficacy. New markers of oxidative damage are measured (8-iso-prostaglandin F2a, protein carbonyls, oxidized and reduced glutathione) to better assess the type of oxidative insults during space flight. The array of nutritional assessment parameters was expanded to include serum folate, plasma pyridoxal 5'-phosphate, and homocysteine to better understand changes in folate, vitamin B6 status, and related cardiovascular risk factors during and postflight. Additionally, stress hormones and hormones that affect bone and muscle metabolism will also be measured (dihydroergocryptine (DHEA), dehydroepiandrosterone (DHEAS), cortisol, testosterone, estradiol). This additional assessment would allow for better health monitoring, and more accurate recommendations to be made for crew rehabilitation.

RESULTS
This experiment is still ongoing. Urine and blood samples for several astronauts have been collected before, during, and after ISS Expeditions. Since the experiment design calls for the combination and comparative analysis of data from all Expeditions, final results are not yet available.

PUBLICATION(S)
There are no publications at this time.

ISS015E10555 — Astronaut Suni Williams, Expeditions 14 and 15 flight engineer, configures her blood samples in the HRF-2 refrigerated centrifuge, preparing to separate the cellular and liquid components of blood to facilitate sample analysis on the ground.

BIOAVAILABILITY AND PERFORMANCE OF PROMETHAZINE DURING SPACE FLIGHT (PMZ)

Principal Investigator(s): Lakshmi Putcha, Ph.D., NASA Johnson Space Center, Houston
Expeditions: 11, 13, 15, 16

Research Area Neurological and Vestibular Systems

Promethazine (PMZ) is a pharmaceutical treatment for the space motion sickness that is experienced by space shuttle astronauts. This investigation measured the amount of medication that is absorbed and available to the body after taking it to estimate the intensity of drug-related side effects that may affect the performance of astronauts during space flights.

For this investigation, two sessions were conducted: a ground-based and an in-flight session. The in-flight session involved astronauts who take Promethazine for motion sickness during their mission. To keep track of these astronauts, Actiwatches were worn to monitor sleep and activity; saliva samples were provided to measure drug levels in the body; and the Karolinska Sleepiness Score (KSS) was used to estimate sleepiness side effects. These data was compared to the postflight data that were collected 30 days after return from their mission. Results from this study will be used to determine how Promethazine is handled by the body, and how severe the performance-impacting side effects of the drug are on orbit and on Earth.

RESULTS
There are no results at this time.

PUBLICATION(S)
There are no publications at this time.

EFFECTS OF EVA AND LONG-TERM EXPOSURE TO MICROGRAVITY ON PULMONARY FUNCTION (PuFF)

Principal Investigator(s):   John B. West, University of California-San Diego, La Jolla, Calif.
Expeditions 3–6

Research Area Physiological Studies: Cardiovascular and Respiratory Systems

This experiment examined the effect of long-term exposure to microgravity and EVA on pulmonary function by studying crewmembers before and after they performed EVAs. It examined whether pulmonary function was affected by long-term exposure to noxious gases or to particulate matter that may accumulate in the atmosphere of the ISS.

There is a large difference in pressure between the inside of ISS and in the spacesuit that is used for EVAs. The effects of this difference in pressure pose a significant risk of decompression sickness (DCS)—known in the diving world as "the bends"—for spacewalkers, including bubble formation within the blood. Even if the symptoms of DCS do not occur, venous gas microbubbles can alter pulmonary function, increasing the risk of forming a venous embolism.

ISS006348005 — Astronaut Donald R. Pettit, Expedition 6 NASA ISS science officer, uses a camera during a spacewalk on Jan 15, 2003.

Each Pulmonary Function in Flight (PuFF) session consisted of five noninvasive tests with the crew breathing only cabin air. The tests measured the pulmonary system's ability to exchange gases, the amount of air that is inspired and expired as a function of time, and the maximum pressure of the air that is inhaled and exhaled. The analysis looked for markers that indicate that the lungs have been weakened from exposure to microgravity, or that the body's ability to exchange and distribute gases has been disrupted.

PuFF hardware, including a manual breathing valve and flow meter, was attached to the HRF gas analyzer system for metabolic analysis physiology (GASMAP) hardware, physiological signal conditioners, and the HRF computer. GASMAP measured the volume of gases that were inspired and expired, frequency of respiration, and ambient barometric pressure.

RESULTS

Excellent quality data were collected from the crews on all four Expeditions, comprising eight crewmembers who performed EVAs. Prisk et al. (2005, 2006) reported no significant changes in pulmonary gas exchange 1 day after an EVA was performed. The researchers report a small increase in metabolic rates in the astronauts on the day following the EVA. The authors note that testing was not ideal because acute effects resulting from EVAs could not be measured immediately after performing the spacewalk due to EVA protocols and logistics, but remain confident in their primary conclusion. However, the small effect observed on the day following EVA suggests that current denitrogenation protocols prevent the decompression stress that is associated with EVA from causing any major lasting disruption to gas exchange in the lung.

ISS006E07133 —Donald R. Pettit sets up PuFF hardware in preparation for an HRF experiment in the U.S. *Destiny* laboratory on station during Expedition 6.

PUBLICATIONS(S)

Prisk GK, Fine JM, Cooper TK, West JB. Pulmonary gas exchange is not impaired 24 h after extravehicular activity. *Journal of Applied Physiology.* 2005; 99(6):2233–8.

Prisk GK, Fine JM, Cooper TK, West JB. Vital Capacity, Respiratory Muscle Strength and Pulmonary Gas Exchange during Long-Duration Exposure to Microgravity. *Journal of Applied Physiology.* 2006; 101:439–447.

Renal Stone Risk During Space Flight: Assessment and Countermeasure Validation (Renal Stone)

Principal Investigator(s): Peggy A. Whitson, NASA Johnson Space Center, Houston
Expeditions 3–6, 8, 11–14

Research Area Physiological Studies: Bone and Muscle

The loss of calcium from bone that is combined with decreased fluid intake in flight increases the probability for kidney stone formation during and after flight. Development of a kidney (or renal) stone in an astronaut can have serious consequences since it cannot be treated in flight as it would be on the ground. To better understand the risks to astronauts on long-duration space flights, quantification of renal stone formation potential and recovery is required. This experiment studied the potential development of renal stones in space crews and the efficacy of a pharmaceutical countermeasure.

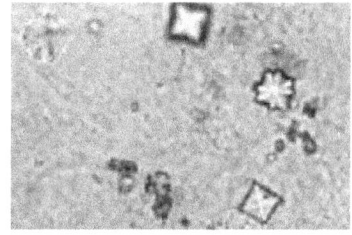

This micrograph shows calcium oxalate crystals in urine. These small crystals can develop to form renal stones.

Potassium citrate (K-cit) is a proven ground-based treatment for patients who are suffering from renal stones. In this study, from 3 days before launch and continuing through 14 days after landing, each crewmember ingested either two K-cit tablets or two placebos daily. Urine samples were collected during 24-hour periods at three points during the crewmembers' mission, once at the beginning, midway point, and at the end. In addition to taking pills and collecting urine samples, crewmembers maintained handwritten logs of their daily food and fluid intake, exercise, and medication during the time of the urine collections. These log books acted as a backup to the barcode reader records that are part of the inventory management system with which crews typically record food intake and medication.

Ultimately, these data will not only help long-duration space flight crews, but also will aid those on Earth in understanding how renal stones form in otherwise healthy persons. This should also provide insight into stone-forming diseases on Earth.

Results
Urine samples were collected before, during, and after flights, as was dietary information from crewmembers. Since the experiment design calls for the combination and comparative analysis of data from all Expeditions, final results are pending.

Publication(s)
There are no publications at this time.

ISS011E13500 — ISS commander Sergei Krikalev with Renal Stone sample collection hardware during ISS Expedition 11.

Sleep-Wake Actigraphy and Light Exposure During Space Flight-Long (Sleep-Long)

Principal Investigator(s): Charles A. Czeisler, M.D., Ph.D., Brigham and Women's Hospital, Harvard Medical School, Boston, Mass.
Expeditions 14–18, ongoing

Research Area Behavior and Performance Studies

The success and effectiveness of human space flight depends on the ability of crewmembers to maintain a high level of cognitive performance and vigilance while operating and monitoring sophisticated instrumentation. Astronauts during long-duration space flights, however, commonly experience sleep disruption and may experience misalignment of circadian phase. Both of these conditions are associated with insomnia, and impairment of alertness and cognitive performance.

There is little information on the effect of long-duration space flight on sleep and circadian rhythm organization. Sleep-Wake Actigraphy and Light Exposure During Spaceflight-Long (Sleep-Long) uses state-of-the-art ambulatory technology to monitor sleep-wake activity patterns and light exposure in crewmembers who are aboard ISS. Subjects wear a small, lightweight activity- and light-recording device (Actiwatch) for the duration of their mission. The sleep-wake activity and light-exposure patterns that are obtained in flight are compared with baseline data that are collected on Earth before and after space flight. These data should help us better understand the effects of space flight on sleep as well as aid in the development of effective countermeasures for long-duration space flight.

Results
Data collection for this investigation is ongoing. Final results will be analyzed for completion of data collection from the last test subject.

Publication(s)
There are no publications at this time.

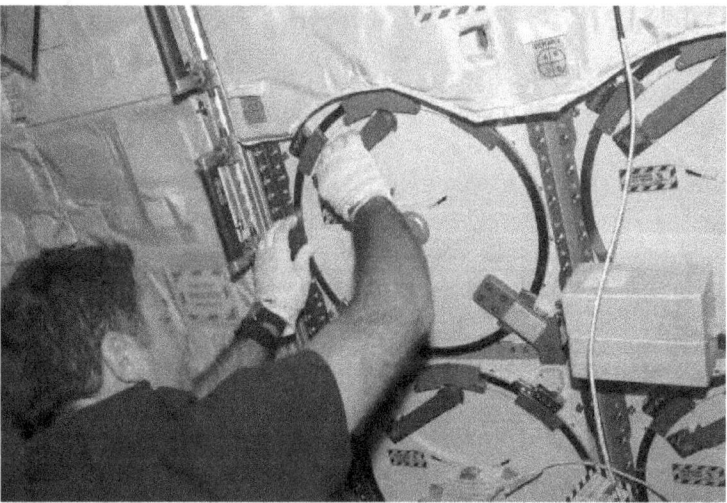

ISS014E05119 — The Sleep-Long Actiwatch is visible on the left arm of astronaut Michael Lopez-Alegria, Expedition 14 commander. The Actiwatch monitors light and activity patterns of crewmembers.

Sleep-Wake Actigraphy and Light Exposure During Space Flight-Short (Sleep-Short)

Principal Investigator(s): Charles A. Czeisler, M.D., Ph.D., Brigham and Women's Hospital, Harvard Medical School, Boston, Mass.
Expeditions 11, 13–18, ongoing

Research Area Behavior and Performance Studies

S104E5114 — Astronaut, Janet Kavandi on STS 104 wearing an Actiwatch on her right wrist for recording activities.

The success and effectiveness of human space flight depends on the ability of crewmembers to maintain a high level of cognitive performance and vigilance while operating and monitoring sophisticated instrumentation. Astronauts during short space flights, however, commonly experience sleep disruption and may experience misalignment of circadian phase. Both of these conditions are associated with insomnia, and impairment of alertness and cognitive performance.

Relatively little is known of the prevalence or cause of space-flight-induced insomnia in short-duration missions. This experiment uses state-of-the-art ambulatory technology to monitor sleep-wake activity patterns and light exposure in crewmembers who are aboard the space shuttle. Subjects wear a small, lightweight activity- and light-recording device (Actiwatch) for the duration of their mission. The sleep-wake activity and light exposure patterns that are obtained in flight are compared with baseline data that are collected on Earth before and after space flight. The data that are collected should help us better understand the effects of space flight on sleep as well as aid in the development of effective countermeasures for short-duration space flight.

Results
Data collection for this investigation is ongoing. Final results will be analyzed for completion of data collection from the last test subject.

Publication(s)
There are no publications at this time.

STABILITY OF PHARMACOTHERAPEUTIC AND NUTRITIONAL COMPOUNDS (Stability)

Principal Investigator(s): Lakshmi Putcha, Ph.D., NASA Johnson Space Center, Houston; Scott M. Smith, Ph.D., NASA Johnson Space Center, Houston
Expeditions 13–18

Research Area Radiation Studies

Data gathered from past space shuttle missions suggest that some of the medications that are packed in the shuttle medical pack degrade even after relatively brief periods (less than 20 days) of space flight. The observed degradation includes both physical and chemical characteristics of medicine formulations. The degradation was sufficient to influence FDA-stipulated shelf-life for these formulations and may result in a loss of potency. Physical and chemical instability of medications could render treatments with degraded drugs ineffective for assurance of optimal crew health during long-duration space exploration missions. An evaluation of subjective data on medications that are used by crewmembers during space flight indicates that 8% of all treatments that have been administered in the Space Shuttle Program were reported ineffective. Pharmaceutical instability may modify effectiveness and safety, and is one possible cause of the ineffectiveness of treatments. Degradation of food products may also render them ineffective in providing health and energy sustenance. The stability of medications and foods that are used by the crew must be adequate to facilitate safe exploration of space in the future. The Stability of Pharmacotherapeutic and Nutritional Compounds (Stability) investigation evaluates mission-critical medications and foods to understand issues relating to loss of potency for medicines and to nutritional adequacy of foods in space.

Four identical Stability kits were delivered to ISS in Jul 2006 during the STS-121/ULF1.1 mission. The first kit was returned to Earth during the STS-121/ULF1.1 mission. The second kit was returned after 11 months of exposure during the STS-117/13A mission in Jun 2007. The third kit was returned after 1 year and 7 months of exposure during the STS-122/1E mission in Feb 2008. The fourth kit was returned on STS-128/ULF2 in Nov 2008.

Pharmaceuticals
Results from the Stability investigation offer an assessment of the stability and chemical integrity of medications in adverse environments that are encountered during space missions. This information will also assist in identifying susceptible medications that may require alternate methods of preparation, dispensing, and storage to improve stability and minimize loss of potency during space missions. Estimates of shelf-life and potency from Stability data can be used for the selection, modification, and development of drug dosage forms and storage techniques for the ISS, moon, and Mars missions. The data will also enhance our understanding of the conditions that contribute to pharmaceutical instability and help define mitigation strategies that assure adequate drug stability for human exploration of space.

Nutrients
Nutrients are vital for every cellular process in the body, both on Earth and in space. It is evident that the status of certain vitamins in the body is altered during long-duration space missions, but it is not known whether vitamin metabolism is altered, the vitamins in the food supply have been degraded, or the food supply contains enough of each nutrient throughout the mission. For example, other studies have shown that folate status is decreased after 3- to 4-month missions. It is possible that high levels of ionizing radiation or long-term storage could destroy nutrients in foods or decrease their bioavailability. Either of these would lead to impaired nutritional status in crews who are consuming these foods.

Stability results from nutritional products could provide important information about the susceptibility of vitamins, amino acids, and fatty acids in the space food system to adverse environmental factors that are encountered during space missions. These results will assist in determining whether modified requirements are necessary to avoid malnutrition in crewmembers. The results will also indicate whether improved packaging designs will be needed to protect food from irradiation and long-term storage conditions during long-duration space exploration missions. These data will be useful in identifying alternate methods of preparation and storage of these specific food systems to minimize loss of nutritional value during long-duration space exploration missions.

Results
There are no results to report at this time.

Publication(s)
There are no publications at this time.

ISS007E07832 — Expedition 7 Science Officer Ed Lu prepares to add garlic paste to a food packet while preparing a meal in the galley area of the *Zvezda* service module. A can of green peas and eating utensils are visible on the table in front of him. Preserving food for long-duration Exploration missions is important to maintaining the safety and health of the crew.

SUBREGIONAL ASSESSMENT OF BONE LOSS IN THE AXIAL SKELETON IN LONG-TERM SPACE FLIGHT (Subregional Bone)

Principal Investigator(s): Thomas F. Lang, University of California, San Francisco, Calif.
Expeditions 2–8

Research Area Physiological Studies: Bone and Muscle

Bone loss is one of the known risks of exposure to reduced gravity—a risk that increases with the length of stay in that environment. Although healthy bone can repair damage done to itself, researchers are yet unsure how much bone is replaced after crewmembers return to Earth. Is bone mass recovered 1 year after flight? Is there a difference in the subregional distribution of bone prior to flight and 1 year after flight? Subregional Bone measured the amount of bone that was lost during space flight and recovered postflight in an effort to answer these questions.

Subregional Bone hardware consisted of several devices that were used before and after flight. Dual-energy X-ray absorptiometry (DEXA) provided a two-dimensional measurement of the entire bone mass of the hip, spine, and heel. These measurements were compared to quantitative computed tomography (QCT), which examined cortical (the bone's dense outer layer) and trabecular (the bone's inner, spongy-looking layer) bone separately and three-dimensionally to determine the extent of bone loss in the hip and spine. QCT measurements allow researchers to determine whether loss is localized in a subregion of the bone. DEXA and QCT measurements were also compared to quantitative ultrasound (QUS) of the heel to evaluate ultrasound as a possible alternative to X-ray measurements.

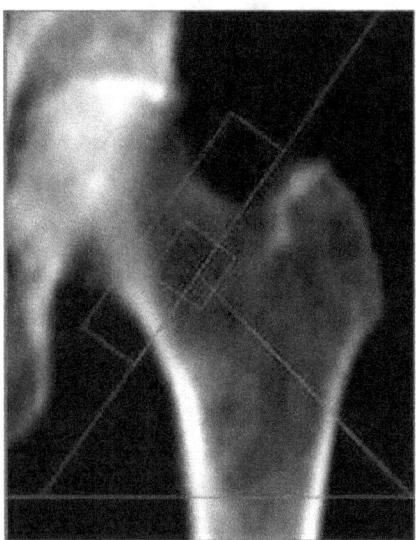

DEXA scan of a human hip

RESULTS

This experiment determined the distribution of bone loss in the spine and hip in long-duration space flight using QCT and assessed how bone is recovered after return. One of the first Bioastronautics research investigations to begin on ISS, this study recruited 16 subjects between Expedition 2 and Expedition 8. The first publication in the *Journal of Bone and Mineral Research* (Lang et al. 2004) included eight subjects who had been back long enough to measure their bone density 1 year postflight. On ISS, bone mineral density was lost at an average rate of about 0.9% per month in the lumbar spine and 1.4% per month in the femoral neck. For comparison, a post-menopausal woman experiences losses of bone mineral on the order of 1% per year. The experiment provides insight into the process of bone loss because it is the first study to differentiate the loss in the cortical bone (the outer part of the bone) and the trabecular bone (the inner parts of the bone). For example, losses of mass in the cortical bone of the hip averaged around 1.6–1.7% per month whereas losses in the trabecular bone averaged 2.2–2.5% per month. In subsequent publications, Lang et al. (2006a and 2006b) measured the femoral neck cross section and calculated volumetric bone loss and recovery after 1 year on the ground after space flight. The authors report that bone mass and structure of the astronauts' femurs recovered—but not fully—after 1 year back on Earth. Astronauts experienced an average 11% decline in femoral bone loss during space flight. While bone mass and volume increased back on Earth, the volumetric bone mass density did not fully recover (proximal femur is larger in size, but less mineralized and more porous than bone lost during space flight).

This research is included in a comprehensive discussion of the effects of space flight on bone health (Cavanaugh and Rice, 2007). Lang et al. (2007) discuss their results in a broader context. The large weight-bearing bones (pelvis, hips, legs) suffer greatest bone loss during space flight, and countermeasures and recovery therapies should focus on those areas to better protect against injuries that may be related to diminished bone strength. This research has direct applications to the design of countermeasures and considerations that will be built into surface activities for future exploration missions to Mars: after a lengthy transit (6 months) in microgravity, astronauts will be expected to perform on the surface of Mars and engage in activities that place loads on large bones. This work feeds into ongoing and future experiments that monitor astronauts' biochemical indicators of bone health and bone loss, as well as the design of diagnostic tools that may provide additional means to monitor bone size and density during exploration space flights (Lang et al. 2007).

In a related follow-on study, Sibonga et al. (2007) examined the bone mineral density (BMD) measurements in five regional sites for 45 crewmembers, both U.S. and Russian, who participated on 56 long-duration flights of at least 4 months' duration. The study population showed variable decreases in BMD across the five sites; the key result was that they could calculate an estimate for time required to restore 50% of the loss of bone in each site, and that full recovery would take up to 3 years—much longer than the mission duration.

PUBLICATION(S)

Lang T, LeBlanc A, Evans H, Lu Y, Gennant H, Yu A. Cortical and Trabecular Bone Mineral Loss from the Spine and Hip in Long-duration Spaceflight. *Journal of Bone and Mineral Research.* 19(6):1006–12, 2004.

Lang TF, LeBlanc AD, Evans HJ, Lu Y. Adaptation of the Proximal Femur to Skeletal Reloading After Long-Duration Spcaeflight. *Journal of Bone and Mineral Research.* 21(8);1224-1230, 2006.

Lang, TF, Keyak, JH and LeBlanc, AD, Defining and Assessing Bone Health During and After Spaceflight, in *Bone Loss During Spaceflight*, Cavanaugh and Rice (eds), Cleveland Clinic Press, Cleveland, Ohio, pp. 63–69, 2007.

Cavanaugh PR, Rice AJ. Bone Loss During Spaceflight, Cleveland Clinic Press, Cleveland, Ohio, 297 pp., 2007.

Sibonga JD, Evans HJ, Sung HG, Spector ER, Lang, TF, Oganov VS, Bakulin AV, Shackelford LC, LeBlanc AD. Recovery of spaceflight-induced bone loss: Bone mineral density after long-duration missions as fitted with an exponential function. *Bone*, 41 (2007) 973–978.

Sibonga JD, Evans HJ, Spector ER, Maddocks MJ, Smith SA, Shackelford LC, LeBlanc AD. Bone Health During and After Spaceflight, in *Bone Loss During Spaceflight* (Cavanaugh and Rice, eds), Cleveland Clinic Press, Cleveland, Ohio, pp. 45–51, 2007.

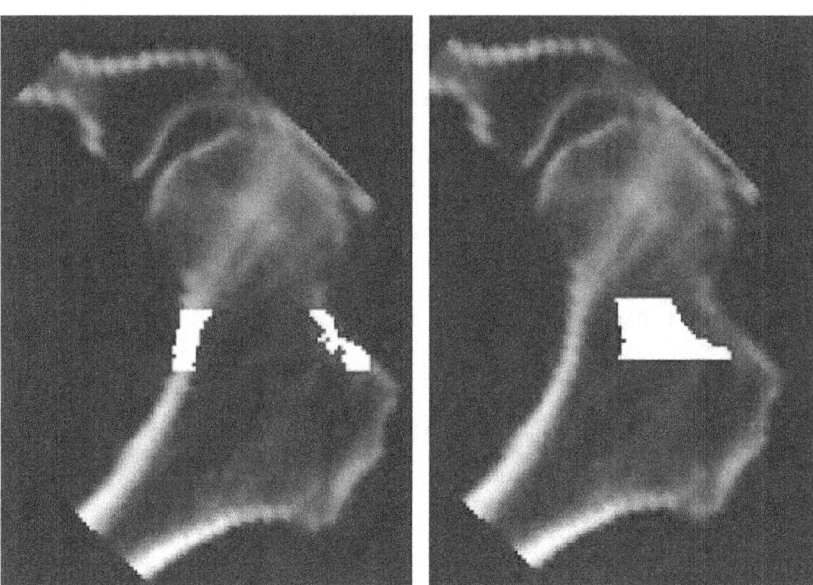

One-mm-thick sections through the mid frontal plane of the hip, showing regions of evaluation in white superimposed on a false color image of the computed tomography (CT) data. The image on the left shows the cortical region of the femoral neck, and the image on the right shows the trabecular bone regions.

Surface, Water, and Air Biocharacterization (SWAB) – A Comprehensive Characterization of Microorganisms and Allergens in Spacecraft

Principal Investigator(s):   Duane L. Pierson, Ph.D., NASA Johnson Space Center, Houston
Expeditions 13–16, ongoing

Research Area Immune System and Integrated Studies

ISS0515E07586 — Astronaut Sunita L. Williams, Expeditions 14 and 15 flight engineer, conducts a SWAB air sampling in the *Destiny* laboratory of the ISS.

During long-duration space flight, spacecraft build up a diverse array of microorganisms that directly interacts with the crew. Most microorganisms are harmless or even beneficial to the crew; however, the presence of medically significant organisms appearing in this environment could adversely affect crew health and performance during long-duration missions. The primary goal of the Surface, Water, and Air Biocharacterization (SWAB) – A Comprehensive Characterization of Microorganisms and Allergen in Spacecraft experiment is to use advanced technologies to better understand the types of organisms that the crew could encounter and their sources, and to assess the potential risks.

This study of microorganisms, allergens, and microbial toxins in the spacecraft environment was initiated to ensure the health, safety, and performance of crewmembers during flight. All previous methods evaluating spacecraft ecology have used culture-based methodology, thus many organisms have been omitted from isolation, including medically significant organisms such as the pathogen *Legionella* (the bacteria that causes Legionnaire's disease). Likewise, culturable bacteria and fungi are the only potential allergens that have been studied; the more potent allergens, such as dust mites, have never been analyzed in spacecraft environments. This study uses modern molecular biology, advanced microscopy, and immunochemical techniques to examine air, surface, and water samples for bacteria and fungi, pathogenic protozoa, allergens, and microbial toxins.

To accomplish this goal, new collection techniques have been developed to improve the quality of the sample that is being returned from ISS for analysis. Air samples are being collected through a novel gelatin filter to improve collection efficiency. These filters can retain particles as small as viruses. Water and surface samples have been designed to improve DNA recovery, using a DNA preservative that is composed of a mixture of sodium dodecyl sulfate (SDS) and ethylenediamine-tetraacetic acid (EDTA) in Tris buffer.

Analysis of the in-flight samples focuses around molecular techniques. These include bacterial fingerprinting, bacterial and fungal ribosomal identification, and quantitative PCR to identify and enumerate specific genes in environmental samples. The identification of specific genes is critical in the assessment of microorganisms for particular characteristics, including the production of microbial toxins. The samples that are returned from flight will also be evaluated using denaturing gradient gel electrophoresis (DGGE), which is a technique that allows for the identification of bacteria without any amplification of organisms with growth on media. This technique holds the potential to increase the number of different identified species by 100-fold.

Results
Sample collection for the surface and air samples has been completed, and the samples have been returned to Earth for analysis. Water sampling will take place during future missions.

Publication(s)
There are no publications at this time.

Organ Dose Measurement Using a Phantom Torso (Torso)

Principal Investigator(s): Gautam D. Badhwar, NASA Johnson Space Center, Houston
Expedition 2

Research Area Radiation Studies

One of the most critical risks to humans in space is radiation exposure. Outside the protection of Earth's atmosphere, space crews are exposed to a wide range of particles, including neutrons, that are not normally a threat on Earth. Exposure to radiation that is found in LEO and beyond can cause cataracts, cancer, damage to reproductive organs and the nervous system, and changes in heredity.

The Organ Dose Measurement Using the Phantom Torso (Torso) employed a model human head and torso (Rando phantom), which was imbedded with more than 350 detectors (thermo-luminescent detectors) and five silicon diode detectors, over five depths to measure absorbed dose to specific organs during shuttle flight. A tissue-equivalent proportional detector and a charged-particle directional spectrometer (CPDS) were placed within 1.5 feet of the torso during these ISS measurements. This was the first NASA experiment to simulate doses at discrete locations within the body.

The tissue-equivalent proportional counter (TEPC) consisted of a spectrometer and cylindrical detector with which to measure external radiation doses. The TEPC measured radiation dose and dose equivalent in complex radiation fields (fields containing a mixture of particle types). The CPDS measures particle energy and direction inside ISS. Both the TEPC and the CPDS remained within 1 to 1.5 feet (30.48–45.72 cm) of Torso during its operation on station.

Results

Torso results were combined with results from various experiments on previous missions to validate NASA's organ dose database for astronauts. Preliminary results suggest that organ dose and dose equivalent can be projected to a ±25% accuracy using a combination of dosimetry and radiation transport models. This accuracy envelope is greatly improved relative to the current accuracy of organ-specific cancer risk projections, which are estimated at ±500%. Further analyses and incorporation of these radiation results into operational planning for exploration is ongoing.

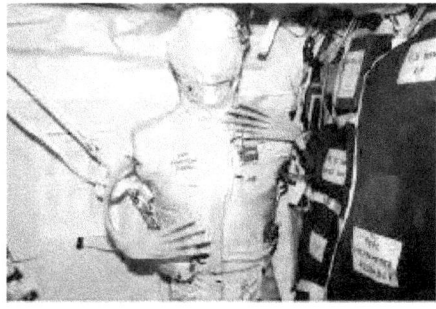

ISS002E5952 — Image of Torso on station during Expedition 2.

Overall, the dose rates that were measured in Torso were in good general agreement with other measured values and with the models that were used to predict these values. The largest differences that were observed between measured data and the simulations were 15%. In addition, a model that considers orbital altitude, attitude, and solar cycle emissions agreed within 25% of the measured data. It was determined that the majority of radiation energy that was deposited in human tissues (~80%) was due to galactic cosmic radiation. This is due to spacecraft material providing effective attenuation of the protons that are trapped in the Earth's magnetic field. The data indicated an average radiation quality factor (a measurement of how damaging a type of radiation is to tissue) of 2.6; these quality factors do not appreciably change with depth in the body. Finally, this experiment indicated that the contribution to both skin and organ doses from secondary neutrons is not negligible (Expedition 2 Postflight Report).

In a follow-on analysis, Cucinotta et al. (2008) report results from post-mission biodosimetry assessments of chromosomal damage in lymphocyte cells from 19 ISS astronauts. These results were compared with space radiation transport models, irradiation of preflight blood samples, and results from the phantom torso experiments. The ISS missions who were sampled include those who flew on the earliest missions near the solar maximum, and conclude with the Increment 15 astronauts, who flew near the solar minimum. During this time frame, although 67 SPEs occurred, the extended solar maximum (particular to this solar cycle) decreased the galactic cosmic ray (GCR) levels. Average effective doses for a 6-month stay on the ISS were 72 mSv. At least 80% of the organ dose equivalents come from GCRs. Another important result shows that the models are predictive within about 10%. The authors conclude that many uncertainties about space radiation remain—both levels and types of radiation, and effects inside the spacecraft. Continued research and analyses are required.

PUBLICATION(S)

Cucinotta FA, Wu H, Semones EJ, Gibbons F, Atwell W, Shavers MR, George KL, Badhwar GD. Astronauts organ doses form 1962–2002: space validation with phantom torso experiments (Abstract), Proceedings of the Bioastronautics Investigators' Workshop, Galveston, Texas. Jan 2003.

Cucinotta FA, Kim M-H, Willingham V, George K. Physical and Biological Organ Dosimetry Analysis for International Space Station Astronauts. *Radiation Research*, 170, 127–138, 2008.

Effect of Microgravity on the Peripheral Subcutaneous Veno-arteriolar Reflex in Humans (Xenon-1)

Principal Investigator(s): Anders Gabrielsen, Danish Aerospace Medical Center of Research National University Hospital (Rigshospitalet), Copenhagen, Denmark
Expeditions 3–5

Research Area Physiological Studies: Cardiovascular and Respiratory Systems

When the body's legs are lower in relationship to the heart, the body triggers what is called a local veno-arteriolar reflex, where small subcutaneous (below the surface of the skin) blood vessels constrict, forcing blood from the feet toward the head. If this reflex is not properly triggered or if blood circulation is impeded, the blood pressure drops, causing dizziness and, possibly, fainting. This effect is called orthostatic intolerance. Due to a number of possible reasons—reduced fluid volume, muscle atrophy, neurovestibular adaptation—astronauts suffer from orthostatic intolerance during entry and landing and for a few days postflight, potentially interfering with their ability to perform entry and landing tasks and prolonging their recovery period. Xenon-1 tested the local veno-arteriolar reflex in an effort to understand the source of, and ways to combat, postflight orthostatic intolerance.

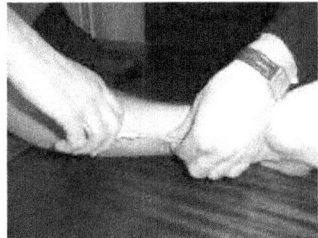

Injection of the tracer into a subject's leg.

Prior to and following Expeditions 3, 4, and 5, station crewmembers were placed on a gurney as a small amount of Xenon-133, which is a radioactive isotope that is dissolved in sterile saline solution, was injected into the subcutaneous tissue of their lower legs. Arterial blood pressure was recorded by a continuous pressure device on the crewmember's index finger. This measurement, which was taken with the Xenon-1 detector unit, was used to trace the movement of the Xenon tracer following injection. As the measurements were taken, the Xenon memory box recorded and displayed the counting rate.

Results

The last group of subjects for this experiment returned after Expedition 5. Data from all subjects were collected successfully. The Gabrielsen and Norsk (2007) findings show veno-arteriolar reflex reduced subcutaneous blood flow by $37 \pm 9\%$ before flight and by $64 \pm 8\%$ following landing, with no statistical difference between the two responses. The mean arterial pressures and heart rates in supine astronauts were very similar before and after flight. These data indicate that this reflex is not attenuated by weightlessness, and suggest that the veno-arteriolar reflex is not a contributor to postflight orthostatic intolerance.

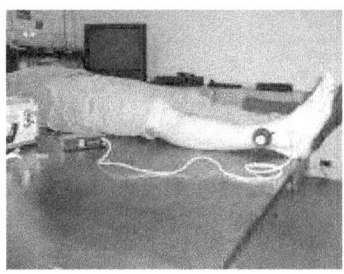

A Xenon detector unit is attached to each crewmember's leg, slightly above the ankle, as shown here.

Publications

Gabrielsen A, Norsk P. Effect of spaceflight on the subcutaneous venoarteriolar reflex in the human lower leg, *J. Appl. Physiol.* 103:959-962 (2007).

Observing the Earth and Educational Activities

The tradition of Earth observations from orbit was born in 1962 when Project Mercury astronaut John Glenn packed an Ansco Autoset 35 mm camera, which was bought at a local drug store, to take photographs of the Earth during the first NASA orbital mission. These images changed our view of ourselves and our relationship to the Earth. Even with the many satellites now orbiting the Earth, ISS continues to provide unique views of our planet.

ISS provides a unique platform for inspiring students to excel in mathematics and science. Station educational activities have had a positive impact on thousands of students by involving them in station research, and by using the station to teach them the science and engineering that are behind space exploration.

Observing the Earth and Educational Activities Performed on the International Space Station, Grouped by Discipline

Educational Activities
ARISS (Amateur Radio on the International Space Station)
CSI-01 (Commercial Generic Bioprocessing Apparatus Science Insert-01)
CSI-02 (Commercial Generic Bioprocessing Apparatus Science Insert-02)
DREAMTIME (DreamTime)
EPO (Education Payload Operations)
EPO-Demos (Education Payload Operation-Demonstrations)
EPO-Educator (Education Payload Operations-Educator)
EPO-Kit C (Education Payload Operations-Kit C: Plant Growth Chambers)
EarthKAM (Earth Knowledge Acquired by Middle School Students)
Education-SEEDS (Space-Exposed Experiment Developed for Students)
SEM (Space Experiment Module)
Observing the Earth
CEO (Crew Earth Observations)
CEO-IPY (Crew Earth Observations-International Polar Year)

AMATEUR RADIO ON THE INTERNATIONAL SPACE STATION (ARISS)

Principal Investigator: Frank Bauer, NASA Headquarters, Washington, D.C.
Expeditions 2–17, ongoing

Research Area Educational Activities

Ever since the Amateur Radio on the International Space Station (ARISS) hardware was first launched aboard Space Shuttle *Atlantis* on STS-106 and transferred to ISS, it has been regularly used education outreach. With the help of Amateur Radio Clubs and HAM radio operators, astronauts and cosmonauts aboard the ISS have been speaking directly with large groups of the general public, showing teachers, students, parents, and communities how amateur radio energizes students about science, technology, and learning. The overall goal of ARISS is to get students interested in mathematics and science by allowing them to talk directly with the crews who are living and working aboard the ISS.

In preparation for a radio contact, students research the ISS and learn about radio waves and amateur radio, among other topics. Before their scheduled contact with the ISS crew, they prepare a list of questions on topics that they have researched, many of which have to do with career choices and science activities aboard station. Depending on the amount of time and complexity of the questions, between 10 and 20 questions can be asked during a single session. During the sessions, the ISS passes over a school or another location that receives a signal from station and relays that signal on to the participating school; there is typically a 5- to 8-minute window for students to make contact with the crews aboard ISS.

While usually only a handful of students can ask questions due to the limited time available, hundreds of other students frequently are listening in to the school event from their school classrooms or auditorium, so that each of these events typically reaches hundreds of students.

RESULTS
ARISS has been instrumental in using amateur radio to connect teachers and students to the crew of the ISS, sparking an interest in science and mathematics for many students around the world.

PUBLICATION(S)
Wright RL. Remember, We're Pioneers! The First School Contact with the International Space Station. Proceedings of the AMSAT-NA 22nd Space Symposium, Arlington, Va. 2004 Oct 8–10.

Cunningham C (N7NFX). NA1SS, NA1SS, This is KA7SKY Calling….. Proceedings of the AMSAT-NA Space Symposium, Arlington, Va. 2004 Oct 8–10.

Palazzolo P (KB3NMS). Launching Dreams: The Long-term Impact of SAREX and ARISS on Student Achievement. Proceedings of the AMSAT-NA Space Symposium, Pittsburgh, Penn. 2006 Oct 26–28.

ISS014E18307 — Astronaut Sunita L. Williams, Expeditions 14 and 15 flight engineer, talks with students at the International School of Brussels in Belgium during an ARISS session in the *Zvezda* service module.

Crew Earth Observations (CEO)

Principal Investigator(s): Sue Runco, NASA Johnson Space Center, Houston
Expeditions 1–17, ongoing

Research Area Observing the Earth

ISS crewmembers use commercial and professional handheld cameras with a suite of lenses (from wide angle to an 800mm lens equivalent) to take Earth observation photographs that support research in a wide variety of Earth Science subdisciplines. Scientists on the ground train the crew in basic areas of Earth system science and provide to the crew a daily list of targets of the greatest scientific interest. Crewmembers takes photographs as time is available and during their leisure time. These digital photographs are downlinked, their location identified, and both images and meta-data are assimilated into a public database. The images are used as educational and research tools, as well as historical records of global environmental change, special geological and weather events, and the growth and change of human-made features, such as cities. CEO can be conducted from any available window on the space station, but is conducted primarily from the windows in the Russian *Zvezda* service module and the nadir-viewing, optical-quality window in the U.S. *Destiny* laboratory module.

Results

ISS provides a unique opportunity to capture a variety of sites on Earth by providing repeated overflight passes of the Earth with different lighting and viewing angles. Through CEO, ISS crewmembers share their view of the Earth with the public and take pictures of some of the most dramatic examples of change on the Earth's surface. These sites have included major deltas in south and east Asia, coral reefs, cities, alpine glaciers, volcanoes, large mega-fans (major fan-shaped river deposits), and features on Earth, such as impact craters, that are analogs to structures on other planets. Astronauts also record dynamic events: in 2004 and 2005, station astronauts took key photographs of the four Florida hurricanes, the Dec 2004 tsunami, and Hurricanes Katrina and Wilma. Other notable images capture volcanic eruptions on Mt Etna (Sicily) and the Cleveland volcano (Alaska), dust storms, smog, forest fires in the western U.S., and the Kolka glacier collapse in Russia.

From Expedition 1 through Dec 2007, ISS crewmembers took more than 300,000 images of Earth, almost half of the total number of images that were taken from orbit by astronauts since the first Mercury missions. Scientists and the public around the world have access to CEO images that were captured by astronauts on ISS through the Gateway to Astronaut Photography of Earth Web site (http://eol.jsc.nasa.gov). Between 400,000 and 1,000,000 digital photographs of Earth taken from the CEO collection are downloaded by the public each month. The Web site also features an "Image of the Week" and searchable access to all the photographs. Scientific analyses using CEO data have been published in scientific journals in a wide variety of disciplines. A few highlights of these publications are summarized here.

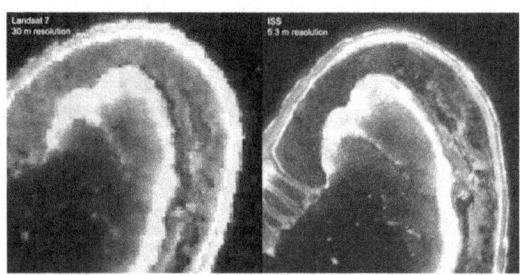

This comparison of Landsat 7 ETM+ (left) and digital photography from station (ISS002E6372, right) of a coral reef atoll illustrates how astronaut photography can be an important source of supplemental data to studies that use satellite imagery. Landsat's 30-meter spatial resolution does not provide enough information about the reef lagoon structure. Detailed photographs of reefs that were taken from ISS are being used around the world. The high spatial resolution of the images (about 5 meter/pixel) make them well suited for comparison to what is seen by divers in the water.

Spatial resolution is a measure of the smallest object that can be resolved by the sensor, or the size of the area on the ground represented by each pixel determined by geometric properties of the altitude of the spacecraft, lens magnification, size of the original image, and look angle. To achieve maximum potential spatial resolution, a camera system must capture information at sufficient speed to eliminate the effects of relative ground motion. Using handheld motion compensation, ISS crewmembers have achieved a spatial resolution of less than 6 meters in photographs of Earth from ISS. The ISS provides great potential as a remote-sensing platform that is capable of providing high-resolution imagery of the Earth's surface (Robinson and Evans 2002).

CEO images captured from ISS of Pacific Ocean atolls (islands consisting of a circular coral reef surrounding a lagoon) allowed for an assessment of spatial resolution on estimates of landscape parameters of the atolls. Data gathered indicated that landscape parameter estimates were fairly accurate regardless of spatial resolution changes from 5 to 30 meters. This study of ISS imagery showed that spatial resolution, as well as spectral resolution, is of equal importance when studying these formations (Andréfouët et al. 2003). The most detailed images of Fangatau Atoll, which were taken from ISS, were used to measure the biomass of the giant clam fishery at Fangatau Atoll with accuracy similar to that obtained from aerial photography (Andréfouët et al. 2005). Astronaut photographs of reefs in the Indian Ocean have been used as base maps for dive surveys of reef resources in the region (Quod et al. 2002).

Extracting clear water depths from a variety of sources allows the examination and mapping of shallow water from global to local scales. Scientists from the National Oceanic and Atmospheric Administration (NOAA) used four sources of data to map shallow water bathymetry near U.S. coral reef areas. These included the sea-viewing wide field-of-view sensor (SeaWiFS) on board the OrbView 2 Satellite (SeaWiFS allows global mapping within 1-kilometer pixels), the IKONOS satellite (global mapping within 4 meters), the Landsat Satellite (global mapping within 30 meter pixels), and handheld photography by the ISS crew (CEO local mapping within 6 meters). A new technique was applied to the blue and green bands from astronaut photography, allowing construction of a bathymetry map for Pearl and Hermes reef with accuracies similar to that obtained from IKONOS (Stumpf et al. 2003).

High-resolution astronaut photography that was collected from station has provided useful data for urban analysis, especially vegetation measurements. The accuracy of the data that are obtained from the astronaut photographs was similar to the data that were obtained by satellite remote sensors. The high-resolution astronaut photography that was obtained by the CEO investigation gives insights into vegetation density in urban areas (Stefanov and Robinson 2003).

Imagery of cities at night captured during ISS Expedition 6 by Astronaut Don Pettit (example, right) and subsequent astronauts aboard ISS has provided researchers preliminary data for potential applications modeling urban land use and population density. ISS photographs of cities at night are unique because they provide greater spatial resolution than any other source of city light data. Nighttime images of cities that are taken by ISS crewmembers have been used to support the concept of a new satellite sensor for nighttime lighting called Nightsat (Elvidge et al. 2007a; 2007b; Lulla 2003).

ISS006E24987 — ISS Expedition 6 science officer Don Pettit pioneered an approach using a homemade tracking system to track the ground as it moves relative to station; this allowed him to acquire long-exposure images under low-light conditions using very long exposures. This image shows the lights of Argentina's capital city, Buenos Aires.

Imagery that is taken by astronauts has been collected and used to identify and build a global inventory of a new class of landform called megafans. These are large cones of sediments that are deposited by rivers that empty into large continental basins. Megafans are more easily identified using the wide, oblique views that are provided by astronaut photography. The results have been applied to interpreting features on both Earth and Mars. Wilkinson et al. (2005) mapped megafans in South America from imagery that was collected by space shuttle and ISS crews to understand how aquatic organisms might diversify in different river systems in South America. By identification of megafan systems and their processes, which involve switching stream courses, habitat fragmentation and new habitat combinations can be hypothesized that, over time, may enable new speciation in aquatic systems. Another application of megafan landform analysis discusses how these features might provide a new perspective on mineral exploration (Wilkinson 2004). This analysis resulted in a patent (Wilkinson 2006). Finally, mapping megafans on Earth have provided a framework for interpreting similar landforms on Mars, supporting the existence of fluvial systems on Mars (Wilkinson et al. 2008).

CEO observations of large tabular icebergs in the South Atlantic Ocean stimulated new research with relevance to global warming and the break-up of Antarctic ice shelves (Scambos et al. 2005). Large chunks of ice calve off the Ronne Ice Shelf in Antarctica and drift northward towards South Georgia Island in the South Atlantic Ocean. Polar scientists track these icebergs; as they drift into warmer waters, they melt and break up. These icebergs can be used as a proxy for understanding how the ice shelves respond to global warming. Imagery that was collected by station astronauts in 2004 showed, for the first time, meltwater on one of the icebergs. This indicated a different ice profile on the icebergs than previously modeled—one with ramparts along the edge that pond the meltwater inboard. These observations allowed scientists to create new models for the forces that are experienced by these icebergs to help explain how they break up. This work has continued under CEO support of the International Polar Year (CEO-IPY).

In a different application that ties two ISS experiments together, Robinson et al. (2006) describe the benefits of photographing the Earth to the mental wellbeing of astronauts on long-duration missions on ISS. For more information about this study, see the *Interactions* results.

PUBLICATION(S)

Andréfouët S, Robinson JA, Hu C, Salvat B, Payri C, Muller-Karger FE. Influence of the spatial resolution of SeaWiFS, Landsat 7, SPOT and International Space Station data on landscape parameters of Pacific Ocean atolls. *Canadian Journal of Remote Sensing*. 29:210–218, 2003.

Andrefouet S, Gilbert A, Yan L, Remoissenet G, Payri C, Chancerelle Y. The remarkable population size of the endangered clam *Tridacna maxima* assessed in Fangatau Atoll using in situ remote sensing data. *ICES Journal of Marine Science*. 62(6):1037–1048, 2005.

Elvidge CD, Cinzano P, Pettit DR, Arvesen J, Sutton P, Small C, Nemani R, Longcore T, Rich C, Safran J, Weeks J, Ebener S. 2007a. The Nightsat mission concept. *Int. J. of Remote Sensing*, V.28, 2645–2677.

Elvidge CD, Safran TB, Sutton P, Cinzano P, Pettit DR, Arvesen J, Small C. 2007b. Potential for Global Mapping of development via a nightsat mission. *GeoJournal*. 69:45–53.

Kohlmann B., Wilkinson MJ, 2007. The Tarcoles Line: biogeographic effects of the Talamanca Range in lower Central America, *Giornale Italiano di Entomologia*. 12:1–30.

Lulla K. 2003 Nighttime Urban Imagery from International Space Station: Potential Applications for Urban Analyses and Modeling. *Photogrammetric Engineering and Remote Sensing*. 69:941–942, 2003.

Quod J-P, Bigot L, Blanchot J, Chabanet P, Durville P, Nicet J-B, Wendling B. Research and monitoring of the coral reefs of the French islands of the Indian Ocean. Assessment activities in 2002. Mission carried out in Glorieuses. *Réunion: IFRECOR (l'Initiative Française pour les Récifs Corallines)*. 2, 2002. [in French]

Robinson JA, Evans CA. Space Station Allows Remote Sensing of Earth to within Six Meters. *Eos, Transactions of the American Geophysical Union*. 83:185–188, 2002.

Robinson JA, Slack KJ, Olsen V, Trenchard M, Willis K, Baskin P, Ritsher JB. Patterns in Crew-Initiated Photography of Earth From ISS – Is Earth Observation a Salutogenic Experience? Presented at the 57th International Astronautical Congress. Valencia, Spain. Oct 2–6, 2006 IAC-06-A1.1.4.

Stefanov WL, Robinson JA. Vegetation Density Measurements From Digital Astronaut Photography. *International Archives of the Photogrammetry, Remote Sensing, and Spatial Information Sciences*. 34:185–189, 2003.

Scambos T, Sergienko O, Sargent A, MacAyeal D, Fastook J. 2005, ICES at profiles of tabular iceberg margins and iceberg breakups at low latitudes. *Geophys. Research Letters*, Vol 32, L23S09 2005.

Stumpf RP, Holderied K, Robinson JA, Feldman G, Kuring N. Mapping water depths in clear water from space. Proceedings of the 13th Biennial Coastal Zone Conference. Baltimore, Md, Jul 13–17, 2003.

Wilkinson, MJ, NASA Tech Brief 2004: Large Fluvial Fans and Exploration for Hydrocarbons, 2004, http://www.techbriefs.com/content/view/170/34/.

Wilkinson, MJ, Allen, CC; Oheler, DZ; and Salvotore, MR, A new Fluvial Analog for the Ridge-forming unit, Northern Sinus Meridiani, Southwest Arabia Terra, Mars, Lunar and Planetary Science XXXIX (2008).

Wilkinson MJ, Marshall LG, Lundberg JG. River behavior on megafans and potential influences on diversification and distribution of aquatic organisms. *Journal of South American Earth Sciences*. 2006; 21:151–172.

PATENT(S)

M.J. Wilkinson, 2006, Method for Identifying Sedimentary bodies from images and its application to mineral exploration, USPTO Patent # 6,985,606, (http://patft.uspto.gov/netacgi/nph-Parser?Sect1=PTO2&Sect2=HITOFF&p=1&u=%2Fnetahtml%2FPTO%2Fsearch-bool.html&r=1&f=G&l=50&co1=AND&d=PTXT&s1=%22fluvial+fan%22&OS)

ARTICLES ON THE WORLD WIDE WEB

Lu E. Watching the World Go By. Earth Observatory. Oct 22, 2003;On-line posting. http://eol.jsc.nasa.gov/EarthObservatory/Watchingtheworldgoby.htm.

Robinson JA, Evans C. Astronaut Photography: Observing the Earth from the International Space Station. Earth Observatory. Feb 16, 2001; On-line posting http://eol.jsc.nasa.gov/EarthObservatory/EO_Study_Astronaut_Photography.htm.

Pettit D. Auroras Dancing in the Night. Earth Observatory. 2004; On-line posting. http://eol.jsc.nasa.gov/EarthObservatory/AuroraDancingintheNight.htm.

Evans CE, Stefanov W. Cities at Night: The View from Space, Earth Observatory; On-line posting Apr 22, 2008, http://eol.jsc.nasa.gov/EarthObservatory/Cities_at_Night_The_View_from_Space.htm.

Stefanov W. The Art of Science, Earth Observatory, On-line posting Oct 18 2005, http://eol.jsc.nasa.gov/EarthObservatory/The_Art_Of_Science.htm.

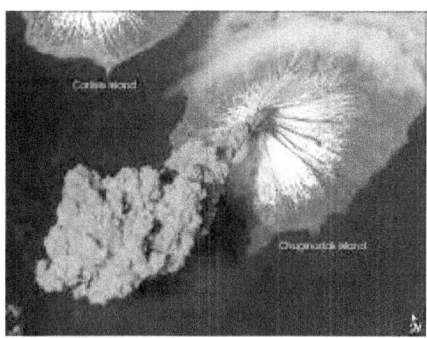

ISS006E44305 — Astronaut Donald R. Pettit, Expedition 6 NASA ISS science officer, runs a drill while looking through EarthKAM mounted on the nadir window in the U.S. *Destiny* laboratory. The device is called a "barn door tracker." The drill turns the screw, which moves the camera and its spotting scope.

ISS013E24184 — ISS flight engineer Jeff Williams was the first to report the volcanic eruption of Cleveland Volcano in Alaska on May 23, 2006. This image shows the eruption cloud moving west-southwest from the volcano summit. The Alaska Volcano Observatory was contacted and was able to monitor the volcanic activity. This image was captured using the Kodak 760c camera equipped with a 800mm lens.

ISS015E2276 — On Aug 13, 2007, while docked to the ISS, the crewmembers of shuttle mission STS-118 and ISS Expedition 15 reported seeing the smoke plumes from widespread fires across Idaho and Montana. The crew photographed and downlinked images of regional views of the smoke. Strong westerly winds were driving the smoke eastward. Tens of thousands of acres were ablaze at this time. This view was taken looking westward toward the horizon. It shows fires in Montana (Gallatin National Forest), to the south in Wyoming (Yellowstone National Park), and to the northwest in Idaho.

ISS010-E-13079 (left) and ISS010-E-13088 (right) were acquired Jan 15, 2005 with a Kodak 760C digital camera using a 400 mm lens.

On Dec 26, 2004, a large (magnitude 9.0) earthquake occurred off the western coast of Sumatra in the Indian Ocean and generated a tsunami that affected coastal regions around the Indian Ocean. The northwestern Sumatra coastline suffered extensive damage and loss of life. These photographs, which were taken from ISS on Jan 15, 2005 (roughly 2 weeks after the tsunami) show some of the damage along the southwestern coast of Aceh Province in the vicinity of the city of Lho' Kruet, Indonesia.

Large areas of bare and disturbed soil (brownish gray) that were previously covered with vegetation are visible along the coastline in the near-nadir (top) image. Embayments in the coastline were particularly hard hit, while adjacent headlands were less affected. The oblique (lower) astronaut photograph was acquired 45 seconds after the near-nadir photograph, and captures sunglint illuminating the Indian Ocean and standing water inland (light gray, yellow). Distortion and scale differences in the images are caused by increased obliquity of the view from the ISS. Arrows on the photographs indicate several points of comparison between the two images. Standing bodies of seawater may inhibit revegetation of damaged areas and act as sources of salt contamination in soil and groundwater.

CREW EARTH OBSERVATIONS-INTERNATIONAL POLAR YEAR (CEO-IPY)

Principal Investigator(s): Donald Pettit, Ph.D., NASA Johnson Space Center, Houston
Expeditions 14–17, ongoing

Research Area Observing the Earth

IPY 2007–2009 is the fourth time in the past 125 years that scientists worldwide combined efforts in observation and exploration of the Earth's Polar Regions. A prime focus of the this IPY is global climate change and the role of the polar regions in understanding climate change. The ISS is participating in the 2007–2009 IPY using the CEO program

The station provides a human observational platform to observe high-latitude features, including sea ice, icebergs, plankton blooms, and atmospheric phenomena such as aurora and polar mesospheric clouds (PMCs). PMCs, which are also known as noctilucent clouds, are thin clouds that are found in the mesosphere. They are the highest known clouds with altitudes around 53 miles (85 km) and are visible only at night when illuminated by sunlight below the horizon. The ISS platform provides excellent observation opportunities for upper-atmospheric features because the crewmembers can observe the upper atmosphere to latitudes as high as 70 to 80°, and their observations are not obscured by clouds.

ISS crewmembers use digital still photography and videos to capture targets that are cataloged and assimilated into the CEO database. The targets are selected according to their relevance to the IPY studies, and are made available to researchers via the CEO Web site at http://eol.jsc.nasa.gov. Additional imaging from satellites and ground observation stations will be compared to the data that are collected by the ISS.

RESULTS

To date, ISS crewmembers have successfully documented the break-up of large tabular icebergs that have calved from the Antarctic ice shelves and drifted northward into the South Atlantic Ocean. Researchers from the National Snow and Ice Data Center have used the imagery from the ISS to examine surface features, including ice margins, cracks, and surface meltwater ponds to better understand the mechanisms and timing of iceberg breakup. Large tabular icebergs can be used to model breakups of the Antarctic ice shelf (Scambos et al. 2005; 2008).

Several sequences of PMCs have been observed and documented by the ISS crews, and imagery of auroras from the ISS and shuttle were collected simultaneously with data from ground stations and meteorological satellites. The integration of these images and data were discussed in the European Optical Meeting in 2007 (Sandahl et al. 2007).

Some of the data have been published and served to the public via NASA's Earth Observatory Web site (http://earthobservatory.nasa.gov).

PUBLICATIONS(S)

Sandahl I, Fuglesang C. The Network for Optical Auroral Research in the Arctic Region, and The Upper Atmospheric Physics Group at the National Institute of Polar Research, Japan. Auroral Observations from Space Shuttle Discovery, 34th Annual European Optical Meeting, Aug 2007.

Scambos T, Ross R, Bauer R, Yermolin Y, Skvarca P, Long D, Bohlander J, Haran T. Calving and ice-shelf break-up processes investigated by proxy: Antarctic tabular iceberg evolution during northward drift; *Journal of Glaciology*, Volume 54(187), 2008; 579–591.

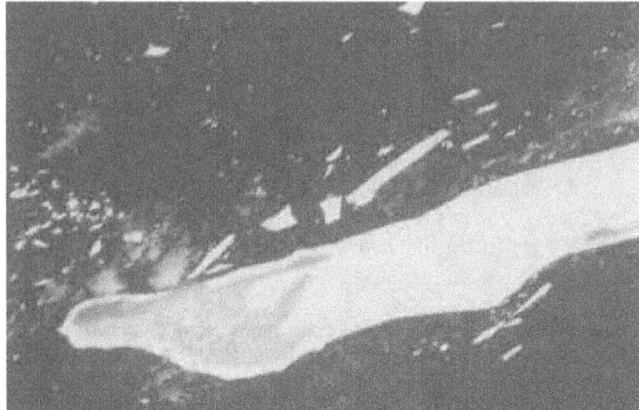

ISS00812558 — Image of iceberg A43B with meltponds on the surface, taken in 2004. This image initiated the construction of a new model for forces that are exerted on large tabular icebergs during their breakup.

STS117e6998 — PMCs observed over central Asia,

ISS006e28961 Auroras as observed from the ISS as it transited the Antarctic region in Feb 2003. Astronauts have unobstructed views of upper atmospheric events around the Earth's polar regions.

COMMERCIAL GENERIC BIOPROCESSING APPARATUS SCIENCE INSERT-01 (CSI-01)

Principal Investigator(s): Louis Stodieck, Ph.D., BioServe Space Technologies, University of Colorado, Boulder, Colo.
Expeditions 14, 15

Research Area Educational Activities

The Commercial Generic Bioprocessing Apparatus Science Insert-01 (CSI-01) was the first in a series of experiments for the K–12 education program from BioServe Space Technologies at the University of Colorado-Boulder. This program provided students learning opportunities based on research that was conducted on the ISS through downlinked data and imagery, which was distributed directly into the classroom via the internet. National Standards-based curriculum materials, including teacher guidebooks, student workbooks, and complementary classroom experiments, were used to ensure the greatest possible benefit to the participating students. The objective of the CSI suite of experiments was to launch small education experiments to be processed in CGBA on an annual basis such that during every academic school year, a "live," on-orbit experiment is available to participating schools.

CSI-01 supported three investigations. The first examined multigenerational, long-term growth of *Caenorhabditis elegans*—a small nematode worm, which is a model organism that is used to perform detailed study of physiological processes that also affect humans. This experiment used two strains of *C. elegans*: wild type (CC1) and a balancer strain (eT1). The eT1 strain was designed to allow the accumulation of mutations without such mutations proving lethal to the organism. This model enabled the study of the biological effects of space radiation. The worms were grown using *C. elegans* maintenance medium (CeMM) and gas exchange sterile chambers, Opticells, which were inside the *C. elegans* habitat (CHab) that was housed in a CGBA. At approximately 1-month intervals, nematodes were automatically transferred from one chamber to another chamber that had fresh CeMM. This was to ensure that the nematode specimens reproduced and propagated for up to 6 months on orbit under nominal, well-defined environmental conditions. This *C. elegans* experiment involved over 5,000 middle school students (grades 6 – 9) who were located in Texas, Arizona, Michigan, Florida, California, New Mexico, Wisconsin, and Montana, as well as several thousand students from Malaysia. The *C. elegans* experiment was part of the Orion's Quest education program (http://www.orionsquest.org/); video of the worms are available on the Web site.

The second investigation provided the opportunity for over 2,000 third-grade students to understand how gravity affects germination and plant development. *Raphanus sativus* (radish plants) and *Medicago sativa* (alfalfa plants) were germinated on orbit in a garden habitat (GHab). The activation for this experiment was coordinated with the participating schools to have their seeds germinate at approximately the same time as the germination of the on-orbit plants. Students examined both root and stem growth of the two plants on Earth and on board the ISS. The seed germination experiment provided the opportunity for schoolchildren to participate in the investigation in conjunction with the Adventures of the Agronauts program (http://www.ncsu.edu/project/agronauts) at North Carolina State University.

The third experiment included orchids, Malaysian red sandalwood, and rosewood tree seeds. These seeds were returned to Earth after approximately 6 months of exposure to the space environment. The orchid seeds were planted alongside control seeds for radiation studies; whereas the Malaysian tree seeds were distributed to students in Malaysia for germination experiments that were performed on Earth.

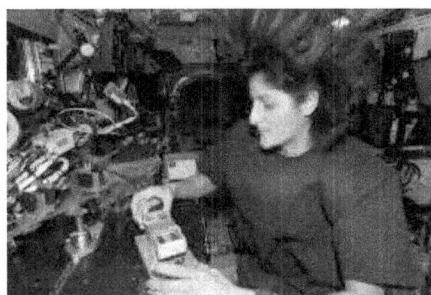

RESULTS

Initial results from Expeditions 14 and 15 indicated that the CSI-01 CHab modules performed properly throughout the 7-month stay on ISS. The CHab modules, which were returned on STS-117/13A, are undergoing analysis; this analysis of the specimens is expected to greatly expand on the results that were already obtained from images and video and analyzed by thousands of students from the U.S. and other countries.

ISS014E20211— Astronaut Sunita L. Williams, Expeditions 14 and 15 flight engineer, is seen here with two GHabs as part of the CSI-01 investigation. The GHabs are placed in the CGBA where their germination will be studied by middle school students.

At first glance, the worms that were recovered from the flight CHabs appeared to be comparable to worms from the equivalent ground control habitats. This would imply that multigenerational propagation (20+ generations over the 6-month flight) of these relatively simple animals can occur normally. A much more detailed study will be required to determine whether long-lasting adaptations to space occurred with the *C. elegans*. It should be noted, however, that this is the first study of its kind to look at long-term adaptations over a large number of generations in such a valuable model organism.

PUBLICATIONS(S)
There are no publications at this time.

COMMERCIAL GENERIC BIOPROCESSING APPARATUS SCIENCE INSERT-02 (CSI-02)

Principal Investigator(s): Louis Stodieck, Ph.D., BioServe Space Technologies, University of Colorado, Boulder, Colo.
Expeditions 15–17, ongoing

Research Area Educational Activities

With the launch of the CGBA-4 on space shuttle mission STS-116/12A.1 on Dec 7, 2006, BioServe initiated a new K–12 education program. Commercial Generic Bioprocessing Apparatus Science Insert-02 (CSI-02) was the second payload that was developed for the program. The program was intended to provide teaching and learning opportunities, primarily targeted to middle school students, that were based on research conducted on orbit and made available through data and imagery downlinked and distributed directly into the classroom through the World Wide Web. National Standards-based curriculum materials, including teacher guidebooks, student workbooks, complementary classroom experiments, etc. are used to ensure the greatest possible benefit to the participating students. For the program, BioServe launched small education experiments that are processed in CGBA-4 on approximately an annual basis such that during every academic school year, a "live," on-orbit experiment is available to participating schools. BioServe Space Technologies partnered with the nonprofit Orion's Quest program (www.orionsquest.org) and the Adventures of Agronaut's program (http://www.ncsu.edu/project/agronauts) at North Carolina State University to implement the educational component of CSI-02.

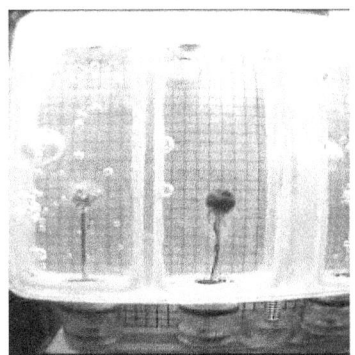

Cobalt chloride is seen growing in this Silicate Garden Habitat during Expedition 17. Image courtesy of BioServe Space Technologies, University of Colorado, Boulder.

CSI-02 is an educational payload that is designed to interest middle school students in STEM by providing the opportunity for these students to participate in near-real-time research conducted on board the ISS. Each experiment was designed to be easily reproducible in the classroom, providing hands-on experience to the students. The seed germination and plant development experiment provided the opportunity for younger students to begin to understand how gravity affects germination and plant development. Small seeds were germinated on orbit in BioServe-developed hardware. The students examined root and stem growth and plant development over a period ranging from a few weeks to 2 months. Classroom kits were available from BioServe for teachers.

The second experiment supported yeast cell growth during space flight. Yeast has been used by researchers for many years as a model organism because there are similar regulatory mechanisms between yeast and mammalian cells. Postflight analysis of the yeast DNA, which is the main component of chromosomes and is the material that transfers genetic characteristics in life forms, is used to determine differences from the ground-based controls. Future plans for the CSI-02 experiment include more detailed scientific and experimental requirements and in-depth analyses for middle and high school students to complete.

The third experiment replicated and extended the chemical garden experiment that was first completed in BioServe hardware for the STARS™ STS-107 program and flight but not recovered due to *Columbia*'s loss in Feb 2003. In this experiment, silicate crystals was activated and grown on orbit. The previous experiment demonstrated that gravity potentially played a significant role in silicate crystal formation. While silicate crystals that are grown on the ground grow in an upwards direction, in space the silicate crystals grow outward and then in a rotational pattern. Growth was very different between flight and ground controls. Students were able to compare silicate crystals that were grown in their own classrooms to those grown in space. Observations were made by using an attached camera and lighting assembly to image all experiments once on orbit.

RESULTS
There are no results to report at this time.

PUBLICATION(S)
There are no publications at this time.

DreamTime (DreamTime)

Principal Investigator(s):   Ben Mason, Dreamtime Holdings Inc., Moffett Field, Calif.
Expedition 3

Research Area Educational Activities

As part of the DreamTime project, a commercial high-definition television (HDTV) system was flown on ISS. When compared to standard television video, high-definition video appears four times sharper, giving a considerably more detailed image. The audio is also improved with HDTV, which records on 5.1 channels vs. the standard two channels in typical stereo systems, in effect providing surround-sound capability. DreamTime was used on ISS to provide these enhanced images and audio for ground-based observers.

Results
In developing the original public-private partnership, NASA had hoped that DreamTime would play a role in developing commercial products that were based on historic activities on ISS. Lacking commercial direction from DreamTime, yet recognizing the historical significance of activities on the station, NASA took the initiative and developed scenarios and created storyboards for the flight crew to record ISS documentary footage of outstanding quality during the mission. The result of this effort returned over 500 minutes of HDTV footage, suitable for commercial purposes, and far exceeding the expected imagery return. The private company that originally sponsored DreamTime was short-lived, and no results were generated. The Bioastronautics Research Program has created the video "Secrets of Science in Outer Space" using some of the DreamTime footage.

Publication(s)
There are no publications at this time.

ISS003-E-5826 — Cosmonaut Vladimir N. Dezhurov, Expedition 3 flight engineer, works with camera equipment in the *Zvezda* service module.

EDUCATION PAYLOAD OPERATIONS (EPO)

Principal Investigator(s): Cynthia McArthur, NASA Johnson Space Center, Houston
Expeditions 4, 5, 7–9

Research Area Educational Activities

Expedition 3 science officer Mike Foale uses small and large magnets to show the pull of the Earth's magnetic field on ISS as part of an EPO event.

The objective of the Education Payload Operations (EPO) investigation was to use toys, tools, and other common items in the microgravity environment of ISS to create educational video and multimedia products that inspire the next generation of engineers, mathematicians, physicists, and other scientists. The products are used for demonstrations and to support curriculum materials distributed across the United States and internationally. The individual EPO projects were designed to explore physical phenomena such as force, motion, and energy. Each Expedition involved different on-orbit activities and themes, as well as different partners, such as museums, universities, and public school districts.

The EPO payloads were small, weighing less than 6.8 kg (15 lbs) each. Whenever possible, the demonstrations use materials and objects that are already available on ISS. Some of the activities cover physical properties, such as Newton's Laws of Motion or Bernoulli's Principle for air pressure, and others are specific to life in space, such as explaining how ISS solar panels work or demonstrating EVAs.

Specific activities are as follows:
- Education demonstration activities (EDAs) showed basic physics, such as in Weight vs. Mass and Center of Mass.
- EDAs illustrated aspects of living in space, such as Tools in Space and Pouring Liquid into a Container.
- International Toys in Space developed a DVD for use in classrooms around the United States that was based on the physics behind a variety of toys.
- Tomatosphere II exposed 1.5 million tomato seeds to the space environment. The seeds have been distributed to classrooms throughout Canada. Students will measure the germination rates, growth patterns, and vigor of growth of the seeds.
- EDAs for use by science museums included a harmonica, puzzles, dexterity puzzles, and a balsa wood Wright Flyer.

RESULTS

EPO has been a successful education program on ISS. By using simple objects and the microgravity environment, NASA is able to produce videos that demonstrate physical properties, such as force, motion, and energy, that may be obscured by gravity on Earth. To date, over 500 videos, DVDs, and video clips have been produced and distributed to science teachers and schools throughout the United States. About 1,500 teachers each year are trained to use the materials in their classrooms. An additional 30.9 million students have had the opportunity to participate in live downlink events where their classmates pose questions of ISS crews on orbit.

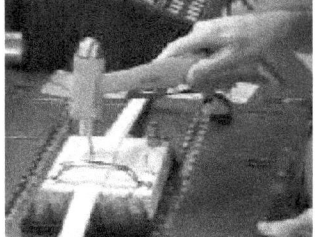

Expedition 8 science officer Mike Foale demonstrates the use of tools on station for an EPO event.

The 1.5 million Tomatosphere-II seeds from Expedition 9 were divided and distributed to 160,000 students in 6,000 classrooms across Canada.

PUBLICATION(S)

National Aeronautics and Space Administration Educational Product. International Toys in Space – Science on the Station DVD. ED-2004-06-001-JSC, 2004.

EDUCATION PAYLOAD OPERATION–DEMONSTRATION (EPO-Demos)

Principal Investigator(s): Matthew Keil, NASA Johnson Space Center, Houston
Expeditions 7–17, ongoing

Research Area Educational Activities

Education Payload Operation-Demonstrations (EPO-Demos) are recorded video education demonstrations that are performed on ISS by crewmembers who are using hardware that is already on board station. EPO-Demos are videotaped, edited, and used to enhance existing NASA education resources and programs for educators and students in grades K–12. EPO-Demos are designed to support the NASA mission to inspire the next generation of explorers.

EPO-Demos are a continuation of education demonstrations that have been conducted by ISS crewmembers since Expedition 4. The products are used for demonstrations and to support curriculum materials that are distributed across the United States and internationally to educators to encourage students to pursue studies and careers in STEM and inspire the next generation of space explorers. Each ISS Expedition involves different on-orbit activities and themes, as well as different partners, such as museums, universities, and public school districts.

Some of the activities cover physical properties, such as Newton's Laws of Motion or Bernoulli's Principle for air pressure, and others are specific to life in space, such as explaining how the ISS solar panels work or demonstrating EVAs.

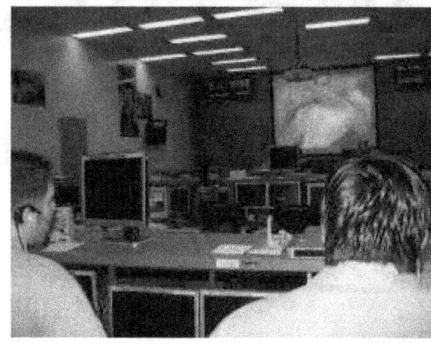

Teaching From Space Office team members in the JSC TeleScience Center supporting an EPO-Demo that is being conducted by astronaut Suni Williams. Image courtesy of Teaching From Space Office, NASA JSC.

RESULTS

EPO-Demos has been a successful education program on ISS. By using simple objects and the microgravity environment, NASA is able to produce videos that demonstrate physical properties, such as force, motion, and energy, that may be obscured by gravity on Earth. Several specific videos demonstrating basic science principles have been created and are available via the NASA Education Project and Central Operations of Resources for Educators (CORE).

PUBLICATION(S) AND VIDEOS

- EPO-Relative_Distance-Earth_Moon_and_Mars.wmv
- Exp_16_EPO_Newtons_First_Law_of_Motion.wmv
- Exp_16_EPO_Newtons_Laws_of_Motion.wmv
- Exp_16_EPO_Newtons_Second_Law_of_Motion.wmv
- Exp_16_EPO_Newtons_Third_Law_of_Motion.wmv
- Inc_10_EPO_Lab_Tour.wmv
- Inc_12_Solar_Arrays.mpg
- Inc_12_Garbage_and_Supplies.mpg
- Inc_12_EPO_Lab_Safety.wmv
- Inc_12_EPO_Trailer.mpg
- Inc_12_Water_Recycling.ram
- Inc_14_EPO-Fitness.wmv

EDUCATION PAYLOAD OPERATIONS-EDUCATOR (EPO-Educator)

Principal Investigator(s): Jonathan Neubauer, NASA Johnson Space Center, Houston
Expedition 15

Research Area Educational Activities

The EPO-Educator payload supports the Educator Astronauts—full-time astronauts with experience teaching in K–12 classrooms—in their mission in orbit. The Educator Astronaut Program (EAP) is one of NASA's pathfinder education initiatives. The main objective of the EAP is inspiring students to look at engineering, science, technology, and mathematics as future career paths.

EPO-Educator uses investigations that are focused on inquiry-based learning. These investigations are performed by students in ground-based studies. Students are also involved in design challenges that are directly tied to the investigation. Part of this payload is developing activities for teachers, creating lesson plans for students K–12, and providing a Web site that will contain the data-collection information.

JSC2007E03712_250 — Barbara Morgan, Educator Astronaut, is pictured with children during a demonstration at Space Center Houston.

RESULTS

Through the EPO-Educator investigation, nearly 1 million students in grades K–12 have participated in the NASA Engineering Design Challenge Lunar Plant Growth Chamber. This engineering design challenge asks students to design, build, and evaluate a plant growth chamber for future missions to the moon. Students learn about the engineering design process and how to conduct a scientific experiment. In conjunction with the engineering design challenge, approximately 10 million cinnamon basil seeds were flown in space. As part of a comprehensive suite of education activities, the seeds are being delivered to students and educators across the country. To get involved in this activity and to see video captured during EPO-Educator, visit www.nasa.gov/education/plantchallenge.

PUBLICATION(S)

There are no publications at this time.

EDUCATION PAYLOAD OPERATIONS-KIT C (EPO-Kit C)

Principal Investigator(s): Jonathan Neubauer, NASA Johnson Space Center, Houston
Expedition 15

Research Area Educational Activities

The excitement of space exploration is a sure way to catch the attention of students of all ages, and space biology is one of many sciences that is critical to understanding the space flight environment. However, many systems used in the past for space-to-classroom biology activities have required extensive crew time and material resources, making space-linked education logistically and financially difficult. The new Educational Payload Operations-Kit C (EPO-Kit C) aims to overcome some of the obstacles to space-linked education and outreach by dramatically reducing the resources that are required to conduct educational activities in plant space biology with a true space flight component.

EPO-Kit C was originally developed as the Astro Garden, a miniature hobby garden that was designed for growing flowers, herbs, and small vegetable plants on orbit. The kit required minimal resources, thus allowing its use for educational opportunities. Illumination is achieved via environmental lighting on the space shuttle or space station, and watering is conducted manually via drink bags. Stowage of the entire kit requires less than 700 cm^3; the garden itself could fit in a large pants pocket.

The EPO-Kit C experiment supported the Educator Astronaut in her mission in orbit. Launched in 2004, the EAP exemplifies NASA's commitment to inspiring and motivating students and educators on a national scale. The program does this through a series of activities and initiatives that is based on astronaut training and the excitement of space flight.

On orbit, crewmembers captured video of the transfer of two, small, collapsible growth chambers for EPO-Kit C. The video, which included a discussion of the growth chambers by the crewmembers, will be used during Phase I and Phase II of the national engineering design challenge. The video will be distributed to education organizations to be incorporated into education products for students in grades K–12. Crewmembers also conducted a 20-day on-orbit plant growth investigation using basil and lettuce seeds. The plant growth inside the growth chambers were documented with still digital imagery.

EPO-Kit C aimed to inspire students to look at engineering, science, technology and mathematics as future career paths. The EPO-Kit C investigations focused on inquiry-based learning; ground-based investigations were performed by students. Students were also involved in design challenges that were directly tied to the payload. Part of this payload also developed activities for teachers, creating lesson plans for students K–12 and providing a Web site that will contain the data-collection information.

Basil plants grown from seeds, on Earth, in a simple plant growth chamber (closed). Image courtesy of NASA Johnson Space Center.

Results

The EPO-Kit C, 20-day on-orbit plant growth investigation was successful. The cinnamon basil seeds germinated in the microgravity environment and had some growth during the short investigation. Toward the end of the experiment, the plants appeared to have received more water than needed, causing them to slowly deteriorate. When the 20 days were up, the growth chambers were collapsed and prepared for their trip home on STS-120.

Through the EPO-Kit C investigation, nearly 1 million students in grades K–12 have participated in the NASA Engineering Design Challenge Lunar Plant Growth Chamber. This engineering design challenge asks students to design, build, and evaluate a plant growth chamber for future missions to the moon. Students learn about the engineering design process and how to conduct a scientific experiment. To get involved in this activity and to see video captured during EPO-Kit C and EPO-Educator, visit www.nasa.gov/education/plantchallenge.

Publication(s)

There are no publications at this time.

EARTH KNOWLEDGE ACQUIRED BY MIDDLE SCHOOL STUDENTS (EarthKAM)

Principal Investigator(s): Sally Ride, University of California at San Diego, La Jolla, Calif.
Expeditions 2, 4–17, ongoing

Research Area Educational Activities

The EarthKAM equipment is mounted in one of the windows of the ISS service module.

Earth Knowledge Acquired by Middle School Students (EarthKAM) is a NASA-sponsored education program that enables thousands of students to photograph and examine Earth from the unique perspective of space. The purpose of EarthKAM is to integrate the excitement of ISS with middle-school education. EarthKAM invites schools from around the world to take advantage of this exceptional educational opportunity. In addition to the many schools in the United States, schools from 12 countries have also participated in the program.

Middle-school students learn about spacecraft orbits and Earth photography, and then target and request their desired images by tracking the orbit of the station, referencing maps and atlases, and checking weather. Their requests are then collected and compiled by students at the University of California, San Diego, Calif. With help from representatives at NASA Johnson Space Center in Houston compiled requests are uplinked to a computer on board ISS. This computer records the requests and transmits them to the digital camera, which takes the desired images and transfers them back to the computer. The images are then downlinked to EarthKAM computers on the ground. Within hours, the EarthKAM team makes the photographs available on the World Wide Web for easy access by participating schools as well as the general public. Schools then explore the images in support of national, state, and local education standards. Students learn to recognize and research features in the images, place the images in global context using maps and atlases, and make connections with the topics and subjects that they are studying.

RESULTS

As of Dec 2007, more than 89,000 students from over 1,400 schools worldwide, as well as members of the general public, have used EarthKAM to investigate every corner of the globe. Images taken by the participating schools are posted on the EarthKAM Web site at http://www.earthkam.ucsd.edu/ for use by the public and participating classrooms around the world. Started in 1996, EarthKAM has conducted 30 missions on ISS and offers more than 24,000 photos of the Earth. No other NASA program gives students such direct control of an instrument that is flying on a spacecraft that is orbiting Earth, and, as a result, students assume an unparalleled personal ownership in the study and analysis of their Earth photography.

Edward T. Lu, Expedition 7 flight engineer, participates in EarthKAM training in the Space Vehicle Mockup Facility at NASA Johnson Space Center.

EarthKAM students determine photo locations by referencing maps and atlases, and by matching these to a display of the upcoming orbit tracks for ISS to generate their image requests.

PUBLICATION(S)
Hurwicz M. Case Study: Attack of The Space Data – Down To Earth Data Management At ISS EarthKAM. *New Architect*. 38, Aug 1, 2002.

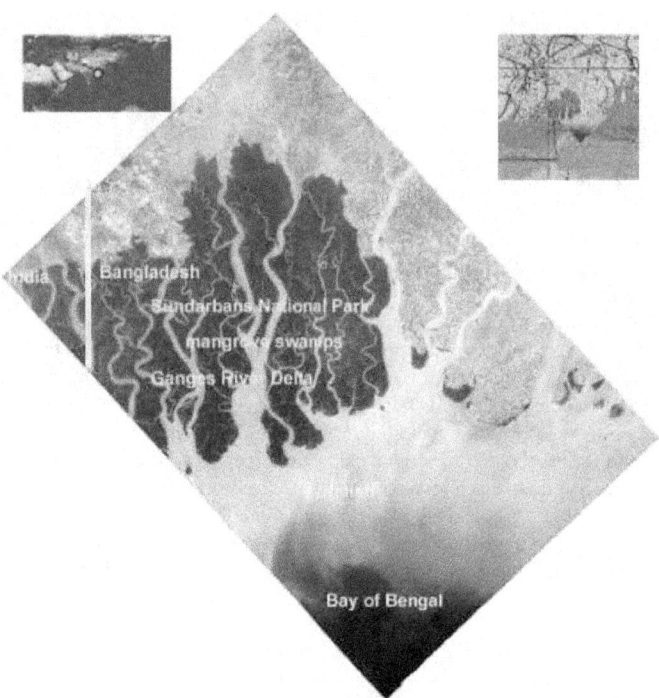

The above image was acquired by ISS EarthKam, annotated, and captioned by students. The caption produced by the students is as follows:

South of the confluence of the Ganges and Brahmaputra Rivers (not shown in this image) and north of the Bay of Bengal lies the vast Ganges Delta, which is about 220 miles (350 kilometers) wide. As the rivers empty, they carry large quantities of sediment into the Bay of Bengal. Parts of this delta—the world's largest—lie in both Bangladesh and the State of West Bengal, India. The very dark part of the delta is the Sundarbans, a vast wildlife preserve and abundant mangrove swamp that is the largest remaining habitat of the Bengal tiger. The entire low-lying region is plagued almost yearly by severe storm surges and powerful low-pressure cyclones (the monsoons) that arrive from the Bay of Bengal. Summer monsoon causes flooding, heavy damage to crops and shelters, and loss of human life. The Ganges River is the most sacred river of India. Many Hindus come to get healed, to wash away their sins, or to die at the river if they are ill or elderly. When a person dies, they are taken to the "purifying stream" and are then dipped in the Ganges before being cremated. The Hindus then scatter the ashes across the river.

Expedition 10 science officer and commander Leroy Chiao in front of the EarthKAM camera holding a student greeting from Wissahickon Middle School, Ambler, Penn.

SPACE-EXPOSED EXPERIMENT DEVELOPMENT FOR STUDENTS (Education-SEEDS)

Principal Investigator(s): Howard Levine, NASA Kennedy Space Center, Cape Canaveral, Fla., and Fredrick Smith, NASA Johnson Space Center, Houston
Expedition 1

Research Area Educational Activities

During the Education-Space-Exposed Experiment Development for Students (SEEDS) experiment, eight pouches of soybean and corn seeds flew on ISS and germinated under either dark or lighted conditions. A grid along the side of the pouch allowed the crew to determine the amount of growth without opening the pouches. In addition, microgravity-exposed seeds were distributed to schools in Fall 2001 and students conducted germination experiments comparing them with seeds that had not flown in space.

RESULTS

The Education-SEEDS investigation, which was part of the Jason XI mission, was the first plant experiment to be performed on station. This experiment studied the effects of microgravity and light on the germination of corn and soybean seeds.

The corn seedlings that were exposed to light appeared to show phototropism (or growth towards light). The shoots grew toward the light and were green, demonstrating chlorophyll synthesis (the creation of the green pigment that is used in photosynthesis). The corn seedlings that were not exposed to light did not turn green and did not grow towards the light. The soybean seedlings that were grown in the light were slightly greener than the seedlings that were grown in the dark. The phototropic effect was more evident in the corn seedlings than in the soybean seedlings. On Earth, gravity influences the roots of plants to grow in a downward direction (gravitropism). While on orbit, the seedlings grew in a microgravity environment. Whether grown in light or dark, the corn roots grew in random directions. The roots of the soybean seeds also grew in random directions (Levine et al. 2001).

Examination of the seeds after their stay on ISS revealed that the nutritional and epidermal layers of the space-exposed seeds were more porous than those of the ground-based control seeds. This might allow nutrients to disperse through the seeds more quickly and explain the faster germination and growth rates that were observed in the space-exposed seeds.

Simple space flight experiments that are suitable for ISS can have significant science impact in the classroom. A total of 750,000 students across the U.S. participated in the experiments, growing corn and soybean seeds in their classrooms to compare with the results from station, and participating in live broadcasts.

PUBLICATION(S)

Levine HG, Norwood KLL, Tynes GK, Levine LH. Soybean and Corn Seed Germination in Space: The First Plant Study Conducted on Space Station Alpha. Proceedings of the 38th Space Congress, Cape Canaveral, Fla. 181–187, May 2001.

Expedition 1 commander Bill Shepherd is shown with a SEEDS pouch. Additional pouches are hanging from the ISS "wall" above the "Astronauts At Work" sign.

Space Experiment Module (SEM)

Principal Investigator(s): Ruthan Lewis, NASA Goddard Space Flight Center, Greenbelt, Md.
Expeditions 10, 11, 13, 14

Research Area Educational Activities

ISS010E12594 0— This image shows an integrated SEM used during Expedition 10; the photograph was taken by ISS commander Leroy Chiao.

The Space Experiment Module (SEM) provided high school students with an opportunity to conduct research on the effects of microgravity, radiation, and space flight on various materials. Research objectives for each experiment were determined by students but generally included hypotheses on changes in selected materials due to the space environment. This was achieved by providing students space capsules that contained passive test articles for flight. These capsules were clear, sealable polycarbonate vials, 1 inch in diameter and 3 inches in depth. The vials were packed in satchels (20 per satchel) that contain special formed foam layers for flight.

Students selected the items that were contained inside the vials. Some of the items included seeds, such as corn, watermelon, cucumber, beans, peas, and several other vegetables. Additional items included materials, such as wool, Kevlar, silk, ultraviolet beads, chicken bones, copper, plastic, dextrose, yeast, over-the-counter medications, human hair, mineral samples, light bulbs, and brine shrimp eggs. Many students tested seed growth after microgravity exposure; other students tested how materials protect against radiation exposure and survival rates of microscopic life forms.

Results

Eleven schools and 300 students developed experiments for SEM Satchel 001. The satchel was launched during ISS Expedition 10 in Dec 2004 and returned to Earth on Space Shuttle *Discovery* (STS-114) in Aug 2005. The sample vials were returned to the students for analysis.

Publication(s)

There are no publications at this time.

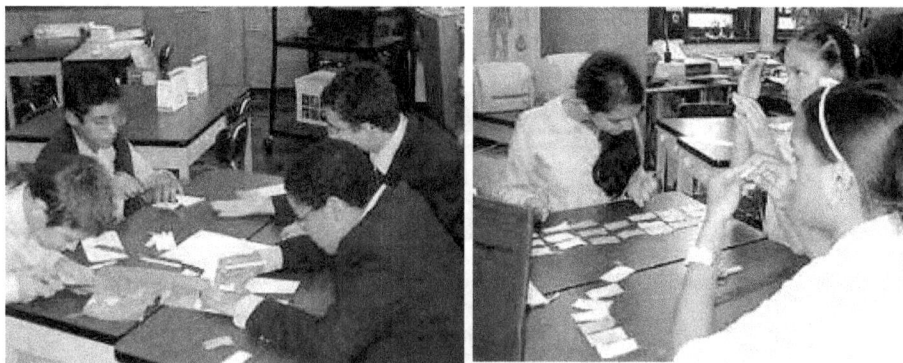

Students from Mott Hall School, New York City, N.Y. wrap fabrics around radiation dosimeters as they prepare samples for the Wearable Radiation Protection Experiment that was part of SEM-2.

RESULTS FROM ISS OPERATIONS

Although not part of a formal investigation or payload on ISS, medical, environmental, and engineering data that are collected as part of the operation of ISS are an important source of information for scientific study. We include and summarize the results of six areas of operations that have generated valuable scientific data that enable scientists and engineers to better define problems and understand the space environment: crewmember-initiated science, environmental monitoring, medical monitoring of crewmembers, station development test objectives, supplementary medical objectives, and lessons learned from the operation of station that are relevant to future mission designs.

CREWMEMBER-INITIATED SCIENCE

This science encompasses activities that are initiated by the crew. These activities are performed at the discretion of the astronauts on board ISS using simple materials that would not impact ISS operations. These activities often take the form of simple demonstrations of life or physical phenomena in microgravity; videos of the activities are downlinked to Earth for use in informal education materials.

EDUCATIONAL ACTIVITIES

Everyday life aboard the ISS provides topics that can be used as springboards for classroom activities. Crewmembers and collaborators with certain payloads have creatively worked with ground-based crews to produce educational activities and materials that grow from normal operations aboard the ISS.

ENVIRONMENTAL MONITORING OF ISS

Environmental monitoring research has been performed on all ISS Expeditions and will continue to be performed on future station missions to ensure the health of the spacecraft as well as of the crew. While the monitoring—both inside and outside the spacecraft—has a hard purpose to support daily life aboard the ISS, the data that are collected are extremely useful for addressing a wide set of questions, such as understanding the evolution of contamination, discerning the detectable limits of contaminants, quantifying anecdotal assessments of environmental conditions that impact the ISS crew, and more.

MEDICAL MONITORING OF ISS CREWMEMBERS

Medical monitoring of ISS crewmembers includes tests before, during, and after space flight to follow the effects of space flight on their health and to ensure that they receive proper medical care. Medical data collection in support of baseline monitoring of crew health is separate from, but supports, the hypothesis-driven medical experiments that are sponsored by the Human Research Program.

SUPPLEMENTARY MEDICAL OBJECTIVE

Supplemental medical objectives are designed to use data that are collected by another investigation to answer new questions that are posed by investigators concerning the effects of space flight on crew health. These small studies can provide the data that are required for defining a full, hypothesis-driven experiment.

STATION DEVELOPMENT TEST OBJECTIVE

Station Development Test Objectives (STDOs) are designed to refine or expand ISS models, assess future ISS capabilities or improvements, and test capabilities for future Exploration systems. The objects of SDTOs are often elevated to operational status after successful demonstration on orbit. Data that are collected from the SDTOs can provide new scientific and engineering insights into aspects of ISS operations or technologies that are relevant to future space flight missions.

EXPLORATION LESSONS LEARNED FROM THE OPERATION OF ISS

Constructing and operating ISS serves as a test bed for new technologies and techniques in support of the crew exploration vehicle (CEV) and lunar mission hardware design and development.

Results from ISS Operations investigations by Sub-discipline.

Crewmember-initiated Science
Saturday Morning Science (Science of Opportunity)
Educational Activities
Education-Solar Cells (Education – How Solar Cells Work)
In-flight Education Downlinks
Environmental Monitoring of ISS
ISS Acoustics (International Space Station Acoustic Measurement Program)
Medical Monitoring of ISS Crewmembers
Clinical Nutrition Assessment (Clinical Nutrition Assessment of ISS Astronauts)
Environmental Monitoring (Environmental Monitoring of the International Space Station)
Supplementary Medical Objective
PFE-OUM (Periodic Fitness Evaluation with Oxygen Uptake Measurement)
Station Development Test Objective
SoRGE (Soldering in Reduced Gravity Experiment, SDTO 17003-U)
Exploration Lessons Learned from the Operation of ISS
Plasma Interaction Model (Analysis of International Space Station Plasma Interaction)
ZPM (International Space Station Zero-Propellant Maneuver (ZPM) Demonstration)
Miscellaneous Science Results from ISS Operations

CLINICAL NUTRITION ASSESSMENT OF ISS ASTRONAUTS (Clinical Nutrition Assessment)
Principal Investigator(s): Scott M. Smith, Ph.D., NASA Johnson Space Center, Houston
Expeditions 1–14

Research Area Medical Monitoring of ISS Crewmembers

To provide nutritional recommendations to crewmembers for long-duration space travel, we need to better understand how nutritional status and general physiology are affected by the microgravity environment. Dietary intake during space flight has often been inadequate, and this can greatly compromise nutritional status. Data from both short- and long-duration space flights provide evidence that energy intake is typically 30 to 40% below World Health Organization recommendations, but energy expenditure is typically unchanged or even increased. This imbalance may explain some of the observed negative changes in overall nutritional status during flight. However, blood concentrations of some nutrients, such as vitamin D, continue to be low even when astronauts receive supplements during flight. The space environment itself results in physiologic changes that can alter nutritional status. For example, changes in iron metabolism are closely associated with blood chemistry alterations during space flight. Similarly, increased levels of radiation and oxidative stress during flight likely contribute to decreased antioxidant status and genetic damage during or after space flight.

ISS012E12635 — ISS science officer Bill MacArthur during Expedition 12, is shown checking out the space linear acceleration mass measurement device (SLAMMD) hardware of HRF-2. Measuring the mass of a crewmember in space is difficult because mass does not equal weight in the absence of gravity.

There are six components to the Clinical Nutrition Assessment research program. These are:
- The food system provides a 6- to 10-day menu cycle; before each mission, crewmembers participate in food-tasting sessions, and dietitians plan menus using crew choices that best fulfill the defined nutritional requirements for space flight.
- During flight, crewmembers were asked to record their dietary intake once per week using a Food Frequency Questionnaire (FFQ) that was designed for use with the space flight food system. The FFQ was designed to obtain a near-real-time estimate of intakes of energy, protein, water, sodium, calcium, and iron, as well as to collect information about vitamin supplement use and any crew comments. The questionnaire inputs from the astronauts were transmitted to the ground, and results were calculated and reported to the flight surgeon within 24 to 48 hours.
- Body mass was measured preflight, postflight, and in-flight, while body composition was determined preflight and postflight using laboratory measurements.
- Blood and urine samples were collected preflight and postflight for analysis of whole blood, plasma, serum, and various analytes; blood pH and ionized calcium levels were measured in-flight using finger sticks.
- Biochemical analysis of the blood and urine samples was performed at NASA Johnson Space Center using standard laboratory methods.
- Statistical analysis of the analytical results were performed to detect differences in nutritional status from preflight to postflight.

Body mass and body composition analyses were performed using an X-ray absorptiometer that was equipped with a fan beam densitometer. In-flight body mass measurements require two astronauts: one to take measurements and the other to record the data. The measurements were obtained using a body-mass measuring device; this device exerts a known force on the body and measures body acceleration. Body mass was then calculated. Standard medical finger sticks and capillary tubes were used to obtain blood samples for input into an analyzer unit that was on board. A single astronaut performed the blood collection and analysis procedure. The FFQ program, which was loaded on station computers, was used to input food consumption information on a weekly basis.

RESULTS
Results have been compiled and analyzed for ISS crewmembers. Intake of energy (relative to World Health Organization standards) was observed to generally decrease over time during missions. However, when dietary counseling was provided to a single astronaut during flight, adequate energy intake was maintained throughout the

mission. Body weight, total bone mineral content, and bone mineral density decreased during flight. Antioxidant capacity decreased during flight, leading to increased susceptibility to genetic damage from radiation. Vitamin D concentration in crew bone was decreased, and bone resorption was increased, by long exposure to microgravity. The relative concentrations of other blood and urine analytes preflight and postflight were variable and subject to several confounding factors that limit conclusions as to the particular effects of space flight (Smith, 2005, 2008). The results of this study formed the basis for the nutrition and repository experiments, which are currently being operated on ISS.

Additionally, these data have been used by other medical researchers who are performing experiments on ISS. Cross-cutting results on bone health in space were reported in Cavanaugh and Rice, 2007 (see Smith et al. 2007; Hall et al. 2007).

PUBLICATION(S)
Hall PS. Past and Current Practice in Space Nutrition, in P Cavanagh and AJ Rice (eds.), Bone Loss During Spaceflight: Etiology, Countermeasures, and Implications for Bone Health on Earth. Cleveland Clinic Press, Cleveland, Ohio (2007), pp. 125–132.

Smith S, Zwart SR, Block G, Rice BL, Davis-Street JE. The nutritional status of astronauts is altered after long-term space flight aboard the International Space Station. *Journal of Nutrition*. 2005; 135(3):437–443.

Smith S, Zwart, SR, Block G, Rice BL, Davis-Street JE. The Nutritional Status of Astronauts is Altered After Long-Term Space Flight Aboard the International Space Station, in P Cavanagh and AJ Rice (eds.), Bone Loss During Spaceflight: Etiology, Countermeasures, and Implications for Bone Health on Earth. Cleveland Clinic Press, Cleveland, Ohio (2007), pp. 133–147.

Smith SM, Zwart SR. Nutrition issues for space exploration. *Acta Astronautica*. 2008; 63:609–613.

EDUCATION-HOW SOLAR CELLS WORK (Education-Solar Cells)
Principal Investigator(s):   Christopher J. Ferguson, NASA Johnson Space Center, Houston
Expedition 13

Research Area Educational Activities

The Teaching From Space (TFS) Office at NASA Johnson Space Center worked closely with astronaut Christopher Ferguson, pilot of STS-115 (*Atlantis*); Lockheed Martin Corporation; Oklahoma State University (OSU), and the Student Observation Network to provide classroom versions of solar cells and learning activities to NASA Explorer Schools. These activities have been designed to engage students through the STS-115 primary mission objective, deployment of a new solar array on ISS. The original suggestion for this activity was brought to the TFS Office by Ferguson.

In his free time during the STS-115 mission, Ferguson demonstrated how solar cells work in front of a video camera. Ferguson discussed, in detail, how solar cells provide energy; open circuit voltage; and power measurement with resistors. With the aid of light-emitting diode (LED) lights, a visual demonstration of solar flux was performed. Following completion, the video will be edited for use in the classroom.

Lockheed Martin Corporation donated solar cells to this project for educational purposes. Engineering students at OSU linked the solar sells together in packs for use in the space shuttle demonstration and for distribution to the schools. The NASA Student Observation Network created lessons and activities that will be provided to the NASA Explorer Schools and the Aerospace Education Specialists when teaching this lesson.

RESULTS
There are no results to report at this time.

PUBLICATION(S)
There are no publications at this time.

Dan Hern, OSU Masters Student in Aviation and Space Education, works on solar cell hardware that will fly aboard STS-115. Image courtesy of OSU.

ENVIRONMENTAL MONITORING OF THE INTERNATIONAL SPACE STATION
(Environmental Monitoring)

Principal Investigator(s): NASA Johnson Space Center, Houston
Expeditions 1–18

Research Area Medical Monitoring of ISS Crewmembers

To successfully live and work in the environment of space, the ISS environment must be monitored to ensure the health of the crew that is living and working there. Astronauts can be more sensitive to air pollutants because of the closed environment. Pollutants in this environment are magnified in ISS because the exposure is continuous.

Sources of physical, chemical, and microbiological contaminants include humans and other organisms, food, cabin surface materials, and experiment devices. One hazard is the offgassing of vapors from plastics and other items on ISS; although this is a small hazard, the accumulation of these contaminants in the air can prove dangerous to crew health. The air sampling systems on ISS periodically checks the air for potential hazards. Advanced HEPA filters and periodic filter cleanings have been successful in keeping harmful vapors out of the air. Other significant contaminants that pose hazards to the crew are microbial growth, both bacterial and fungal; air, water, and surface sampling by the crew in conjunction with periodic cleaning keep the microbial levels on ISS in check.

ISS006E45319 — Cosmonaut and flight engineer for ISS Expedition 6, Nikolai Budarin, is posing near the Potok 150MK air-decontamination equipment in the *Zvezda* service module.

The volatile organic analyzer (VOA) is an atmospheric analysis device on ISS that uses a gas chromatograph and ion mobility spectrometer to detect, identify, and quantify a selected list of volatile organic compounds (i.e., ethanol, methanol, and 2-propanol) that are harmful to humans at high levels in a closed environment such as ISS.

To monitor microbial levels on ISS, crewmembers use devices called grab sample containers, dual absorbent tubes, and swabs to collect station air, water and surface samples and send them to Earth for detailed analysis and identification every 6 months. These data provide controllers on Earth detailed information about the type of microbial contaminants that are on board ISS. The controllers can then give direction to the crew on sanitation if increased microbial growth is identified. The crew keeps microbes under control on ISS through periodic scheduled sanitation of the station.

Missions to the moon and Mars will increase the length of time that astronauts live and work in closed environments. To complete future long-duration missions, the crews must remain healthy in closed environments; hence, future spacecraft must provide sensors to monitor environmental health and accurately determine and control the physical, chemical, and biological environment of the crew living areas and their environmental control systems.

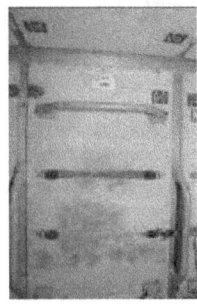

ISS010E11563 — An example of microbial contamination that developed on interior panel 406 aboard ISS. Following workouts, the crewmembers placed their clothes on this panel to dry out. Attempts to clean this panel with supplies aboard ISS failed, and the panel had to be replaced.

RESULTS

During one study of the ISS atmosphere, 12 bacterial strains were isolated and fingerprinted from the ISS water system. These bacteria consisted of common strains and were encountered at levels below 10,000 colony-forming units/10 cm^2, which is well below the minimum of bacteria needed to cause illness. These data represent the beginning of station habitation and indicate that the lessons learned from previous *Mir* and *Skylab* missions were implemented and have been effective in keeping station a safe place in which to live and work (Castro et al. 2004).

Other studies performed an in-depth microbial examination of the drinking water in various stages (from the NASA Kennedy Space Center, Cape Canaveral, Fla. to the ISS ports). These studies have revealed that the NASA policy for biocide treatment has effectively removed pathogenic microbes that are traveling to space (La Duc et al. 2004; Plumlee et al. 2002; Plumlee et al. 2003). Studies on station air quality found that the active (VOA) and passive (HEPA filters) controls that are in place on ISS are effective in controlling trace contaminants of volatile organic compounds on space station (James 2003; Perry 2003).

In another analysis, 39 mold species were identified on dust in the ISS that was collected from the HEPA filters (Vesper et al. 2008). Because some molds pose health risks, including infections and allergic reactions, and others break down organic substances that could compromise parts of the ISS hardware, understanding the mold populations on the ISS is important.

PUBLICATION(S)

Castro VA, Thrasher AN, Healy M, Ott CM, Pierson DL. Microbial Characterization during the Early Habitation of the International Space Station. *Microbial Ecology*. 2004; 47:119–126.

La Duc MT, Sumner R, Pierson D, Venkat P, Venkateswaran K. Evidence of pathogenic microbes in the International Space Station drinking water: reason for concern? *Habitation*. 2004; 10:39–48.

Plumlee D, Mudgett PD, Schultz JR. Chemical Sampling and Analysis of ISS Potable Water: Expeditions 1–3. 32nd International Conference on Environmental Systems, San Antonio, Texas. 2002; SAE Technical Paper 2002-01-2537.

Plumlee D, Mudgett P, Schultz J. ISS Potable Water Sampling and Chemical Analysis: Expeditions 4 & 5. 33rd International Conference on Environmental Systems, Vancouver, Canada. Jul 2003; SAE Technical Paper 2003-01-2401.

Perry J, Peterson B. Cabin air quality Dynamics on Board the International Space Station. 33rd International Conference on Environmental Systems, Vancouver, Canada. Jul 2003; SAE Technical Paper 2003-01-2650.

James J, Limero T, Beck S, Martin B, Covington P, Boyd J, Peters R. Toxicological Assessment of the International Space Station Atmosphere with Emphasis on Metox Canister Regeneration. 33rd International Conference on Environmental Systems, Vancouver, Canada. Jul 2003; SAE Technical Paper 2003-01-2647.

Vesper SJ, Wong W, Kuo CM, Pierson DL. Mold species in dust from the International Space Station identified and quantified by mold-specific quantitative PCR. *Research in Microbiology*. (2008).

INTERNATIONAL SPACE STATION IN-FLIGHT EDUCATION DOWNLINKS
(In-flight Education Downlinks)

Principal Investigator(s): Rene Flores, NASA Johnson Space Center, Houston
Expeditions 1–22, ongoing

Research Area Educational Activities

The NASA TFS Office is responsible for facilitating the flight of educational activities on the space shuttle and the ISS. The TFS Office activities focus on demonstrating science, mathematics, engineering, technology, or geography principles in a microgravity environment. Most activities involve video recording of the demonstrations and/or still photographic documentation of a crewmember.

The TFS Office also coordinates In-flight Education Downlinks, which provides students and educators with the opportunity to learn about science and the ISS by speaking to the crew in orbit. Prior to the event, students study the ISS and on-board science activities and develop questions to ask the crew. Since In-flight Education Downlinks began, thousands of students have had the opportunity to participate in a downlink event where ISS crewmembers answered their questions and performed simple educational demonstrations. Through educational broadcasting, more than 30 million students have been able to watch the live interviews. Through In-flight Education Downlinks, students and educators can communicate live with the ISS crew.

Usually two education downlinks occur each month. Members of the formal and informal education communities, NASA centers and education programs, and the ISS International Partners host these events. The hosts make downlinks part of a comprehensive education package that supports national and state education standards and initiatives. Live in-flight education downlinks, which have one-way video (from ISS) and two-way audio, are broadcast live on NASA Television.

RESULTS

As of Apr 2007, over 30 million students have participated in In-flight Education Downlink activities, which include a Channel One (network broadcast to schools) downlink to schools across the western U.S. that reached millions of students during Expedition 6 and a U.S. Department of Education downlink that reached millions of students around the country during Expedition 12.

The following statements were written by seventh-grade students in Phelps, Ky., after participating in an In-flight Education Downlink event during Expedition 10:
- "I have decided to become more educated and become some sort of scientist."
- "I have been influenced – I want to get an education in a field of science or technology."
- "Small towns have big technology too."
- "I didn't know about all of the good jobs we could get when we grow up. My parents say they didn't have these opportunities when they were growing up but we do."

Expedition 13 crewmembers on board ISS answered questions that students asked in four languages during the International Education Week event that was held on Nov 14, 2006, at the U.S. Department of Education. Image courtesy of the NASA Educational Technology Services.

PUBLICATION(S)
There are no publications at this time.

From 210 miles above Earth, Expedition 14 ISS commander Michael Lopez-Alegria thrilled an audience of students, faculty, staff, and their kids Apr 5, 2007. Pictured are 8-year-old Mark and 9-year-old Julianne, who asked Lopez-Alegria what it's like to float and what do stars look like from the space. Lopez-Alegria did one better. He tumbled himself like a whirling dervish and then told Julianne that when they are on the far side of Earth, away from the sun, there are so many billions-of-stars that it's more like seeing a whitish background with little black specks! U.S. Navy photograph taken by Javier Chagoya.

INTERNATIONAL SPACE STATION ACOUSTIC MEASUREMENT PROGRAM (ISS Acoustics)

Principal Investigator(s): Christopher Allen, NASA Johnson Space Center, Houston; Samuel A. Denham, Ph.D., The Boeing Company, Huntsville, Ala.; S. Reynold Chu, Ph.D., Lockheed Martin, Houston
Expeditions 0, 1–18, ongoing

Research Area Environmental Monitoring of ISS

The ISS is a noisy place. To better characterize the acoustic environment as it changes with assembly, the International Space Station Acoustic Measurement Program (ISS Acoustics) has collected data from before a permanent crew occupied the ISS. The experiment uses one B&K Type 2260 sound level meter (SLM) and three Ametek Mark I audio (acoustic) dosimeters to monitor the ISS acoustic noise environment.

There are two types of SLM activities: the SLM survey of ISS and acoustic engineering evaluation. An SLM survey is performed once every 2 months to measure the acoustic spectral levels at specified locations. An acoustic engineering evaluation is performed to diagnose acoustic abnormalities; investigate crew complaints, and evaluate effectiveness of newly installed noise-reduction measures.

Noise exposure levels are measured by crew-worn dosimeters and dosimeters that are deployed at fixed locations to determine work, sleep, and 24-hour noise exposure levels. Periodic reports of SLM surveys and dosimeter measurements are generated for noise trend tracking, acoustic diagnosis, and development of noise abatements/acoustic remedies.

ISS005E12372 - Flight engineer Sergei Treschev takes measurements with an SLM in the U.S. *Destiny* laboratory during Expedition 5.

ISS007E08990 – Expedition 7 science officer Ed Lu poses for a photograph in the *Zvezda* service module while wearing the audio dosimeter.

RESULTS

The ISS presents a significant acoustics challenge considering all of the modules and equipment that make it an on-orbit laboratory and home with long-duration crew occupation. The acoustic environment on board station has become one of the highest crew habitability concerns. The acoustics mission support function, which includes training, Mission Control support, and data analysis, is necessary to monitor crew exposure and ensure that the crewmembers' hearing is not at risk. Without accurate on-orbit data, all preventative ground efforts are rendered ineffective. Mission monitoring and support is critical to the control and mitigation of acoustic noise on station. ISS Acoustics preserves crewmembers' hearing and provides for a safe, productive, and comfortable noise environment.

The Acoustics Office at NASA Johnson Space Center performs valuable management oversight over acoustic activities. The NASA Johnson Space Center acoustic team provides beneficial support of modules, payloads, and government-furnished equipment requirements definition, design, and development and consultation, and applies proactive efforts to help hardware providers achieve compliance. The acoustic team also manages predictions for flight readiness and on-orbit measurements, and maintains a database of measurements; it also distributes reports and assessments of the data. It is important that the ISS noise be in compliance with current specifications. This is important to ensure acceptable crew communications, health, and well-being. Data that were collected from the Space Shuttle Program and *Mir* Program indicate that levels at or close to 70 dBA should be considered ISS daily exposure limits. These limits are justified in view of crew experience, especially considering the variability in crewmember physiological and psychological response to noise (Goodman 2003).

ISS Acoustics develops measures to safeguard the crewmembers' hearing, ensures there are workarounds for excessively noisy areas or mission events, and provides for a secure, productive, and comfortable noise environment.

This is aided by module noise monitoring, noise abatement, and restricting crew noise exposure during a mission (Pilkinton 2003).

PUBLICATION(S)

Goodman JR. International Space Station Acoustics. Noise Conference, Cleveland, Ohio. Jun 23–25, 2003.

Pilkinton GD. ISS Acoustics Mission Support. Noise Conference, Cleveland, Ohio. Jun 23–25, 2003.

ISS003E329018 — Expedition 3 commander Frank L. Culbertson holds the SLM in the U.S. laboratory *Destiny*. The SLM is part of the Acoustic Countermeasures Kit (ACK).

Periodic Fitness Evaluation with Oxygen Uptake Measurement (PFE-OUM)

Principal Investigator(s): Sean K. Roden M.D., NASA Johnson Space Center, Houston; Filippo Castrucci, M.D., European Astronaut Centre, Cologne, Germany
Expeditions 13–16

Research Area Supplementary Medical Objective

ISS crewmembers routinely perform ground-based exercise tests using a metabolic gas analysis before and after space flight. During these tests, oxygen uptake measurements (OUMs) are made are used to determine a person's aerobic capacity. Measurement of aerobic capacity allows exercise physiologists and flight doctors to assess crew health and fitness and accurately prescribe exercise countermeasures for use on board the ISS.

During space flight, ISS crewmembers perform a monthly periodic fitness evaluation (PFE) starting on flight day 14, and repeated every 30 days. While performing this test, a crewmember's heart rate and blood pressure are recorded using the blood pressure/electrocardiograph (BP/ECG). Heart rate data along with the prescribed Cycle Ergometer with Vibration Isolation System (CEVIS) workloads allow ground personnel to estimate changes in crewmember aerobic capacity by comparing the heart rate response to increases in workload performed during preflight evaluations. There are several limitations to this approach of measuring oxygen capacity that the PFE-OUM will eliminate.

To date, OUMs have not been possible during the PFE because hardware that is capable of measuring oxygen uptake has not been available aboard ISS. Now the Pulmonary Function System (PFS) provides the ability to perform the required OUMs. The PFS is a development in the field of respiratory physiology instrumentation by ESA. The PFS consists of the Photoacoustic Analyzer Module (PAM), the Pulmonary Function Module (PFM), and the Gas Delivery System (GDS), which is capable of a wide range of respiratory and cardiovascular measurements. The PFS was initially launched to the ISS aboard STS-114/LF1, followed by a hardware upgrade that was launched on the Russian Progress cargo module 21 in Apr 2006. All hardware pertaining to the OUM activity is currently stored in the HRF Rack 2 and was checked out during Expedition 12.

ISSE01356862 — NASA ISS science officer Jeff Williams assisting flight engineer-2 Thomas Reiter performing his PFE-OUM on the CEVIS during ISS Expedition 13.

This evaluation was performed in two phases. The first phase assessed the feasibility of making OUMs during the PFE. The data gave investigators and technical personnel the necessary information to determine whether the PFS is able to accurately perform OUMs aboard the ISS within the current timelines that are designated for the PFE.

The second phase of this evaluation compares the current methods of estimating ISS crewmembers' aerobic fitness with the more direct method of performing OUMs. The data that were collected during this phase along with data that were collected during phase one are being analyzed using the current method of estimating changes in aerobic fitness (i.e., workload and heart rate observations alone) and direct analysis of the heart rate data in association with OUMs.

Results
There were some initial instrument calibration problems when the experiment was first used aboard the ISS.

This experiment is still ongoing, and results are pending.

Publication(s)
There are no publications at this time.

ANALYSIS OF INTERNATIONAL SPACE STATION PLASMA INTERACTION (Plasma Interaction Model)

Principal Investigator(s): Ronald Mikatarian, Boeing, Houston
Expeditions 11–16

Research Area Exploration Lessons Learned from the Operation of ISS

This experiment was designed to create a model to predict the voltage difference between the ISS and the plasma background. The ionospheric plasma interacts with the ISS solar arrays and conducting surfaces, causing excess charge to be accumulated, thus creating the potential difference. This model will be used to predict the ISS floating potentials to assess vehicle and EMU dielectric breakdown.

The plasma interaction model (PIM) collected measurements of the ionospheric plasma around the ISS using the floating potential probe (FPP), which was mounted outside on an ISS truss until it was jettisoned in late 2005. Data were also collected using the Incoherent Scatter Radar (ISR) data, which is a ground-based technique for the study of the Earth's ionosphere and its interactions with the upper atmosphere, the magnetosphere, and the interplanetary medium (solar wind). These measurements helped to validate the models that used magnetic field inputs from the International Geomagnetic Reference Field (IGRF) and the plasma environmental parameters from the International Reference Ionosphere (IRI), which is an empirical standard model of the ionosphere that is based on all available data sources. Joint investigations with ISR principal investigators at Millstone Hill (U.S.) and Irkutsk (Russia) have been conducted. Plasma data from a previously launched satellite, Dynamics Explorer-2 (which was launched Aug 3, 1981 and collected data through Feb 19, 1983) can also be used as input into the PIM. This plasma database will be increased as more plasma data become available, i.e., from the CHAMP [challenging minisatellite payload] satellite and ISS floating potential measurement unit (FPMU). Other inputs to the PIM include vehicle coordinates, configuration of solar arrays, conducting area, and the plasma contactor unit (PCU) currents and current-to-voltage relationship.

RESULTS

The PIM has been used to predict the charge build-up and associated dangers for future ISS solar array configurations. Ground data were collected from various locations around the world. Additionally, historical data—e.g., the IRI 2001 model, FPP data, and the Dynamics Explorer-2 satellite data—are available. Using these data, the PIM has been able to characterize the peak voltage levels for the various ISS stage builds. (Reddell et al. 2006).

PUBLICATION(S)

Reddell B, Alred J, Kramer L, Mikatarian R. Analysis of ISS Plasma Interaction. Proceedings of the 44th AIAA Aerospace Sciences Meeting and Exhibit, Reno, Nev. Jan 9–12, 2006; AIAA 2006-865.

SATURDAY MORNING SCIENCE (Science of Opportunity)

Principal Investigator(s): Donald Pettit, Ph.D., NASA Johnson Space Center, Houston
Expedition 6

Research Area Crewmember-initiated Science

Science of Opportunity was the brainchild of Expedition 6 NASA ISS science officer Don Pettit. On ISS, crewmembers have access to world-class laboratory facilities in the unique environment of microgravity; the topics for Saturday Morning Science were therefore spawned by living and working in the microgravity environment.

Science of Opportunity, which was dubbed "Saturday Morning Science" by Expedition 6 ISS commander Ken Bowersox, was done at the discretion of the ISS astronauts. The experiments used simple materials that would not impact ISS operations. During Expedition 6, a number of scientific principles were demonstrated through Saturday Morning Science. The value of this science is the ability to provide observation-based insights for the reduced-gravity environment. Some of the many experiments that were performed are described below.

Noctilucent Cloud Observations in the Southern Hemisphere
Noctilucent clouds, which are clouds that occur in the polar regions in the upper atmosphere (or at about 80 km), appear as a thin but distinct cloud layer that is well above the visible part of the atmosphere. This ISS crewmembers have an excellent vantage point for observing these phenomena. Pettit's observations included photographs, space-craft position, date and time, and approximate viewing direction. They were compiled into a data set that laid the groundwork for the ISS participation in the IPY in 2007.

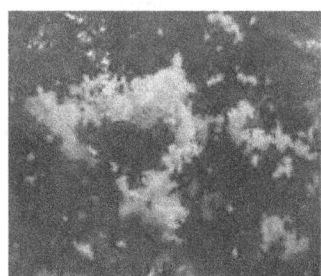

Water Observations
An analog to planetary accretion processes was demonstrated using a plastic bag, sugar, tea grains, and water. The particles (e.g., tea leaves) were suspended in water before they were manipulated and photographed. The "planetary" accretion (increase in the mass of object by the collection of surrounding "interstellar gases and objects" by gravity) of the particles in microgravity was observed.

As part of the Water Observations demonstrations, these clumps of 1- to 6-mm salt (sodium chloride (NaCl)) particles in air were formed on ISS Expedition 6.

Studying Water Films
In microgravity, thin films were surprisingly robust and could withstand numerous mechanical durability tests without breaking. Blowing on the film created ripples that quickly dampened when the perturbations ceased. Oscillating the loop through tens of centimeters with a period of about 2 seconds distorted the film with patterns like seen in a soft rubber membrane when driven by a sound oscillator.

Growing Plants in Zero-g
Using materials readily available aboard ISS (ear plugs, underwear, toilet paper, drinking straw, basil and tomato seeds), basil and tomato seeds were germinated and sprouted on ISS.

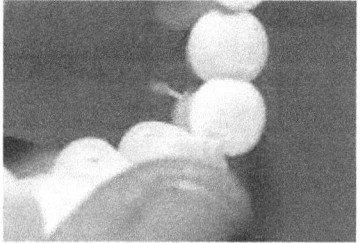

ISS006E20853 – Close-up view of earplugs strung together by NASA ISS science officer Don Pettit to create a sprouter for basil and tomato seeds.

Symphony of Spheres
A 3-inch-diameter sphere of water, which was an analog to a pond, was injected with a bubble of air (~1.5 inch diameter). Several more droplets of air were injected into the air bubble. Ripples moving outward from a disturbance on a pond (e.g., a rock thrown into a pond) were demonstrated in microgravity using a pulse that was applied to the sphere of water. The resulting ripples moved in an outward direction around the

Water bubble injected with many air bubbles aboard ISS during Expedition 6.

spherical surface. The ripples met on the opposite side of the water sphere, 180 degrees away from the pulse. As the ripples met, a spurt of water was shot out, which collapsed back down, sending a ripple back the other way. In this demonstration, the surface waves damped out, leaving the spherical body waves to continue for 5 or 10 minutes.

Results

Several articles have been published on Saturday Morning Science. Although simple, this experiment often demonstrated phenomena that had not been seen in microgravity.

An article that was published in *Sky and Telescope* in Oct 2003 titled "Shooting the Heavens from Space" discusses the amazing views and images that are captured by astronauts as they live and work in space.

In Apr 2004, another article was published in *Sky and Telescope* titled "Building Planets in Plastic Bags" that was based on Water Observations that were performed as a Saturday Science Demonstration. The demonstration involved observing how salt grains clumped in water while in microgravity. During this demonstration, Pettit unknowingly demonstrated middle-stage planetary accretion. A summary paper of the demonstration, which was published in *Lunar and Planetary Science* in 2004, concluded that although the demonstration lacked formal controls to identify the exact clumping mechanism, the mechanism is obviously electrostatic. Future investigations of this phenomenon will use realistic materials such as rock dust.

Publication(s)

Grunsfeld JM. Shooting the Heavens from Space. *Sky and Telescope*. 2003; 128–132.

Love SG, Pettit DR. Fast, Repeatable Clumping of Solid Particles in Microgravity. Lunar and Planetary Science XXXV. 2004; 1119.

Tytell D. Building Planets in Plastic Bags. *Sky and Telescope*. Apr 13, 2004 [http://www.skyandtelescope.com/news/3308986.html?page=1&c=y].

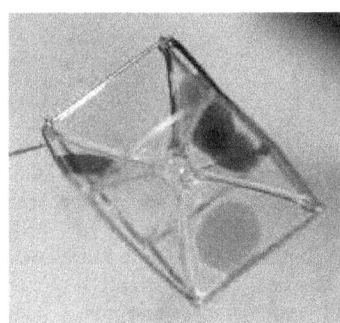

ISS006E18432 — View of surface tension demonstration during Saturday Morning Science, using water that is being held in place by a metal loop. Food coloring has been added to the water for demonstration purposes only.

Soldering in Reduced Gravity Experiment, SDTO 17003U (SoRGE)

Principal Investigator(s): Peter M. Struk, Ph.D., NASA Glenn Research Center, Cleveland, Ohio;
Richard D. Pettegrew, Ph.D., National Center for Space Exploration Research, Cleveland, Ohio
Expedition 14

Research Area Station Development Test Objective

Current electronics maintenance procedures aboard the ISS call for the replacement of failed hardware instead of repair. This strategy relies on re-supply flights from Earth to provide the replacement units. Since this logistical support may not be easily available for future exploration missions beyond LEO, repairing electronics at the lowest component level will potentially ease the logistical burden by minimizing the upmass and volume of required spares.

Soldering in Reduced Gravity Experiment (SoRGE) involves soldering small electronic components in reduced gravity aboard the ISS. SoRGE uses the soldering kit that is currently aboard the ISS and seeks to quantitatively examine the effects of microgravity on the soldering process. This experiment is the next step in a systematic study of soldering in microgravity, following normal gravity studies and reduced-gravity studies aboard NASA's reduced-gravity aircraft.

Earlier studies showed two primary differences between joints that were produced in normal gravity and microgravity, including external geometric differences, and a greatly increased amount of voids in the simulated microgravity samples (about three times the normal Earth gravity in reduced gravity). The changes in geometric shape are due to the dominance of surface tension in microgravity, where body forces are not acting on the mass of molten solder. The increase in void fraction during microgravity is due to the lack of buoyant forces on flux and water vapor that are present in the molten solder and inherent to the soldering process. These entrapped gasses, the presence of which is due to the need for flux in the soldering process and from water vapor evolving from the circuit board itself during the soldering process, are normally transported to the joint surface and eliminated while the joint is molten during normal gravity soldering operations.

ISS015E06764 — This image shows the samples that were used during the SoRGE investigation. SoRGE examined how the microgravity environment affects soldering joints.

The amount of porosity in soldered samples is anticipated to be even greater when conducted in the microgravity environment of ISS when compared with the aircraft results. This is because the "noisy" acceleration environment aboard the aircraft (from atmospheric turbulence, airframe vibrations, etc.) is believed to have provided some residual buoyant forces that may have helped to drive out some of the bubbles prior to the solidification of the joint. Strategies for mitigating the formation of voids in the solder joint will also be tested during SoRGE. Certain techniques showed considerable promise during extensive ground and aircraft testing, but must be proven in the true microgravity environment of ISS before use during future long-duration missions to the moon and Mars.

After flight operations aboard the ISS, the samples (as well as the video recording of the soldering operations) were returned to the investigators for analysis using both nondestructive CT-scanning techniques that use X ray, and industry-standard internal metallographic examination methods.

Results
Data are still undergoing analysis.

Publication(s)
There are no publications at this time.

INTERNATIONAL SPACE STATION ZERO-PROPELLANT MANEUVER (ZPM) DEMONSTRATION

Principal Investigator(s): Nazareth Bedrossian, Ph.D., Charles Stark Draper Laboratory Incorporated, Houston
Expeditions 14, 15

Research Area Exploration Lessons Learned from the Operation of ISS

To maintain its orbit and perform necessary attitude maneuvers, the ISS is equipped with thrusters and control moment gyroscopes (CMGs); i.e., spinning wheel momentum-storage devices that are powered by solar energy. Although small attitude adjustments can be accomplished with CMGs, large-angle rotations that are carried out with the flight software require more momentum than CMGs can provide, resulting in saturation. For this reason, thrusters are used to rotate the ISS. But, thrusters consume precious propellant and their plume can contaminate and stress the solar arrays.

Naval Postgraduate School Professor of Mechanical and Astronautical Engineering I. Michael Ross briefs NPS Space Systems Academic Group students on NASA's use of his optimal control software to maneuver the ISS cost-free, without the need to use thrusters and expend valuable fuel. U.S. Navy photo by Javier Chagoya.

To perform rotations, the ISS flight software uses an eigenaxis trajectory, which is the shortest-distance kinematic path. Most spacecraft use this approach as it is simple to implement in flight software. According to Euler's rotation theorem, any two orientations are related by a common axis, the eigenaxis, about which rotation by a specific angle, the eigenangle, accomplishes the transition from one orientation to the other. To follow the eigenaxis trajectory, the attitude control system must overcome inertial and environmental dynamics, such as torques due to gravity or aerodynamics. As a result, CMGs reach momentum capacity even if maneuvering at a slow rate.

The zero-propellant maneuver (ZPM) concept is based on developing a special non-eigenaxis attitude trajectory that takes advantage of nonlinear system dynamics to complete the maneuver without the need to use thrusters. The attitude trajectory modulates the attitude-dependent environmental torques that are acting on the ISS to maintain the CMGs within their capacity limits. ZPMs can be used to perform rotational state transitions (attitude, rate, and momentum), which can be either a maneuver between prescribed states and/or an attitude maneuver that is used to reset the CMGs. The ZPM attitude trajectory is generated by formulating the ISS maneuver as an optimal control problem. It is solved using DIDO, which is an optimal control software package that was developed at the Naval Postgraduate School. ZPM will also reduce propellant use for spacecraft maneuvers by means of only thrusters.

RESULTS

The ZPM concept was successfully demonstrated on ISS. On Nov 5, 2006 and Mar 3, 2007, the ISS was rotated 90 degrees and 180 degrees, respectively, without using any propellant.

The 90-degree maneuver of ISS Stage 12A was completed in 2 hours and reached 70% of CMG momentum capacity (Bedrossian, AIAA, 2007). The 180-degree maneuver of ISS Stage 12A.1 was completed in 2 hours and 47 minutes and reached 76% of CMG momentum capacity (Bedrossian, International Symposium on Space Flight Dynamics, 2007) The same 180-degree maneuver was performed with thrusters on Jan 2, 2007, consuming 50.8 kilograms or 112 pounds of propellant. At an estimated cost of $10,000 per pound, the 180-degree maneuver with ZPM saved %1,120,000 dollars (Kang 2007).

The flight results were documented and compared to predicted performance. The data that were documented included attitude, momentum, and gimbal rates during the maneuver. Flight reconstruction was performed to resolve discrepancies between predicted and flight results.

The impact of this new technology is to substantially reduce ISS lifetime propellant use, and avoid solar array contamination and loads. Future applications that can also be performed non-propulsively include maneuvering the ISS to unload accumulated CMG momentum, recovering attitude control when CMGs are saturated, and recovering attitude control in the event of a tumbling ISS. Since ZPM will also reduce propellant consumption for maneuvers that use thrusters, it can also be put into service for moon, Mars and beyond missions where propellant savings are even more valuable than for ISS and will allow for increased payload and provisions.

PUBLICATION(S)

Kang W, Bedrossian N. Pseudospectral Optimal Control Theory Makes Debut Flight, Saves NASA 1M dollars in Under Three Hours. *Society for Industrial and Applied Mathematics News*. 2007, Sep; 40(7).

Bedrossian N, Bhatt S, Lammers M, Nguyen L. Zero Propellant Maneuver Flight Results for 180deg ISS Rotation. 20th International Symposium on Space Flight Dynamics. 2007, Sep; NASA CP-2007-214158.

Bhatt S. Optimal Reorientation of Spacecraft Using Only Control Moment Gyroscopes. Master's Thesis, Department of Computational and Applied Mathematics, Rice University, Houston. 2007.

Bedrossian N, Bhatt S, Lammers M, Nguyen L, Zhang Y. First Ever Flight Demonstration of Zero Propellant Maneuver Attitude Control Concept. AIAA GN and C Conference, Hilton Head, SC. 2007, Aug. AIAA 2007-6734.

ENVIRONMENTAL MONITORING OF THE INTERNATIONAL SPACE STATION (Environmental Monitoring): MISCELLANEOUS RESULTS

Principal Investigator(s)   NASA Johnson Space Center, Houston
Expeditions 1–18

Research Area Medical Monitoring of ISS Crewmembers, Exploration Lessons Learned from the Operation of ISS

Several papers have been published reporting the engineering and operational ramifications resulting from the data that were collected from the ISS space environment. The results fall into a couple of broad categories:
- radiation environmental effects on ISS
- external contamination of ISS
- other environmental assessments, including thruster plume contamination and condensate venting

Many of the external analyses were performed with the assistance of the Image Science & Analysis Group at NASA Johnson Space Center, at the request of the ISS Mission Evaluation Room. Results are maintained at NASA Web site http://isag.jsc.nasa.gov/. Periodic reports from the external surveys, as well as analyses of other conditions relating to the weathering of the ISS (MMOD strikes, etc.). are also maintained on this Web site.

RESULTS

Radiation environmental effects on ISS

Koontz (2005) reported on in-situ measurements demonstrating the structural shielding of the ISS for ionizing radiation, both natural and induced, and including single event effects (SEEs). While the altitude of the ISS (350–400 km) is well below the worst-case ionizing radiation environment that the structures were designed for, the relatively high latitude that takes in the SAA and the contribution of SEEs creates a potential evironmental source for ISS crews. Data that were collected from dosimeters indicate that ISS has been performing well; the total ionizing dose (TID) is well below the threshold for degradation in performance. While different parts of ISS receive different doses, the overall design and operational controls have maintained an environment that is safe from the negative effects of ionizing radiation. Models have overpredicted the rate of ionizing-radiation-induced failures by a few factors.

External contamination ISS

The ISS is subject to several sources of contamination, including component outgassing and breakdown, water dumps, thruster plumes from both the service module and visiting spacecraft, orbital debris, and micrometeorites. Several papers provide data that model these different contamination sources on the ISS. Certain surfaces of the ISS are particularly sensitive to contamination; knowledge of the station environment leads to mitigation strategies and feeds back into materials and mission designs for future missions.

Regular photographic surveys of the condition of the ISS are conducted with visiting shuttle flights. Changes are documented, including deposits that are related to external contamination from thruster plumes (see Schmidl et al, 2006), offgassing, MMOD impacts, changes in configuration due to thermal cycling, and other damage resulting from operations. Schmidl et al. (2004) report that a main contaminant includes deposits of fuel oxidizer reaction products (FORPs), including the carcinogenic N-nitrosodimethylamine. A concern is secondary contamination inside the crew cabin by EVA suits coming in contact with contamination on the outside of the spacecraft. Mitigation operations are now in place to control potential contaminations as a result of EVAs. Models were also created to predict concentrations of the substance with distance from service module thrusters.

In addition, operations document the degradation of materials form the space environment , including paint peeling on the service module, coloration changes, and degradation of the coating on pressurized module 2 (PM2) retroreflectors. The data are used to help predict component life cycles, provide data that are relevant to performance, create operational rules for spacecraft and EVAs, and feed into future mission design considerations. NASA's Image Science & Analysis Lab produces regular reports of their findings. ISS engineers use the data for the various studies that are reported here.

Thruster plume contamination
Liquid droplets of partly reacted propellants are produced by thruster exhaust plumes from visiting spacecraft. Both the chemical nature and the physical acceleration of the drops can produce pitting, mechanical erosion, and other damage on surfaces. For critical surfaces such as solar arrays, the models translated into flight rules for positions of the arrays during thruster firings (Alred at al. 2003; Pankop et al. 2004).

Condensate venting
An experiment was performed in 2001 to document a water dump and venting event with video for analysis. The venting was characterized from the imagery, including the extent of the cone of the vent, number of particles outside the cone, and velocities of ice particles. Because the particles can be large (2 mm) and have significant velocities (30–50 ft/sec), they have the potential for creating damage on station hardware. When the video was analyzed, some of the ice particles were observed bouncing off of the nearby ISS hardware. Results agreed well with models; during the nominal phase of the vent, larger particles stayed within a cone with a 20-degree half-angle (Schmidle et al. 2002; Schmidle 2001). These data were used to develop an operational constraint on the location of the solar arrays and photovoltaic radiators to mitigate potential damage.

PUBLICATION(S)
Alred J, Boeder P, Mikatarian R, Pnakop C, Schmidl W. Modeling of the thruster plume induced erosion. 9th International Symposium on Materials in Space, ESTEC Noordwijk, Netherlands, Jun 2003.

Boeder P, Mikatarian R, Lorenz MJ, Koontz S, Albyn K, Finckenor M. Effects of Space Environment Exposure on the Blocking Force of Silicone Adhesive., in Protection of Materials and Structures from the Space Environment. Edited by Jacob I. Kleiman, Integrity Testing Laboratory Inc. Markham, Toronto, Canada. ISBN 978-1-4020-4319-2 (e-book). Series: Space Technology Proceedings, Springer, Dordrecht, 2006, p.295

Pankop C, Alred J, Boeder P. Mitigation of thruster plume-induced erosion of ISS sensitive hardware, in Protection of Materials and Structures from the Space Environment. Edited by Jacob I. Kleiman, Integrity Testing Laboratory Inc. Markham, Toronto, Canada. ISBN 978-1-4020-4319-2 (e-book). Series: Space Technology Proceedings, Springer, Dordrecht, 2006, p.71

Koontz S, Boeder PA, Pankop C, Reddell B. The Ionizing Radiation Environment on the International Space Station: Performance vs. expectations for avionics and materials,, in Radiation Effects Data Workshop, 2005, IEEE, 11-15 July 2005, p 110-116..

Soares C, Mikatarian R, Scharf R, Miles E. International Space Station 1A/R - 6A External Contamination Observations and Surface Assessment. Int. Symposium on Optical Science and Technology, SPIE 47th Annual Meeting, Jul 7–11 2002, Seattle, Wash.

Schmidle W, Alred J, Mikatarian R, Soares C, Miles E, Howorth L, Mishina L, Murtazan R. Characterization of On-orbit US Lab Condensate Vacuum Venting, Int Astronautical Congress, 10–19 Oct 2002, Houston. IAF-02-T.P.06.

Schmidl W, Mikatarian R, Lam C-W, West B, Buchanan V, Dee L., Baker D., Koontz S. Fuel Oxidizer Reaction Products (FORP) Contamination of Service Module and Release of N-nitrosodimethylamine in a humid environment from crew EVA suits contaminated with FORP, in Space Technology Proceedings, V. 6, *Protection of Materials and Structurs from the Space Environment*, Springer Netherlands, 2006, p. 193-204

Schmidl W, Visentine J, and Mikatarian R. Mitigation of Damage to the International Space Station (ISS) from water dumps, in Space Technology Proceedings, V. 6, Protection of Materials and Structurs from the Space Environment, Springer Netherlands, 2006, pp. 93-105.

SPACE STATION SCIENCE BENEFITING LIFE ON EARTH

It is difficult to track all of the science research, applications, and innovations that have been "touched" by research aboard the ISS. In the best examples, data from the ISS have provided critical insight or results—perhaps ancillary to the actual hypothesis that is being tested—that influenced the trajectory of subsequent research on the ground.

In almost all cases, experiments on ISS are a small part of an overall research program. Even highly targeted research demonstrations stem from a broad base of earlier investigations and findings. Research aboard ISS is preceded by years of related research, supported by ongoing research on the ground, and may continue for years after data are downlinked or the experimental samples or hardware are returned from the ISS. Along the way the science that falls under the ISS research umbrella has spawned many innovations, patents, and real-world applications of techniques or findings.

Each experiment that is part of the U.S. integrated science portfolio is tracked in the ISS Program Science database (http://www.nasa.gov/mission_pages/station/science/), and each payload entry includes a section describing how the work is relevant to exploration and to life on Earth. We have not captured all of the spin-offs or "Aha!" ideas that have included space station research in their developmental pathway; neither have we fully tracked scientific and technical innovations that resulted from research on board ISS. But many experiments have direct applications to Earth-based processes. We are providing below a few examples of some of the success stories from ISS research, and anticipate this chapter to grow in coming years as results from ISS experiments feed new innovations in scientific and technical circles.

SUCCESS STORY 1
MISSE and exposure of materials to the space environment: beneficial uses of AO
Perhaps the most prolific suite of experiments to date is included in the MISSE payloads. Hundreds of researchers and organizations from industry, other agencies, and academia have tested materials and components in the space environment. One part of this research is understanding and quantifying the damaging effects of highly reactive AO on different materials, and calibrating ground tests with results from space exposure. While preparing for the first MISSE experiments, researchers at the NASA Glenn Research Center performed AO erosion experiments on the ground that led to new ways in which to harness the corrosive properties of AO.

Art Restoration
The NASA Glenn Research Center researchers realized that AO could be used in beams to remove organic contamination on nonorganic materials such as organic coatings on old paintings that use metal-oxide pigments. The concept for the process for noncontact removal of organic coatings from painting surfaces was patented in 1996 (#5,560,781) and 1997 (#5,693,241) (http://www.uspto.gov). The technique was first demonstrated to remove soot particles from a painting that had been in a fire. The technique was later applied in a high-profile restoration of the Andy Warhol painting "The Bathtub" that was damaged with lipstick after

Work being performed on the Andy Warhol painting "Bathtub" (1961).

someone kissed the painting (left). Subsequently, the technique has been tested on a variety of art media, textures, and contaminants, and licensed as a technique for art restoration. The links below describe the art restoration success story. The researchers, Bruce Banks and Sharon Miller, have received numerous awards and citations for their Atomic Oxygen Restoration System.

http://rt.grc.nasa.gov/main/featured-technologies/atomic-oxygen-restoration/
http://discovermagazine.com/2001/jan/featrocketsci

The left photo was taken after the Cleveland Museum of Art's staff attempted to clean and restore it using acetone and methylene chloride. The right photo is after cleaning by the AO technique.

Cleaning and texturing of surgical implants
AO bombardment has been used to clean and texture medical surfaces such as bones that are used for surgical implants. The same corrosive properties of AO that can damage materials that are used in spacecraft can also be used for beneficial applications. For example, AO can be used to remove harmful organic contaminants such as bacteria or their metabolic products that remain on surgical implants, therefore reducing the potential for inflammation. It can also be used to provide nano-scale texturing on artificial surfaces such as plastics used in transplants, enabling places for cells (e.g., bone cells) to adhere.

Mata A., Su X., Fleischman AJ, Roy S, Banks B, Miller S, Midura RJ. Osteoblast Attachment to a Textured Surface in the Absence of Exogenous Adhesion Proteins. IEEE Transactions in Nanobioscience. 2:287–295, 2003.

New glucose testing equipment
The microscopic texturing properties of directed AO beams have been applied as a tool to texture polymer surfaces that are used in medical equipment. Specifically, a technique has been developed and patented to use AO to etch surfaces at the tip of optical fibers that are used for rapid measurement of glucose or other analytes in very small blood samples. The technique has been patented and commercially applied by a company (Light-Pointe) in a glucose sensor.

Banks B, "Energetic Atomic and Ionic Oxygen Textured Optical Surfaces for Blood Glucose Monitoring,"
http://patft.uspto.gov/netacgi/nph-Parser?Sect1=PTO2&Sect2=HITOFF&p=1&u=%2Fnetahtml%2FPTO%2Fsearch-bool.html&r=1&f=G&l=50&co1=AND&d=PTXT&s1=%22Energetic+Atomic+Ionic+Oxygen+Textured+Optical+Surfaces+Blood+Glucose+Monitoring%22.TI.&OS=TTL/

SUCCESS STORY 2
MEPS and new methods for delivering drugs
The HRP supports many experiments that provide fundamental knowledge about the human body or test techniques that can be useful to the practice of medicine on Earth. Work that is related to one experiment, MEPS, has produced exciting results and several patents.

The MEPS-II experiment on ISS, which is a follow-on to earlier shuttle experiments, demonstrated an automated system that produced tiny (roughly the size of blood cells) liquid-filled micro-balloons in microgravity that are capable of delivering drugs to targeted locations in the human body. Experiments on ISS included encapsulation of several different anti-cancer drugs, magnetic triggering particles, and encapsulation of genetically engineered DNA. The experiment system brought together two immiscible liquids in such a way that allowed surface tension forces (rather than fluid shear) to dominate at the interface of the fluids. The microgravity environment on ISS was vital to the development of these capsules; the experiment results provided insight about the formulations and conditions (e.g., fluid shear limitations and interfacial behaviors) that are required to produce microcapsules of different drugs and particles.

Microencapsulation of drugs using protocols that were developed by research aboard the space shuttle and ISS have demonstrated great promise for cancer therapies and delivery of drugs for other diseases. For example, microcapsules that were made by NASA and that are targeted at inhibiting growth of human prostate tumors were successfully demonstrated in clinical trials (Le Pivert et al. 2004). Microgravity demonstration of co-encapsulation of multiple drugs has enabled new engineering strategies for the production of microcapsules on Earth. Other microcapsules have now been made for the treatment of deep tissue infections and clotting disorders, and to provide delivery of genetically engineered materials for potential gene therapy approaches (Morrison et al. 2003).

Several NASA patents have issued from the space research, and more are pending. Commercial companies have licensed some of the MEPS technologies and methods; collaborations with academic institutions are under way to develop new applications (see Morrison et al. 2003).

Le Pivert P, Haddad RS, Aller A, Titus K, Doulat J, Renard M, Morrison DR. Ultrasound Guided, Combined Cryoablation and Microencapsulated 5-Fluorouracil, Inhibits Growth of Human Prostate Tumors in Xenogenic Mouse Model Assessed by Fluorescence Imaging. *Technology in Cancer Research and Treatment.* 2004; 3(2):135–42.
Morrison DR, Haddad RS, Ficht A. Microencapsulation of Drugs: New cancer therapies and improved drug delivery derived from microgravity research. Proceedings of the 40th Space Congress, Cape Canaveral, Fla. Apr 2003.

SUCCESS STORY 3
AdvAsc and anthrax: air-purifying technology
The-air purifying technology (TiO_2-based ethylene) that was employed in the plant growth chamber that was used in the AdvAsc experiment was incorporated into an airborne pathogen scrubber that is effective against anthrax spores and other pathogens.

Research on the ISS that is related to plant growth and plant biology requires special plant growth chambers that are capable of maintaining a healthy environment and atmosphere, delivering appropriate water and nutrients to the plants, and measuring and monitoring the test conditions for later data analysis. To support plant research in microgravity, the University of Wisconsin-Madison and the Wisconsin Center for Space Automation and Robotics developed plant growth chambers, including one used to grow plants on ISS that is called Advanced Astroculture.™ Several new technologies were folded into these chambers, including a novel air scrubber that employs TiO_2 to remove ethylene from the chamber atmosphere.

A commercial company, KES Science & Technology, Inc., licensed the TiO_2-based ethylene removal technology. The intended application was to remove ethylene from the atmosphere to preserve food perishables such as fruits and vegetables. However, the technology is also extremely effective at removing bacterial pathogens in the air. After the 2001 anthrax attacks, an employee of KES speculated that the ethylene scrubber that uses a catalyst bed that is coated with TiO_2 (a photocatalyst) and ultraviolet lamps could be modified and used as an air purifier. The combination of ultraviolet light and TiO_2 creates super-oxides and hydroxyl radicals (strong oxidizing agents) that completely oxidize organic material. The technology is patented by KES as AiroCide, which is an air purifier that kills 98% of airborne pathogens that pass through it, including *Bacillus anthraci* (anthrax), dust mites, molds, and harmful viruses and bacteria such as Influenza A (flu), *E. coli, Staphylococcus aureas, Streptococcus pyogenes*, and *Mycoplasma pneumoniae*.

Today, KES markets several AiroCide units that are designed to be installed in buildings (e.g., mounted on a ceiling) and are capable of purifying large volumes of air. Commercial applications for the device include mail rooms, conference rooms, hospitals, waiting rooms, day care centers, schools, athletic facilities, and other common areas that are susceptible to bacteria, including homes.

http://www.kesair.com/airocide_and_nasa.htm
http://www.sti.nasa.gov/tto/spinoff2002/er_5.html

Appendix A: Scientific Research by Expedition

Expedition 0 (Sep 8 2000 – Nov 2 2000)

Physical Sciences in Microgravity
- Protein Crystal Growth – Enhanced Gaseous Nitrogen Dewar (PCG-EGN)

Biological Sciences in Microgravity
- Commercial Generic Bioprocessing Apparatus:
    - Kidney Cell Gene Expression (CGBA-KCGE)
    - Synaptogenesis in Microgravity (CGBA-SM)

Human Research and Countermeasure Development for Exploration
- Incidence of Latent Virus Shedding During Space Flight (Latent Virus)

Results from ISS Operation
- International Space Station Acoustics Measurement Program (ISS Acoustics)

EXPEDITION 1 (NOV 2 2000 – MAR 18 2001)

> "... these technical demands that we have—flying in space, keeping humans healthy and able to work up there—have huge side benefits to the way that we live and the style of life we enjoy [on Earth]." – Bill Shepherd Expedition 1 Commander

When the three-man Expedition 1 crew moved into ISS, they began not only a habitation that continues to this day but also became part of a station that was a work-in-progress.

The *Zvezda* service module provided their early living quarters. At this time the ISS also consisted of the *Zarya* module (the Functional Cargo Block) Node 1 (the *Unity* module) and a Soyuz spacecraft (in which the crew had arrived and which supplied their assured crew return capability). Later in Expedition 1, the U.S. *Destiny* laboratory was installed. Yet despite the Spartan living conditions and the sense of working on a platform under construction, the crew made significant advances in experiments:

Technology Development for Exploration
- Middeck Active Control Experiment-II (MACE-II)

Physical Sciences in Microgravity
- Protein Crystal Growth-Enhanced Gaseous Nitrogen Dewar (PCG-EGN)

Human Research and Countermeasure Development for Exploration
- Incidence of Latent Virus Shedding During Space Flight (Latent Virus)

Observing the Earth and Educational Activities
- Crew Earth Observations (CEO)
- Space Exposed Experiment Developed for Students (Education-SEEDS)

Results from ISS Operations
- Clinical Nutrition Assessment of ISS Astronauts (Clinical Nutrition Assessment)
- Environmental Monitoring of the International Space Station (Environmental Monitoring)
- International Space Station In-flight Education Downlinks (In-flight Education Downlinks)
- International Space Station Acoustics Measurement Program (ISS Acoustics)

EXPEDITION 2 (MAR 10 2001 – AUG 20 2001)

> "... I know that the research that we're doing will make us take steps forward that eventually will help us to solve many of the problems that we face here on Earth. ... I think we can only imagine what will happen on board the International Space Station as we continue to do research and explore." – James S. Voss Expedition 2 Flight Engineer

As ISS continued to grow and expand with the installation of new experiment facilities and hardware on the U.S. *Destiny* laboratory, the capacity for research on board station increased accordingly. From the original four experiments that were conducted on Expedition 1, Expedition 2 expanded the research potential vastly to include many more experiments:

- Technology Development for Exploration
    - Active Rack Isolation System-ISS Characterization Experiment (ARIS-ICE)
    - Middeck Active Control Experiment-II (MACE-II)
    - Microgravity Acceleration Measurement System (MAMS)
    - Space Acceleration Measurement System-II (SAMS-II)

Physical Sciences in Microgravity
- Commercial Protein Crystal Growth-High Density (CPCG-H)
- ExPRESS Physics of Colloids in Space (EXPPCS)
- Protein Crystal Growth-Enhanced Gaseous Nitrogen Dewar (PCG-EGN)
- Protein Crystal Growth
    - Single Locker Thermal Enclosure System-Improved Diffraction Quality of Crystals (PCG-STES-IDQC)
    - Single Locker Thermal Enclosure System-Science and Applications of Facility Hardware for Protein Crystal Growth (PCG-STES-SA)
    - Single Locker Thermal Enclosure System-Vapor Equilibrium Kinetic Studies (PCG-STES-VEKS)

Biological Sciences in Microgravity
- Advanced Astroculture™ (AdvAsc)
- Commercial Generic Bioprocessing Apparatus-Antibiotic Production in Space (CGBA-APS)

Human Research and Countermeasure Development for Exploration
- Bonner Bell Neutron Detector (BBND)
- Dosimetric Mapping (DOSMAP)
- Effects of Altered Gravity on Spinal Cord Excitability (H-Reflex)
- Crewmember and Crew-Ground Interaction During International Space Station Missions (Interactions)
- Incidence of Latent Virus Shedding During Space Flight (Latent Virus)
- Subregional Assessment of Bone Loss in the Axial Skeleton in Long-term Space Flight (Subregional Bone)
- Organ Dose Measurement Using the Phantom Torso (Torso)

Observing the Earth and Educational Activities
- Crew Earth Observations (CEO)
- Earth Knowledge Acquired by Middle-School Students (EarthKAM)

Results from ISS Operations
- Amateur Radio on the International Space Station (ARISS)
- Clinical Nutrition Assessment of ISS Astronauts (Clinical Nutrition Assessment)
- Environmental Monitoring of the International Space Station (Environmental Monitoring)
- International Space Station In-flight Education Downlinks (In-flight Education Downlinks)
- International Space Station Acoustics Measurement Program (ISS Acoustics)

EXPEDITION 3 (AUG 12 2001 – DEC 12 2001)

"We need a space station because we need a frontier. We need to keep pushing the human race to expand beyond the current boundaries that we have." – Frank L. Culbertson Jr. Expedition 3 Commander

As with other Expedition crews, the Expedition 3 crew focused on expansion and scientific matters. While various experiments naturally carried over from previous Expeditions—thereby establishing a pattern of building on earlier work that carries forward to this day—the crew also conducted several new experiments as follows:

Technology Development for Exploration
- Active Rack Isolation System-ISS Characterization Experiment (ARIS-ICE)
- Microgravity Acceleration Measurement System (MAMS)
- Materials International Space Station Experiment-1 and -2 (MISSE-1 and -2)
- Space Acceleration Measurement System-II (SAMS-II)

Physical Sciences in Microgravity
- Advanced Protein Crystallization Facility
    - Extraordinary Structural Features of Antibodies from Camelids (APCF-Camelids)
    - Solution Flows and Molecular Disorder of Protein Crystals (APCF-Crystal Growth)
    - Effect of Different Growth Conditions on the Quality of Thaumatin and Asparatyl-tRNA Synthetase Crystals Grown in Microgravity (APCF-Crystal Quality)
    - Crystallization of Human Low Density Lipoprotein (LDL) Subfractions
    - (APCF-Lipoprotein)
    - Testing New Trends in Microgravity Protein Crystallization (APCF-Lysozyme)
    - Crystallization of the Next Generation of Octarellins (APCF-Octarellins)
    - Protein Crystallization in Microgravity Collegan Model (X-Y-Gly) Polypeptides (APCF-PPG10)
    - Crystallization of Rhodopsin in Microgravity (APCF-Rhodopspin)
- Dynamically Controlled Protein Crystal Growth (DCPCG)
- Experiment Physics of Colloids in Space (EXPPCS)

Biological Sciences in Microgravity
- Cellular Biotechnology Operations Support System
    - Human Renal Cortical Cell Differentiation and Hormone Production (CBOSS-01-02-Renal)
    - Use of NASA Bioreactor to Study Cell Cycle Regulation: Mechanism of Colon Carcinoma Metastasis in Microgravity (CBOSS-01-Colon)
    - Evaluation of Ovarian Tumor Cell Growth and Gene Expression (CBOSS-01-Ovarian)
    - PC12 Pheochromocytoma Cells – A Proven Model for Optimizing 3-D Cell Culture Biotechnology in Space (CBOSS-01-PC12)

Human Research and Countermeasure Development for Exploration
- Bonner Bell Neutron Detector (BBND)
- Effects of Altered Gravity on Spinal Cord Excitability (H-REFLEX)
- Crewmember and Crew-Ground Interactions During International Space Station Missions (Interactions)
- Effects of EVA and Long-term Exposure to Microgravity on Pulmonary Function (PuFF)
- Renal Stone Risk During Space Flight: Assessment and Countermeasure Validation (Renal Stone)
- Subregional Assessment of Bone Loss in the Axial Skeleton in Long-term Space Flight (Subregional Bone)
- Effect of Microgravity on the Peripheral Subcutaneous Veno-arteriolar Reflex in Humans (Xenon1)

Observing the Earth and Educational Activities
- Crew Earth Observations (CEO)
- DreamTime (DreamTime)
- Earth Knowledge Acquired by Middle School Students (EarthKAM)

Results from ISS Operations
- Amateur Radio on the International Space Station (ARISS)
- Clinical Nutrition Assessment of ISS Astronauts (Clinical Nutrition Assessment)
- Environmental Monitoring of the International Space Station (Environmental Monitoring)
- International Space Station In-flight Education Downlinks (In-flight Education Downlinks)
- International Space Station Acoustics Measurement Program (ISS Acoustics)

EXPEDITION 4 (7 DEC 2001 – 15 JUN 2002)

> "... I think the culmination of our participation ... will allow us to really make some headway and get some answers to some ... scientific questions." – Carl E. Walz Expedition 3 Flight Engineer

The Expedition 4 three-man crew continued to maintain focus not only on ongoing experiments several of which had already carried through Expeditions 2 and 3 but also turned their focus to new experiments. Or in the words of Carl Walz (above) this crew continued to "make some headway".

Technology Development for Exploration
- Active Rack Isolation System-ISS Characterization Experiment (ARIS-ICE)
- Microgravity Acceleration Measurement System (MAMS)
- Materials International Space Station Experiment-1 and -2 (MISSE-1 and -2)
- Space Acceleration Measurement System-II (SAMS-II)

Physical Sciences in Microgravity
- Commercial Protein Crystal Growth-High Density (CPCG-H)
- Experiment Physics of Colloids in Space (EXPPCS)
- Protein Crystal Growth-Enhanced Gaseous Nitrogen Ewer (PCG-EGN)
- Protein Crystal Growth-Single Locker Thermal Enclosure System
  - Synchrotron-Based Mosaicity Measurement of Crystal Quality (PCG-STES-MM)
  - Science Applications of Facility Hardware for Protein Crystal Growth (PCG-STES-SA)
  - Vapor Equilibrium Kinetics Studies (PCG-STES-VEKS)
- Zeolite Crystal Growth (ZCG)

Biological Sciences in Microgravity
- Development and Function of the Avian Otolith System in Normal and Altered Gravity Environments (ADF-Otolith)
- Avian Development Facility-Skeletal Development in Embryonic Quail (ADF-Skeletal)
- Advanced Astroculture™ (AdvAsc)
- Biomass Production System-Photosynthesis Experiment and Systems Testing (BPS)
- Commercial Generic Bioprocessing Apparatus-Antibiotic Production in Space (CGBA-APS)
- Cellular Biotechnology Operations Support Systems
  - Production of Recombinant Human Erythropoientin by Mammalian Cells (CBOSS-02 Erythropoietin)
  - The Effect of Microgravity on the Immune Function of Human Lymphoid Tissue (CBOSS-02-HLT)
  - Human Renal Critical Cell Differentiation and Hormone Production (CBOSS-01-02-Renal)
- Photosynthesis Experiment and System Testing and Operation (PESTO)

Human Research and Countermeasure Development for Exploration
- Commercial Biomedical Testing Module (CBTM): Effects of Osteoprotegerin on Bone Maintenance in Microgravity
- A Study of Radiation Doses Experienced by Astronauts in EVA (EVARM)
- Effects of Altered Gravity on Spinal Cord Excitability (H-Reflex)
- Crewmember and Crew-Ground Interactions During International Space Station Missions (Interactions)
- Incidence of Latent Virus Shedding During Space Flight (Latent Virus)
- Effects of EVA and Long-term Exposure to Microgravity on Pulmonary Function (PuFF)
- Renal Stone Risk During Space Flight: Assessment and Countermeasure Validation Evaluation (Renal Stone)
- Subregional Assessment of Bone Loss in the Axial Skeleton in Long-term Space Flight (Subregional Bone)
- Effect of Microgravity on the Peripheral Subcutaneous Veno-arteriolar Reflex in Humans (Xenon1)

Observing the Earth and Educational Activities
- Crew Earth Observations (CEO)
- Earth Knowledge Acquired by Middle-School Students (EarthKAM)
- Education Payload Operations (EPO)

Results from ISS Operations
- Amateur Radio on the International Space Station (ARISS)
- Clinical Nutrition Assessment of ISS Astronauts (Clinical Nutrition Assessment)
- Environmental Monitoring of the International Space (Environmental Monitoring)
- International Space Station In-flight Education Downlinks (In-flight Education Downlinks)
- International Space Station Acoustics Measurement Program (ISS Acoustics)

EXPEDITION 5 (JUN 7 2002 – DEC 2 2002)

> "[The] ISS is the one step for future investigation future science experiments ..." – Sergei Y. Treschev Expedition 5 Flight Engineer

As ISS continued to grow and expand so, too, did the scientific workload for the Expedition crews—especially and critically augmented by the installation of the MSG. The Expedition 5 crew arrived on station faced with the largest workload to date of both continuing and new experiments:

Technology Development for Exploration
- Microgravity Acceleration Measurement System (MAMS)
- Materials International Space Station Experiment-1 and -2 (MISSE-1 and -2)
- Space Acceleration Measurement System-II (SAMS-II)

Physical Science in Microgravity
- Protein Crystal Growth-Single Locker Thermal Enclosure System
    - Crystallization of the Integral Membrane Protein Using Microgravity (PCG-STES-IMP)
    - Synchrotron-Based Mosaicity Measurement of Crystal Quality (PCG-STES-MM)
    - Crystallization of the Mitochondrial Metabolite Transport Protein (PCG-STES-MMTP)
    - Engineering a Ribozyme for Diffraction Properties (PCG-STES-RDP)
    - Science and Applications of Facility Hardware for Protein Crystal Growth (PCG-STES-SA)
    - Vapor Equilibrium Kinetics Studies (PCG-STES-VEKS)
- Toward Understanding Pore Formation and Mobility During Controlled Directional Solidification in a Microgravity Environment (PFMI)
- Solidification Using a Baffle in Sealed Ampoules (SUBSA)
- Zeolite Crystal Growth (ZCG)

Biological Science in Microgravity
- Advanced Astroculture (AdvAsc)
- Microencapsulation Electrostatic Processing System (MEPS)
- Plant Generic Bioprocessing Apparatus (PGBA)
- StelSys Liver Cell Function Research (StelSys)

Human Research and Countermeasure Development for Exploration
- Effect of Prolonged Space Flight on Human Skeletal Muscle (Biopsy)
- A Study of Radiation Doses Experienced by Astronauts in EVA (EVARM)
- Crewmember and Crew-Ground Interactions During International Space Station Missions (Interactions)
- Incidence of Latent Virus Shedding During Space Flight (Latent Virus)
- Promoting Sensorimotor Response Generalizability (Mobility)
- Effects of EVA and Long-term Exposure to Microgravity on Pulmonary Function (PuFF)
- Renal Stone Risk During Space Flight: Assessment and Countermeasure Evaluation (Renal Stone)
- Space Flight-induced Reactivation of Latent Epstein-Barr Virus (Epstein-Barr)
- Subregional Assessment of Bone Loss in the Axial Skeleton in Long-term Space Flight (Subregional Bone)
- Effect of Microgravity on the Peripheral Subcutaneous Veno-arteriolar Reflex in Humans (Xenon1)

Observing the Earth and Educational Activities
- Crew Earth Observations (CEO)
- Education Payload Operations (EPO)
- Earth Knowledge Acquired by Middle School Students (EarthKAM)

Results from ISS Operations
- Amateur Radio on the International Space Station (ARISS)
- Clinical Nutrition Assessment of ISS Astronauts (Clinical Nutrition Assessment)
- Environmental Monitoring of the International Space (Environmental Monitoring)
- International Space Station Acoustics Measurement Program (ISS Acoustics)
- International Space Station In-flight Education Downlinks (In-flight Education Downlinks)

EXPEDITION 6 (NOV 25 2002 – MAY 3 2003)

> "... [Experiments] done on space station are designed around the reduction in the gravitational force so that you can see other forces manifest themselves and you can make new observations that are very difficult if not impossible to make any other way." – Donald R. Pettit Expedition 6 Flight Engineer

As station entered its third year of continuous operations, the scientific workload continued to occupy a significant amount of the Expedition crews' time. The Expedition 6 crew worked unremittingly on the scientific mission in face of the changes wrought following the loss of *Columbia*, conducting experiments in:

Technology Development for Exploration
- Microgravity Acceleration Measurement System (MAMS)
- Materials International Space Station Experiment-1 and -2 (MISSE-1 and -2)
- Space Acceleration Measurement System-II (SAMS-II)

Physical Sciences in Microgravity
- Investigating the Structure of Paramagnetic Aggregates from Colloidal Emulsions (InSPACE)
- Protein Crystal Growth-Single Locker Thermal Enclosure System
    - Crystal Growth Model System for Material Science (PCG-STES-MS)
    - Regulation of Gene Expression (PCG-STES-RGE)
    - Science and Applications of Facility Hardware for Protein Crystal Growth (PCG-STES-SA)
    - Vapor Equilibrium Kinetics Studies (PCG-STES-VEKS)
- Zeolite Crystal Growth (ZCG)

Human Research and Countermeasure Development for Exploration
- Effect of Prolonged Space Flight on Human Skeletal Muscle (Biopsy)
- Chromosomal Aberrations in Blood Lymphocytes of Astronauts (Chromosome)
- A Study of Radiation Doses Experienced by Astronauts in EVA (EVARM)
- Space Flight-induced Reactivation of Latent Epstein-Barr Virus (Epstein-Barr)
- Foot/Ground Reaction Forces During Space Flight (Foot)
- Promoting Sensorimotor Response Generalizability (Mobility)
- The Effects of EVA and Long-Term Exposure to Microgravity on Pulmonary Function (PuFF)
- Renal Stone Risk During Space Flight: Assessment and Countermeasure Validation (Renal Stone)
- Subregional Assessment of Bone Loss in the Axial Skeleton in Long-term Space Flight (Subregional Bone)

Observing the Earth and Educational Activities
- Crew Earth Observations (CEO)
- Earth Knowledge Acquired by Middle-School Students (EarthKAM)

Results from ISS Operations
- Amateur Radio on the International Space Station (ARISS)
- Clinical Nutrition Assessment of ISS Astronauts (Clinical Nutrition Assessment)
- Environmental Monitoring of the International Space (Environmental Monitoring)
- International Space Station In-flight Education Downlinks (In-flight Education Downlinks)
- International Space Station Acoustics Measurement Program (ISS Acoustics)
- Science of Opportunity (Saturday Morning Science)

EXPEDITION 7 (APR 28 2003 – OCT 27 2003)

> "... [although] the science program will be reduced ... the most exciting results so far ... will be continued. .. [For example] it does seem possible to reduce or even eliminate ... calcium loss in bones from astronauts." – Edward T. Lu Expedition 7 Flight Engineer

The Expedition 7 crew was the first two-person crew to occupy space station and the first crew to launch from Baikonur on board a Soyuz spacecraft. Despite the loss of one-third of the expected crew complement, this crew continued work to fulfill ISS's scientific goals conducting experiments in:

Technology Development for Exploration
- In Space Soldering Investigation (ISSI)
- Microgravity Acceleration Measurement System (MAMS)
- Materials International Space Station Experiment-1 and -2 (MISSE-1 and -2)
- Space Acceleration Measurement System-II (SAMS-II)

Physical Sciences in Microgravity
- Coarsening in Solid Liquid Mixtures-2 (CSLM-2)
- Investigating the Structure of Paramagnetic Aggregates from Colloidal Emulsions (InSPACE)
- Protein Crystal Growth-Single Locker Thermal Enclosure System
  - Regulation of Gene Expression (PCG-STES-RGE)
  - Science and Applications of Facility Hardware for Protein Crystal Growth (PCG-STES-SA)
- Toward Understanding Pore Formation and Mobility During Controlled Directional Solidification in a Microgravity Environment (PFMI)

Biological Sciences in Microgravity
- Cellular Biotechnology Operations Support Systems: Fluid Dynamics Investigation (CBOSS-FDI)

Human Research and Countermeasure Development for Exploration
- Effect of Prolonged Space Flight on Human Skeletal Muscle (Biopsy)
- Chromosomal Aberrations in Bloody Lymphocytes of Astronauts (Chromosome)
- Hand Posture Analyzer (HPA)
- Crewmember and Crew-Ground Interactions During International Space Station Missions (Interactions)
- Promoting Sensorimotor Response Generalizability: A Countermeasure to Mitigate Locomotor Dysfunction After Locomotive Dysfunction After Long Duration Space Flight (Mobility)
- Subregional Assessment of Bone Loss in the Axial Skeleton in Long-term Space Flight (Subregional Bone)

Observing the Earth and Educational Activities
- Crew Earth Observations (CEO)
- Education Payload Operations (EPO)
- Education Payload Operation – Demonstrations (EPO-Demos)
- Earth Knowledge Acquired by Middle School Students (EarthKAM)

Results from ISS Operations
- Amateur Radio on the International Space Station (ARISS)
- Clinical Nutrition Assessment of ISS Astronauts (Clinical Nutrition Assessment)
- Environmental Monitoring of the International Space (Environmental Monitoring)
- International Space Station In-flight Education Downlinks (In-flight Education Downlinks)
- International Space Station Acoustics Measurement Program (ISS Acoustics)

EXPEDITION 8 (OCT 20 2003 – APR 29 2004)

> "... I think [our scientific mission is] going to be advanced quite significantly ... [That's] a ... bold statement but it's supported by the fact that I have many investigations to carry out on board the station." – C. Michael Foale Expedition 8 Commander and ISS Science Officer

As compared to the scientific workload that was borne by Expedition 7 as the crewmembers accustomed themselves to working with a reduced number, the Expedition 8 crew carried a slightly increased workload in research:

Technology Development for Exploration
- In Space Soldering Investigation (ISSI)
- Microgravity Acceleration Measurement System (MAMS)
- Materials International Space Station Experiment-1 and -2 (MISSE-1 and -2)
- Space Acceleration Measurement System-II (SAMS-II)
- Synchronized Position Hold Engage Reorient Experimental Satellites (Spheres)

Physical Sciences in Microgravity
- Binary Colloidal Alloy Test-3-Binary Alloys (BCAT-3-BA)
- Binary Colloidal Alloy Test-3-Critical Point (BCAT-3-4-CP)
- Binary Colloidal Alloy Test-3-Surface Crystallization (BCAT-3-SC)
- Miscible Fluids in Microgravity-Isothermal (MFMG)
- Protein Crystal Growth-Single Locker Thermal Enclosure System
  - Regulation of Gene Expression (PCG-STES-RGE)
  - Science and Applications of Facility Hardware for Protein Crystal Growth (PCG-STES-SA)
  - Toward Understanding Pore Formation and Mobility During Controlled Directional Solidification in a Microgravity Environment (PFMI)

Biological Sciences in Microgravity
- Cellular Biotechnology Operations Support Systems – Fluid Dynamics Investigation (CBOSS-FDI)
- Yeast-Group Activation Packs (Yeast-GAP)

Human Research and Countermeasure Development for Exploration
- Advanced Diagnostic Ultrasound in Microgravity (ADUM)
- Chromosomal Aberrations in Blood Lymphocytes of Astronauts (Chromosome)
- Foot/Ground Reaction Forces During Space Flight (Foot)
- Hand Posture Analyzer (HPA)
- Crewmember and Crew-Ground Interactions During International Space Station Missions (Interactions)
- Behavioral Issues Associated with Isolation and Confinement (Journals)
- Promoting Sensorimotor Response Generalizability: A Countermeasure to Mitigate Locomotor Dysfunction After Long-Duration Spaceflight (Mobility)
- Renal Stone Risk During Space Flight: Assessment and Countermeasure Validation (Renal Stone)
- Subregional Assessment of Bone Loss in the Axial Skeleton in Long-term Space Flight (Subregional Bone)

Observing Earth and Educational Activities
- Crew Earth Observations (CEO)
- Earth Knowledge Acquired by Middle School Students (EarthKAM)
- Education Payload Operations (EPO)
- Education Payload Operations-Demonstrations (EPO-Demos)

Results from ISS Operations
- Amateur Radio on the International Space Station (ARISS)
- Clinical Nutrition Assessment of ISS Astronauts (Clinical Nutrition Assessment)
- Environmental Monitoring of the International Space (Environmental Monitoring)
- International Space Station Acoustics Measurement Program (ISS Acoustics)
- International Space Station In-flight Education Downlinks (In-flight Education Downlinks)

EXPEDITION 9 (APR 21 2004 – OCT 23 2004)

"... we're hoping that we show on our mission the value of working together the value of teamwork the value of knowing one's job and a good work ethic ..." – Edward Michael (Mike) Fincke Expedition 9 Flight Engineer and ISS Science Officer

For Expedition 9 the crew of Gennady Pedalka and Mike Fincke performed the following investigations:

Technology Development for Exploration
- In Space Soldering Investigation (ISSI)
- Microgravity Acceleration Measurement System (MAMS)
- Materials International Space Station Experiment-1 and -2 (MISSE-1 and -2)
- Space Acceleration Measurement System-II (SAMS-II)
- Serial Network Flow Monitor (SNFM)

Physical Sciences in Microgravity
- Binary Colloidal Alloy Test-3 and -4-Critical Point (BCAT-3-4-CP)
- Binary Colloidal Alloy Test-3-Surface Crystallization (BCAT-3-SC)
- Capillary Flow Experiment (CFE)
- Fluid Merging Viscosity Measurement (FMVM)
- Viscous Liquid Foam – Bulk Metallic Glass (Foam)
- Miscible Fluids in Microgravity (MFMG)-Isothermal
- Protein Crystal Growth-Single Locker Thermal Enclosure System (PCG-STES)
- Regulation of Gene Expression (PCG-STES-RGE)
- Science and Application of Facility Hardware for Protein Crystal Growth (PCG-STES-SA)

Human Research and Countermeasure Development for Exploration
- Advanced Diagnostic Ultrasound in Microgravity (ADUM)
- Effect of Prolonged Space Flight on Human Skeletal Muscle (Biopsy)
- Chromosomal Aberrations in Blood Lymphocytes of Astronauts (Chromosome)
- Crewmember and Crew-Ground Interactions During International Space Station Missions (Interactions)
- Behavioral Issues Associated with Isolation and Confinement (Journals)
- Promoting Sensorimotor Response Generalizability (Mobility)

Observing the Earth and Educational Activities
- Crew Earth Observations (CEO)
- Earth Knowledge Acquired by middle School Students (EarthKAM)
- Education Payload Operations (EPO)
- Education Payload Operations-Demonstrations (EPO-Demos)

Results from ISS Operations
- Amateur Radio on the International Space Station (ARISS)
- Clinical Nutrition Assessment of ISS Astronauts (Clinical Nutrition Assessment)
- Environmental Monitoring of the International Space (Environmental Monitoring)
- International Space Station In-flight Education Downlinks (In-flight Education Downlinks)

EXPEDITION 10 (OCT 15 2004 – APR 25 2005)

> "If we're going to go to the moon and Mars we're going to have to know how ... [and] the space station is an ideal platform for ... that kind of work." – Leroy Chiao Expedition 10 Commander and ISS Science Officer

For the Expedition 10 crew of Salizan Sharipov and Leroy Chiao, the focus was on "knowing how" to work for long durations in space and to continue to develop methods and technologies that would aid the space program in a future return to the moon and exploration of Mars. Accordingly and in keeping with the pattern that had been established by earlier Expeditions, they conducted several experiments:

Technology Development for Exploration
- Dust and Aerosol Measurement Feasibility Test (DAFT)
- In Space Soldering Investigation (ISSI)
- Microgravity Acceleration Measurement System (MAMS)
- Materials International Space Station Experiment-1 and -2 (MISSE-1 and -2)
- Space Acceleration Measurement System-II (SAMS-II)
- Serial Network Flow Monitor (SNFM)

Physical Sciences in Microgravity
- Binary Colloidal Alloy Test-3 and -4-Critical Point (BCAT-3-4-CP)
- Binary Colloidal Alloy Test-3-Surface Crystallization (BCAT-3-SC)
- Miscible Fluids in Microgravity (MFMG)
- Protein Crystal Growth-Single Locker Thermal Enclosure System
    - Regulation Gene Expression (PCG-STES-RGE)
    - Science and Applications of Facility Hardware for Protein Crystal Growth (PCG-STES-SA)

Biological Science in Microgravity
- Cellular Biotechnology Operations Support System-Fluid Dynamic Investigation (CBOSS-FDI)

Human Research and Countermeasure Development for Exploration
- Advanced Diagnostic Ultrasound in Microgravity (ADUM)
- Effect of Prolonged Space Flight on Human Skeletal Muscle (Biopsy)
- Chromosomal Aberrations in Blood Lymphocytes of Astronauts (Chromosome)
- Behavioral Issues Associated with Isolation and Confinement (Journals)
- Promoting Sensorimotor Response Generalizability (Mobility)

Observing Earth and Observation
- Crew Earth Observations (CEO)
- Earth Knowledge Acquired by Middle School Students (EarthKAM)
- Education Payload Operations-Demonstrations (EPO-Demos)
- Space Experiment Module (SEM)

Results from ISS Operations
- Amateur Radio on the International Space Station (ARISS)
- Clinical Nutrition Assessment of ISS Astronauts (Clinical Nutrition)
- Environmental Monitoring of the International Space (Environmental Monitoring)
- International Space Station Acoustics Measurement Program (ISS Acoustics)
- International Space Station In-flight Education Downlinks (In-flight Education Downlinks)

EXPEDITION 11 (APR 16 2005 – OCT 10 2005)

"With our new emphasis as announced by the president [last] year, we're going to be focusing our science on things that will take us farther and longer into space." – John Phillips Expedition 11 Flight Engineer and ISS Science Officer

For the Expedition 11 crew of Sergei Krikalev and John Phillips, the following experiments were the focus:

Technology Development for Exploration
- Materials International Space Station Experiment-1 and -2 (MISSE-1 and -2)
- Materials International Space Station Experiment-5 (MISSE-5)
- Microgravity Acceleration Measurement System (MAMS)
- Space Acceleration Measurement System-II (SAMS-II)
- Serial Network Flow Monitor (SNFM)

Physical Sciences in Microgravity
- Fluid Merging Viscosity Measurement (FMVM)
- Miscible Fluids in Microgravity (MFMG)
- Protein Crystal Growth
    - Single Locker Thermal Enclosure System-Regulation of Gene Expression (PCG-STES-RGE)
    - Single Locker Thermal Enclosure System-Science and Applications of Facility Hardware for Protein Crystal Growth (PCG-STES-SA)

Human Research and Countermeasure Development for Exploration
- Advanced Diagnostic Ultrasound in Microgravity (ADUM)
- Behavioral Issues Associated with Isolation and Confinement: Review and Analysis of Crew Journals (Journals)
- Bioavailability and Performance Effects of Promethazine During Space Flight (PMZ)
- Chromosomal Aberrations in Blood Lymphocytes of Astronauts (Chromosome)
- Effect of Prolonged Space Flight on Human Skeletal Muscle (Biopsy)
- Foot Reaction Forces During Space Flight (Foot)
- Hand Posture Analyzer (HPA)
- Incidence of Latent Virus Shedding During Space Flight (Latent Virus)
- Promoting Sensorimotor Response Generalizability (Mobility)
- Renal Stone Risk During Space Flight: Assessment and Countermeasure Validation (Renal Stone)
- Space Flight-induced Reactivation of Latent Epstein-Barr Virus (Epstein-Barr)
- Sleep-Wake Actigraphy and Light Exposure During Spaceflight-Short (Sleep-Short)

Observing the Earth and Educational Activities
- Crew Earth Observations (CEO)
- Earth Knowledge Acquired by middle School Students (EarthKAM)
- Education Payload Operation-Demonstration (EPO-Demos)
- Space Experiment Module (SEM)

Results from ISS Operations
- Amateur Radio on the International Space Station (ARISS)
- Clinical Nutrition Assessment of ISS Astronauts (Clinical Nutrition Assessment)
- Environmental Monitoring of the International Space Station (Environmental Monitoring)
- International Space Station Acoustic Measurement Program (ISS Acoustics)
- Analysis of International Space Station Plasma Interaction (Plasma Interaction Model)
- International Space Station In-flight Education Downlinks (In-flight Education Downlinks)

EXPEDITION 12 (OCT 3 2005 – APR 8 2006)

"I think that in the future we'll have a good knowledge about scientific things... a new direction for the future development of space science and the exploration of space." – Valery Tokarev Expedition 12 Flight Engineer and ISS Science Officer

The Expedition 12 crew of William McArthur and Valery Tokarev worked on the following:

Technology Development for Exploration
- Dust and Aerosol Measurement Feasibility Test (DAFT)
- Microgravity Acceleration Measurement System (MAMS)
- Maui Analysis of Upper Atmospheric Injections (Maui)
- Materials International Space Station Experiment-5 (MISSE-5)
- Serial Network Flow Monitor (SNFM)
- Space Acceleration Measurement System-II (SAMS-II)

Physical Sciences in Microgravity
- Binary Colloidal Alloy Test-3-Surface Crystallization (BCAT-3-SC)
- Binary Colloidal Alloy-3 and -4-Critical Point (BCAT-3-4-CP)
- Capillary Flow Experiment (CFE)
- Investigating the Structure of Paramagnetic Aggregates from Colloidal Emulsions (InSPACE)
- Toward Understanding Pore Formation and Mobility During Controlled Directional Solidification in a Microgravity Environment (PFMI)

Biological Sciences in Microgravity
- Cellular Biotechnology Operations Support Systems-Fluid Dynamics Investigation (CBOSS-FDI)
- Passive Observatories for Experimental Microbial Systems (POEMS)

Human Research and Countermeasure Development
- Advanced Diagnostic Ultrasound in Microgravity (ADUM)
- Behavioral Issues Associated with Isolation and Confinement: Review an Analysis of Crew Journals (Journals)
- Foot Reaction Forces During Space Flight (Foot)
- Promoting Sensorimotor Response Generalizability Mitigate Locomotor Dysfunction After Long-Duration Space Flight (Mobility)
- Space Flight-induced Reactivation of Latent Epstein-Bar Virus (Epstein-Barr)
- Renal Stone Risk During Space Flight: Assessment (Renal Stone)

Observing the Earth and Educational Activities
- Crew Earth Observations (CEO)
- Earth Knowledge Acquired by middle School Students (EarthKAM)
- Education Payload Operations-Demonstrations (EPO-Demos)

Results from ISS Operations
- Amateur Radio on the International Space Station (ARISS)
- Analysis of International Space Station Plasma Interaction (Plasma Interaction Model)
- Clinical Nutrition Assessment of ISS Astronauts (Clinical Nutrition Assessment)
- Environmental Monitoring of the International Space Station (Environmental Monitoring)
- International Space Station Acoustic Measurement Program (ISS Acoustics)
- International Space Station In-flight Education Downlinks (In-flight Education Downlinks)

EXPEDITION 13 (APR 1 2006 – SEP 28 2006)

> "It's quite important to have a three-person crew obviously; one of the most important aspects is the fact that we can dedicate more [time to] scientific research." – Thomas Reiter Expedition 13 Flight Engineer and ISS Science Officer

The Expedition 13 crew of Pavel Vinogradov, Jeffrey Williams, and Thomas Reiter (Jul 2006 – Dec 2006) performed the following experiments:

Technology Development for Exploration
- Dust and Aerosol Measurement Feasibility Test (DAFT)
- Microgravity Acceleration Measurement System (MAMS)
- Maui Analysis of Upper Atmospheric Injections (Maui)
- Materials International Space Station Experiment-3 and -4 (MISSE-3 and -4)
- Materials International Space Station Experiment-5 (MISSE-5)
- Ram Burn Observations (RAMBO)
- Space Acceleration Measurement System-II (SAMS-II)
- Synchronized Position Hold, Engage, Reorient, Experimental Satellites (SPHERES)

Physical Sciences in Microgravity
- Binary Colloidal Alloy Test-3 and -4-Critical Point (BCAT-3-4-CP)
- Capillary Flow Experiment (CFE)
- Investigating the Structure of Paramagnetic Aggregates from Colloidal Emulsions (InSPACE)
- Toward Understanding Pore Formation and Mobility During Controlled Directional Solidification in a Microgravity Environment (PFMI)

Biological Sciences in Microgravity
- Cellular Biotechnology Operations Support Systems-Fluid Dynamics Investigation (CBOSS-FDI)
- Fungal Pathogenesis, Tumorigenesis, and Effects of Hosts Immunity in Space (Fit)
- Effect of Space Flight on Microbial Gene Expression and Virulence (Microbe)
- Passive Observatories for Experimental Microbial Systems (POEMS)
- Yeast-Group Activation Packs (Yeast-GAP)

Human Research and Countermeasure Development for Exploration
- Anomalous Long-Term Effects in Astronauts' Central Nervous System (ALTEA)
- Chromosomal Aberrations in Blood Lymphocytes of Astronauts (Chromosome)
- Space Flight-induced Reactivation of Latent Epstein-Barr Virus (Epstein-Barr)
- Behavioral Issues Associated with Isolation and Confinement: Review and Analysis of ISS Crew Journals (Journals)
- Incidence of Latent Virus Shedding During Space Flight (Latent Virus)
- Bioavailability and Performance Effects of Promethazine During Space Flight (PMZ)
- Renal Stone Risk During Space Flight: Assessment (Renal Stone)
- Sleep-Wake Actigraphy and Light Exposure During Space Flight-Short (Sleep-Short)
- Stability of Pharmacotherapeutic and Nutritional Compounds (Stability)
- Surface, Water, and Air Biocharacterization (SWAB) – A Comprehensive Characterization of Microorganisms and Allergens in Spacecraft Environment

Observing the Earth and Educational Activities
- Crew Earth Observations (CEO)
- Earth Knowledge Acquired by Middle School Students (EarthKAM)
- Education Payload Operations-Demonstrations (EPO-Demos)
- Space Experiment Module (SEM)

Results from ISS Operations
- Amateur Radio on the International Space Station (ARISS)
- Clinical Nutrition Assessment of ISS Astronauts (Clinical Nutrition Assessment)
- Environmental Monitoring of the International Space Station (Environmental Monitoring)
- International Space Station Acoustic Measurement Program (ISS Acoustics)
- Periodic Fitness Evaluation with Oxygen Uptake Measurement (PFE-OUM)
- Analysis of International Space Station Plasma Interaction (Plasma Interaction Model)
- International Space Station In-flight Education Downlinks (In-flight Education Downlinks)

EXPEDITION 14 (SEP 20 2006 – APR 21 2007)

> "... we're trying to understand better the effects of long-duration spaceflight on humans, because our goal is to extend our presence not just in low Earth orbit but to go back to the moon with some kind of a longer-term presence, and hopefully on to Mars someday. " – Michael E. Lopez-Alegria
> Expedition 14 Commander

For the Expedition 14 crew of Michael Lopez-Alegria, Mikhail Tyurin, Thomas Reiter (Jul 2006 – Dec 2006), and Sunita Williams (Dec 2006 – Jun 2007), the crew performed the following experiments:

Technology Development for Exploration
- Lab-on-a-Chip Application Development-Portable Test System (LOCAD-PTS)
- Maui Analysis of Upper (Maui)
- Materials International Space Station Experiment-3 and -4 (MISSE-3 and -4)
- Microgravity Acceleration Measurement System (MAMS)
- Ram Burn Observations (RAMBO)
- Space Acceleration Measurement System-II (SAMS-II)
- Synchronized Position Hold, Engage, Reorient, Experimental Satellites (SPHERES)
- Space Test Program-H2-Atmospheric Neutral Density Experiment (STP-H2-ANDE)
- Space Test Program-H2-Microelectromechanical System-Bases (MEMS) PICOSAT Inspector (STP-H2-MEPSI)
- Space Test Program-H2-Radar Fence Transponder (STP-H2-RAFT)

Physical Sciences Microgravity
- Capillary Flow Experiment (CFE)

Biological Sciences Microgravity
- Analysis of a Novel Sensory Mechanism in Root Phototropism (Tropi)
- Passive Observatories from Experimental Microbial Systems (POEMS)
- Threshold Acceleration for Gravisensing (Gravi)
- The Optimization of Root Zone Substrates (ORZS) for Reduced-Gravity Experiments Program

Human Research and Countermeasure Development for Exploration
- Anomalous Long-Term Effects in Astronauts' Central Nervous System (ALTEA)
- Space Flight-induced Reactivation of Latent Epstein-Barr Virus (Epstein-Barr)
- Behavioral Issues Associated with Isolation and Confinement: Review and Analysis of ISS Crew Journals (Journals)
- Incidence of Latent Virus Shedding During Space Flight (Latent Virus)
- Test of Midodrine as a Countermeasure Against Post-flight Orthostatic Hypotension-Short-Duration Biological Investigation (Midodrine-SDBI)
- Nutrition Status Assessment (Nutrition)
- Perceptual Motor Deficits in Space (PMDIS)
- Renal Stone Risk During Space Flight (Renal Stone)
- Sleep-Wake Actigraphy and Light Exposure During Space Flight-Long (Sleep-Long)
- Sleep-Wake Actigraphy and Light Exposure During Space Flight-Short (Sleep-Short)
- Stability of Pharmacotherapeutic and Nutritional Compounds (Stability)
- Surface, Water and Air Biocharacterization (SWAB) – A Comprehensive Characterization of Microorganisms and Allergens in Spacecraft
- Test of Reaction and Adaptation Capabilities (TRAC)

Observing the Earth and Educational Activities
- Crew Earth Observations (CEO)
- Crew Earth Observations-International Polar Year (CEO-IPY)
- Commercial Generic Bioprocessing Apparatus Science Insert-01 (CSI-01)

- Earth Knowledge Acquired by Middle School Students (EarthKAM)
- Education Payload Operations-Demonstrations (EPO-Demos)
- Space Experiment Module (SEM)

Results from ISS Operations
- Amateur Radio of the International Space Station (ARISS)
- Clinical Nutrition Assessment of ISS Astronauts (Clinical Nutrition Assessment)
- Education-How Solar Cells Work (Education-Solar Cells)
- Environmental Monitoring of the International Space Station (Environmental Monitoring)
- International Space Station Acoustic Measurement Program (ISS Acoustics)
- Periodic Fitness Evaluation with Oxygen Uptake Measurement (PFE-OUM)
- Analysis of International Space Station Plasma Interaction (Plasma Interaction Model)
- Soldering in Reduced Gravity Experiment (SORGE), SDTO 17003-U
- International Space Station In-flight Education Downlinks (In-flight Education Downlinks)
- International Space Station Zero-Propellant Maneuver (ZPM) Demonstration

EXPEDITION 15 (APR 9 2007 – OCT 21 2007)

> "There are two limiting factors currently to deep-space exploration. One of those is the technological problem, and it can be solved. ... the second limiting factor ... is the prevention of the negative impact of zero gravity, how to be involved in a long-duration space flight while maintaining good health, because station work requires good health and good functioning." – Oleg Kotov Expedition 15 Flight Engineer

The Expedition 15 crew of Fyoder Yurchikhin, Oleg Kotov, Sunita Williams (Dec 2006 – Jun 2007),, and Clay Anderson (Jun 2007 – Oct 2007), performed the following experiments:

Technology and Development for Exploration
- Analyzing Interferometer for Ambient Air (ANITA)
- Elastic Memory Composite Hinge (EMCH)
- Lab-on-a-Chip Application Development-Portable Test System (LOCAD-PTS)
- Microgravity Acceleration Measurement System (MAMS)
- Maui Analysis of Upper Atmospheric Injections (Maui)
- Materials International Space Station Experiment-3 and -4 (MISSE-3 and -4)
- Ram Burn Observations (RAMBO)
- Smoke and Aerosol Measurement Experiment (SAME)
- Space Acceleration Measurement System-II (SAMS-II)
- Synchronized Position Hold, Engage, Reorient, Experimental Satellites (SPHERES)

Physical Sciences in Microgravity
- Capillary Flow Experiment (CFE)

Biological Sciences in Microgravity
- Molecular and plant Physiological Analyses of the Microgravity Effects on Multigeneration Studies of *Arabidopsis thaliana* (Multigen)
- *Streptococcus pneumoniae* Expression of Genes in Space (SPEGIS)
- The Optimization of Root Zone Substrates (ORZS) for Reduced-Gravity Experiments Program

Human Research and Countermeasure Development for Exploration
- Anomalous Long-Term Effects in Astronauts' Central Nervous System (ALTEA)
- Commercial Biomedical Test Module-2 (CBTM-2)
- Cardiovascular and Cerebrovascular Control on Return from ISS (CCISS)
- Cell Culture Module-Immune Response of Human Monocytes in Microgravity (CCM-Immune Response)
- Cell Culture Module-Effect of Microgravity on Wound Repair in Vitro Model of New Blood Vessel Development (CCM-Wound Repair)
- Space Flight-induced Reactivation of Latent Epstein-Barr Virus (Epstein-Barr)
- Behavioral Issues Associated with Isolation and Confinement: Review and Analysis of ISS Crew Journals (Journals)
- Incidence of Latent Virus Shedding During Space Flight (Latent Virus)
- Test of Midodrine as a Countermeasure Against Post-flight Orthostatic Hypotension-Short-Duration Biological Investigation (Midodrine-SDBI)
- Nutritional Status Assessment (Nutrition)
- Perceptual Motor Deficits in Space (PMDIS)
- Bioavailability and Performance Effects of Promethazine During Space Flight (PMZ)
- Sleep-Wake Actography and Light Exposure During Space Flight-Long (Sleep-Long)
- Sleep-Wake Actography and Light Exposure During Space Flight-Short (Sleep-Short)
- Stability of Pharmacotherapeutic and Nutritional Compounds (Stability)
- Surface, Water, and Air Biocharacterization (SWAB) – Comprehensive Characterization of Microorganisms and Allergens in Spacecraft Environment
- Test of Reaction and Adaptation Capabilities (TRAC)

Observing the Earth and Educational Activities
- Crew Earth Observations (CEO)
- Crew Earth Observations-International Polar Year (CEO-IPY)
- Commercial Generic Bioprocessing Apparatus Science Insert-01 (CSI-01)
- Commercial Generic Bioprocessing Apparatus Science Insert-02 (CSI-02)
- Education Payload Operations-Demonstrations (EPO-Demos)
- Education Payload Operations-Educator (EPO-Educator)
- Education Payload Operations-Kit C Plant Growth Chambers (EPO-Kit C)
- Earth Knowledge Acquired by Middle School Students (EarthKAM)

Results from ISS Operations
- Amateur Radio on the International Space Station (ARISS)
- Environmental Monitoring of the International Space Station (Environmental Monitoring)
- International Space Station Acoustic Measurement Program (ISS Acoustics)
- Periodic Fitness Evaluation with Oxygen Uptake Measurement (PFE-OUM)
- Analysis of International Space Station Plasma Interaction (Plasma Interaction Model)
- International Space Station In-flight Education Downlinks (In-flight Education Downlinks)
- International Space Station Zero-Propellant Maneuver (ZPM) Demonstration

ACKNOWLEDGMENTS

We are especially grateful to the ISS crewmembers who have been dedicated to making the most of research activities, even in the years of ISS assembly when there are so many maintenance and construction tasks. We would like to the Associate Administrator for Space Operations and former head of the ISS Program, Bill Gerstenmaier, and the ISS Program Scientist, Don Thomas, for their unwavering support of our efforts to assemble complete information on ISS research and results. We thank the many scientists who have shared information on their station research and supported us in tracking their results and publications. We thank them, too, for reviewing the evolving versions of descriptions in the database and for their patience as we tried to describe their work for a general audience. We want to thank those in the Mission Science group of Lockheed Martin who drew the information together for the first draft of the database, especially Peggy Delaney, Pasha Morshedi, Andy Self, and Wes Tarkington. Our original NASA internal Web site and database structure was developed by Wes Tarkington. The system evolved, and dynamic updates to the NASA Portal Web site were made possible by the efforts of Ryan Elliott of the Engineering & Science Contract Group and Alex Pline, NASA Headquarters, and his colleagues. Last, but not least, we want to thank Sharon Hecht of TES for her editorial help in making this NASA Technical Publication a reality. Her patience with our continued updates and her gentle reminders kept this project on track.

Appendix B: Publications Resulting from Research Aboard the International Space Station

PAYLOAD	CITATION	YEAR	DISCIPLINE
Biological Sciences in Microgravity			
AdvAsc	Zhou W, Durst SJ, DeMars M, Stankovic B, Link BM, Tellez G, Meyers RA, Sandstrom PW, Abba JR. Performance of the Advanced Astroculture™ plant growth unit during ISS-6A/7A mission. SAE Technical Paper Series. 2002; Paper No. 02ICES-267.	2002	Biological Sciences in Microgravity
AdvAsc	Link BM, Durst SJ, Zhou W, Stankovic B. Seed-to-seed growth of Arabidopsis Thaliana on the International Space Station. Advances in Space Research. 2003; 31(10):2237-2243.	2003	Biological Sciences in Microgravity
AdvAsc	Zhou W. Advanced AstrocultureTM Plant Growth Unit: Capabilities and Performances. 35th International Conference on Environmental Systems, Rome, Italy. Jul 11 – 14, 2005.	2005	Biological Sciences in Microgravity
BPS	Frazier CM, Simpson JB, Roberts MS, Stutte GW, Fields NW, Melendez-Andrade J, Morrow RC. Bacterial and fungal communities in BPS chambers and root modules. SAE Technical Paper Series. 2003 Paper No. 03ICES-147.	2003	Biological Sciences in Microgravity
BPS	Iverson JT, Crabb TM, Morrow RC, Lee MC. Biomass Production System Hardware Performance. SAE Technical Paper Series. 2003 Paper No. 03ICES-67.	2003	Biological Sciences in Microgravity
BPS	Musgrave ME, Kuang A, Tuominem LK, Levine LH, Morrow R.C. Seed Storage Reserves and Glucosinolates in *Brassica rapa L*. Grown on the International Space Station. *Journal of the American Society for Horticultural Science*. 2005; 130(6): 848–856.	2005	Biological Sciences in Microgravity
BPS	Morrow RC, Iverson JT, Richter RC, Stadler JJ. Biomass Production System (BPS) Technology Validation Test Results. *Transactions Journal of Aerospace* 2004 1:1061–1070.	2004	Biological Sciences in Microgravity
CBOSS-01-02-Renal	Hammond DK, Elliott TF, Holubec K, Baker TL, Allen PL, Hammond TG, Love JE. Proteomic Retrieval from Nucleic Acid Depleted Space-Flown Human Cells. *Gravitational and Space Biology*. 2006 Jun;19(2).	2006	Biological Sciences in Microgravity
CBOSS-01-Ovarian	Hammond DK, Becker J, Elliot TF, Holubec K, Baker TL, Love JE. Antigenic Protein in Microgravity-Grown Human Mixed Mullerian Ovarian Tumor (LN-1) Cells Preserved in a RNA Stabilizing Agent. *Gravitational and Space Biology*. 2005;18(2):99–100.	2005	Biological Sciences in Microgravity
CGBA-APS	Klaus D, Benoit M, Bonomo J, Bollich J, Freman J, Stodieck LL, McClure G, Lam KS Antiobiotic Production in Space using Automated Fed-Bioreactor System. Conference and Exhibit on International Space Station Utilization – 2001, Cape Canaveral, Fla. Oct 15–18 2001. AIAA 200-1-4921.	2001	Biological Sciences in Microgravity
CGBA-APS	Benoit MR, Li W, Stodieck LS, Lam KS, Winther CL, Roane TM, Klaus DM. Microbial Antibiotic Production Aboard the International Space Station. *Applied Microbiology and Biotechnology*. 2006; 70(4): 403–411.	2006	Biological Sciences in Microgravity
Gravi	Driss-Ecole D, Legue V, Carnero-Diaz E, Perbal G. Gravisensitivity and automorphogenesis of lentil seedling roots grown on board the International Space Station. *Physiologia plantarum*. 2008; Epub.	2008	Biological Sciences in Microgravity

PAYLOAD	CITATION	YEAR	DISCIPLINE
MEPS	Morrison DR, Haddad RS, Ficht A. Microencapsulation of Drugs: New cancer therapies and improved drug delivery derived from microgravity research. Proceedings of the 40th Space Congress, Cape Canaveral, Fla. Apr 2003.	2003	Biological Sciences in Microgravity
MEPS	Le Pivert P, Haddad RS, Aller A, Titus K, Doulat J, Renard M, Morrison DR. Ultrasound Guided, Combined Cryoablation and Microencapsulated 5-Fluorouracil, Inhibits Growth of Human Prostate Tumors in Xenogenic Mouse Model Assessed by Fluorescence Imaging. *Technology in Cancer Research and Treatment*. 2004; 3(2):135–42.	2004	Biological Sciences in Microgravity
Microbe	Wilson JW, Ott CM, Hoener zu Bentrup K, Ramamurthy R, Quick L, Porwollik S, Cheng P, McClelland M, Tsaprailise G, Radabaugh T, Hunt A, Fernandez D, Richter E, Shah M, Kilcoyne M, Joshi L, Nelman-Gonzalez M, Hing S, Parra M, Dumars P, Norwood K, Bober R, Devich J, Ruggles A, Goulart C, Rupert M, Stodieck L, Stafford P, Catella L, Schurr MJ, Buchanan K, Morici L, McCracken J, Allen P, Baker-Coleman C, Hammond T, Vogel J, Nelson R, Pierson DL, Stefanyshyn-Piper HM, Nickerson CA. Space flight alters bacterial gene expression and virulence and reveals a role for global regulator Hfq. Proceedings of the National Academy of Sciences of the United States of America. 2007; 104(41):16299–16304.	2007	Biological Sciences in Microgravity
ORZS	Jones SB, Heinse R, Bingham GE, Or D. Modeling and Design of Optimal Growth Media from Plant-Based Gas and Liquid Fluxes, SAE 2005-01-2949, 2005.	2005	Biological Sciences in Microgravity
PESTO	Stutte GW, Monje O, Anderson S. Wheat (*Triticum Aesativum L.* cv. USU Apogee) Growth Onboard the International Space Station (ISS): Germination and Early Development. *Proceedings of the Plant Growth Regulation Society of America*. 2003; 30:66–71.	2003	Biological Sciences in Microgravity
PESTO	Monje O, Stutte GW, Goins GD, Porterfield DM, Bingham GE. Farming in Space: Environmental and Biochemical Concerns. *Advances in Space Research*. 2003; 31:151–167.	2003	Biological Sciences in Microgravity
PESTO	Paul A, Levine HG, McLamb W, Norwood KL, Reed D, Stutte GW, Wells HW, Ferl RJ. Plant molecular biology in the space station era: Utilization of KSC fixation tubes with RNAlater. *Acta Astronautica*. 2005; 56:623–628.	2005	Biological Sciences in Microgravity
PESTO	Stutte GW, Monje O, Goins GD, Tripathy BC. Microgravity effects on thylakoid, single leaf, and whole canopy photosynthesis on dwarf wheat. *Planta*. 2005; 1–11.	2005	Biological Sciences in Microgravity
PESTO	Monje O, Stutte G, Chapman D. Microgravity does not alter plant stand gas exchange of wheat at moderate light levels and saturating CO_2 concentration. *Planta*. Jun 2005; Online.	2005	Biological Sciences in Microgravity
PESTO	Stutte GW, Monje O, Hatfield RD, Paul AL, Ferl RJ, Simone CG. Microgravity effects on leaf morphology, cell structure, carbon metabolism and mRNA expression of dwarf wheat. *Planta*. 2006; 224(5):1038–1049.	2006	Biological Sciences in Microgravity
PGBA	Heyenga AG, Stodieck L, Hoehn A, Kliss M, Blackford C. Approaches in the Design of a Space Plant Cultivation Facility for *Arabidopsis thaliana*. 34th International Conference On Environmental Systems (ICES), Colorado Springs, Colo. Jul 2004; SAE Paper 2004-01-2459 (New designs influenced by lessons learned).	2004	Biological Sciences in Microgravity
PGBA	Heÿenga G, Kliss M, Blackford C. (2005): The Performance of a Miniature Plant Cultivation System Designed for Space Flight Application, SAE-paper: 2005-01-2844. In *International Conference Environmental Systems (ICES) Proceedings*, Rome, Italy.	2005	Biological Sciences in Microgravity
POEMS	Roberts MS, Reed DW, Rodriguez JI. 2005. Passive Observatories for Experimental Microbial Systems (POEMS): Microbes Return to Flight. 35th Annual International Conference Environmental Systems (ICES). SAE 05ICES-214.	2005	Biological Sciences in Microgravity

PAYLOAD	CITATION	YEAR	DISCIPLINE
Human Research and Countermeasure Development for Exploration			
ADUM	Fincke EM, Padalka G, Lee D, van Holsbeeck M, Sargsyan AE, Hamilton DR, Martin D, Melton SL, McFarlin K, Dulchavsky SA. Evaluation of Shoulder Integrity in Space: First Report of Musculoskeletal US on the International Space Station. *Radiology*. 234(2):319–322, 2005.	2005	Human Research and Countermeasure Development for Exploration
ADUM	Chiao L, Sharipov S, Sargsyan AE, Melton S, Hamilton DR, McFarlin K, Dulchavsky SA. Ocular examination for trauma; clinical ultrasound aboard the International Space Station. *Journal of Trauma*. 2005; 58(5):885–889.	2005	Human Research and Countermeasure Development for Exploration
ADUM	Sargsyan AE, Hamilton DR, Jones JA, Melton S, Whitson PA, Kirkpatrick AW, Martin D, Dulchavsky SA. FAST at MACH 20: Clinical Ultrasound Aboard the International Space Station. *The Journal of Trauma, Injury, Infection, and Critical Care*. 2004; 58(1):35–39.	2005	Human Research and Countermeasure Development for Exploration
ADUM	Kwon D, Bouuffard JA, van Holsbeeck M, Sargsyan AE, Hamilton DR, Melton SL, Dulchavsky SA. Battling Fire and Ice: Remote guidance ultrasound to diagnose injury on the International Space Station and the ice rink. *Am J. Surgery* 93 (2007) 417–420.	2007	Human Research and Countermeasure Development for Exploration
ADUM	Foale CM, Kaleri AY, Sargsyan AE, Hamilton DR, Melton S, Margin D, Dulchavsky SA. Diagnostic instrumentation aboard ISS: just in time training for non-physician crewmembers. *Aviation, Space and Environmental Medicine* 2005; 76:594–598.	2005	Human Research and Countermeasure Development for Exploration
ADUM-Ultrasound	Sargsyan AE, Hamilton DR, Jones JA, Melton S, Whitson PA, Kirkpatrick AW, Martin D, Dulchavsky SA. FAST at MACH 20: Clinical Ultrasound Aboard the International Space Station. *The Journal of Trauma, Injury, Infection, and Critical Care*. 2004; 58(1):35–39.	2004	Human Research and Countermeasure Development for Exploration, Results from ISS Operations
ADUM-Ultrasound	Sargsyan A, Hamilton DR, Melton SL, Young J. The International Space Station Ultrasound Imaging Capability Overview for Prospective Users. NASA/TP-2006-213731 2006.	2006	Human Research and Countermeasure Development for Exploration, Results from ISS Operations
ADUM-Ultrasound	Kirkpatrick, AW, Jones, J., Hamilton, DR, Melton, S, Beck, G, Savvas, N, Campell, M, Dulchavsky S. Trauma Sonography for Use in microgravity, *Aviation, Space and Environ. Medicine* 78, No 4, Sec II A38-A42, Apr 2007.	2007	Human Research and Countermeasure Development for Exploration, Results from ISS Operations

PAYLOAD	CITATION	YEAR	DISCIPLINE
BBND	Koshiishi H, Matsumoto H, Koga K, Goka T. Evaluation of Low-Energy Neutron Environment inside the International Space Station. Technical Report of Institute of Electronics, Information, and Communications Engineers. 2003; SANE2003-79:11-14. [*Japanese*]	2003	Human Research and Countermeasure Development for Exploration
CBTM	Gridley DS, Nelson GA, Peters LL, Kostenuik PJ, Bateman TA, Morony S, Stodieck LS, Lacey DL, Simske SJ, Pecaut MJ. Genetic models in applied physiology: selected contribution: effects of spaceflight on immunity in the C57BL/6 mouse. II. Activation, cytokines, erythrocytes, and platelets. *Journal of Applied Physiology*. 2003; 94(5):2095–2103.	2003	Human Research and Countermeasure Development for Exploration
CBTM	Pecaut MJ, Nelson GA, Peters LL, Kostenuik PJ, Bateman TA, Morony S, Stodieck LS, Lacey DL, Simske SJ, Gridley DS. Genetic models in applied physiology: selected contribution: effects of spaceflight on immunity in the C57BL/6 mouse. I. Immune population distributions. *Journal of Applied Physiology*. 2003; 94(5):2085–2094.	2003	Human Research and Countermeasure Development for Exploration
CBTM	Harrison BC, Allen DL, Girten B, Stodieck LS, Kostenuik PJ, Bateman TA, Morony S, Lacey DL, Leinwand LA. Skeletal muscle adaptations to microgravity exposure in the mouse. *Journal of Applied Physiology*. 2003; 95(6):2462–2470.	2003	Human Research and Countermeasure Development for Exploration
CBTM	Bateman TA, Morony S, Ferguson VL, Simske SJ, Lacey DL, Warmington KS, Geng Z, Tan HL, Shalhoub V, Dunstan CR, Kostenuik PJ. Molecular therapies for disuse osteoporosis. *Gravitational and Space Biology Bulletin*. 2004; 17:83–89.	2004	Human Research and Countermeasure Development for Exploration
CBTM-2	Gridley DS, Slater JM, Luo-Owen X, Rizvi A, Chapes SK, Stodieck LS, Ferguson VL, Pecaut MJ. Spaceflight effects on T lymphocyte distribution, function and gene expression. *J Appl Physiol*. 2008 Nov 6.	2008	Human Research and Countermeasure Development for Exploration
DOSMAP	Reitz G, Beaujean R, Dachev Ts, Deme S, Luszik-Bhadra M, Heinrich W, Olko P. Dosimetric Mapping. Conference and Exhibit on International Space Station Utilization, Cape Canaveral, Fla. Oct 15 – 18, 2001; AIAA-2001-4903.	2001	Human Research and Countermeasure Development for Exploration
DOSMAP	Reitz G, Beaujean R, Benton E, Burmeister S, Dachev Ts, Deme S, Luszik-Bhadra M, Olko P. Space radiation measurements on-board ISS – the DOSMAP experiment. *Radiation Protection Dosimetry*. 2005; 116 (1-4): 374–379.	2005	Human Research and Countermeasure Development for Exploration
Foot	Cavanagh PR, Maender C, Rice AJ, Gene KO, Ochia RS, Snedeker JG. Lower-Extremity Loading During Exercise on the International Space Station. Transactions of the Annual Meeting of the Orthopaedic Research Society. 2004; 0395.	2004	Human Research and Countermeasure Development for Exploration
Foot	Pierre MC, Genc KO, Litow M, Humphreys B, Rice A, Maender CC, Cavanagh PR. Comparison of Knee Motion on Earth and in Space: An Observational Study. *Journal of Neuroengineering and Rehabilitation*. 2006; 3:8.	2006	Human Research and Countermeasure Development for Exploration

PAYLOAD	CITATION	YEAR	DISCIPLINE
H-Reflex	Watt DG, Lefebvre L. Effects of altered gravity on spinal cord excitability. First Research on the International Space Station. Conference and Exhibit on International Space Station Utilization, Cape Canaveral, Fla. Oct 15 – 18, 2001; AIAA 2001-4939.	2001	Human Research and Countermeasure Development for Exploration
H-Reflex	Watt DG. Effects of altered gravity on spinal cord excitability (final results). Proceedings of the Bioastronautics Investigators' Workshop, Galveston, Texas. Jan 2003.	2003	Human Research and Countermeasure Development for Exploration
H-Reflex	Watt DG. Effects of prolonged exposure to microgravity on H-reflex loop excitability. Proceedings of the 14th IAA Humans in Space Symposium, Banff, Alberta. May 2003.	2003	Human Research and Countermeasure Development for Exploration
Interactions	Kanas N, Ritsher J. Leadership Issues with Multicultural Crews on the International Space Station: Lessons learned from Shuttle/Mir. *Acta Astronautica*. 2005; 56:932–936.	2005	Human Research and Countermeasure Development for Exploration
Interactions	Kanas N, Salnitskiy V P, Ritsher JB, Gushin VI, Weiss DS, Saylor S, Marmar C. Human interactions in space: ISS versus Shuttle/Mir. 56th International Astronautical Congress. Fukuoka, Japan. Oct 17–21, 2005; IAC-05-AI.5.02: 8.	2005	Human Research and Countermeasure Development for Exploration
Interactions	Ritsher JB, Kanas N, Gushin VI, Saylor S. Cultural differences in patterns of mood states on board the International Space Station. 56th International Astronautical Congress. Fukuoka, Japan. Oct 17–21, 2005; IAC-05-AI.5.03: 4 pp.	2005	Human Research and Countermeasure Development for Exploration
Interactions	Clement J, Ritsher JB. Operating the ISS: Cultural and leadership challenges. 56th International Astronautical Congress. Fukuoka, Japan. Oct 17–21, 2005; IAC-05-AI.5.05: 11 pp.	2005	Human Research and Countermeasure Development for Exploration
Interactions	Ritsher JB, Kanas N, Salnitskiy VP, Gushin VI, Saylor S, Weiss DS, Marmar C. Cultural and Language Backgrounds of International Space Station Program Personnel. Presented at the 57th International Astronautical Congress. Valencia, Spain. 2006; IAC-06-A1.1.3.Oct 2–6, 2006.	2006	Human Research and Countermeasure Development for Exploration
Interactions	Kanas NA, Ritsher JB, Saylor SA. Do Psychological Decrements Occur During the 2nd Half of Space Missions? Presented at the 57th International Astronautical Congress. Valencia, Spain. Oct 2–6, 2006; IAC-06-A1.1.02i.	2006	Human Research and Countermeasure Development for Exploration
Interactions	Clement JL, Ritsher JB, Kanas N, Saylor S. Leadership Challenges in ISS Operations: Lessons Learned from Junior and Senior Mission Control Personnel. Presented at the 57th International Astronautical Congress. Valencia, Spain. Oct 2–6, 2006; IAC-06-A1.1.6.	2006	Human Research and Countermeasure Development for Exploration

PAYLOAD	CITATION	YEAR	DISCIPLINE
Interactions	Kanas NA, Salnitskiy VP, Ritsher JB, Gushin VI, Weiss DS, Saylor SA, Kozerenko OP, Marmar CR Psychosocial interactions during ISS missions *Acta Astronautica* 2007; 60:329–335.	2007	Human Research and Countermeasure Development for Exploration
Interactions	Kanas NA, Salnitskiy VP, Boyd JE, Gushin VI, Weiss DS, Saylor SA, Kozerenko OP, Marmar CR. Crewmember and mission control personnel interactions during International Space Station missions. *Aviation Space and Environmental Medicine*. 2007; 78(6): 601–607.	2007	Human Research and Countermeasure Development for Exploration
Latent Virus	Pierson DL, Stowe RP, Phillips TM, Lugg DJ, Mehta SK. Epstein-Barr Virus Shedding by Astronauts During Space Flight. *Brain, Behavior, and Immunity*. 2004; 19:235–242.	2004	Human Research and Countermeasure Development for Exploration
Latent Virus	Mehta SK, Cohrs RJ, Forghani B, Zerbe G, Gilden DH, Pierson DL. Stress-induced Subclinical Reactivation of Varicella Zoster Virus in Astronauts. *Journal of Medical Virology*. 2005; 72:174–179.	2005	Human Research and Countermeasure Development for Exploration
PuFF	Prisk GK, Fine JM, Cooper TK, West JB. Pulmonary gas exchange is not impaired 24 h after extravehicular activity. *Journal of Applied Physiology* 2005; 99(6):2233–8.	2005	Human Research and Countermeasure Development for Exploration
PuFF	Prisk GK, Fine JM, Cooper TK, West JB. Vital Capacity, Respiratory Muscle Strength and Pulmonary Gas Exchange during Long-Duration Exposure to Microgravity. *Journal of Applied Physiology*. 2006; 101:439–447.	2006	Human Research and Countermeasure Development for Exploration
Subregional Bone	Lang T, LeBlanc A, Evans H, Lu Y, Gennant H, Yu A. Cortical and Trabecular Bone Mineral Loss from the Spine and Hip in Long-duration Spaceflight. *Journal of Bone and Mineral Research*. 2004; 19(6):1006–12.	2004	Human Research and Countermeasure Development for Exploration
Subregional Bone	Lang TF, LeBlanc AD, Evans HJ, Lu Y. Adaptation of the Proximal Femur to Skeletal Reloading After Long-Duration Spcaeflight. *Journal of Bone and Mineral Research*. 2006; 21(8);1224–1230.	2006	Human Research and Countermeasure Development for Exploration
Subregional Bone	Cavanagh P, Rice AJ (eds), Bone Loss During Spaceflight: Etiology, Countermeasures, and Implications for Bone Health on Earth. Cleveland Clinic Press, Cleveland, Ohio (2007), 297 pp.	2007	Human Research and Countermeasure Development for Exploration
Subregional Bone	Lang TF, Keyak JH, LeBlanc AD. Defining and Assessing Bone Health During and After Spaceflight, in Cavanagh P, Rice AJ (eds), Bone Loss During Spaceflight: Etiology, Countermeasures, and Implications for Bone Health on Earth. Cleveland Clinic Press, Cleveland, Ohio (2007), pp. 63–69.	2007	Human Research and Countermeasure Development for Exploration

PAYLOAD	CITATION	YEAR	DISCIPLINE
Subregional Bone	Lang TF, LeBlanc A, Evans H, Lu Y. Geometric adaptation of the proximal femur to skeletal reloading after long-duration spaceflight. *J Bone Miner Res* 2006: 21:1224–1230.	2006	Human Research and Countermeasure Development for Exploration
Subregional Bone	Sibonga JD, Evans HJ, Sung HG, Spector, ER, Land, TF, Oganov VS, Bakulin AV, Shackelford LC, LeBlanc AD. Recovery of spaceflight-induced bone loss: Bone mineral density after long-duration mission as fitted with an exponential function. *Bone* 41 (2007) 973–978.	2007	Human Research and Countermeasure Development for Exploration
Subregional Bone	Sibonga JD, Evans HJ, Spector ER, Maddocks MJ, Smith SA, Shackelford LC, LeBlanc AD. Bone Health During and After Spaceflight, in Cavanagh P, Rice AJ (eds), Bone Loss During Spaceflight: Etiology, Countermeasures, and Implications for Bone Health on Earth, Cleveland Clinic Press, Cleveland, Ohio (2007), pp.45-51.	2007	Human Research and Countermeasure Development for Exploration
Torso	Cucinotta FA, Kim M-H, Willingham V, George K. 2008, Physical and Biological Organ Dosimetry Analysis for International Space Station Astronauts, *Radiation Research* 170, 127–138 (2008).	2008	Human Research and Countermeasure Development for Exploration
Xenon-1	Gabrielsen A, Norsk P. Effect of spaceflight on the subcutaneous venoarteriolar reflex in the human lower leg. *J Appl Physiol* 103: 959–962 (2007).	2007	Human Research and Countermeasure Development for Exploration
Observing the Earth and Educational Activities			
ARISS	Wright RL. Remember, We're Pioneers! The First School Contact with the International Space Station. Proceedings of the AMSAT-NA 22nd Space Symposium, Arlington, Va., Oct 8–10, 2004.	2004	Observing the Earth and Educational Activities
ARISS	Cunningham C (N7NFX). NA1SS, NA1SS, This is KA7SKY Calling….. Proceedings of the AMSAT-NA Space Symposium, Arlington, Va. 2004, Oct 8–10, 2004.	2004	Observing the Earth and Educational Activities
ARISS	Palazzolo P (KB3NMS). Launching Dreams: The Long-term Impact of SAREX and ARISS on Student Achievement. Proceedings of the AMSAT-NA Space Symposium, Pittsburgh, Penn. 2006, Oct 26 – 28.	2006	Observing the Earth and Educational Activities
CEO	Quod J-P, Bigot L, Blanchot J, Chabanet P, Durville P, Nicet J-B, Wendling B. Research and monitoring of the coral reefs of the French islands of the Indian Ocean. Assessment activities in 2002. Mission carried out in Glorieuses. Réunion: IFRECOR (l'Initiative Française pour les Récifs Coralliennes). 2002; 2. [*French*]	2002	Observing the Earth and Educational Activities
CEO	Robinson JA, Evans CA. Space Station Allows Remote Sensing of Earth to within Six Meters. Eos, Transactions of the American Geophysical Union. 2002; 83:185–188.	2002	Observing the Earth and Educational Activities
CEO	Andréfouët S, Robinson JA, Hu C, Salvat B, Payri C, Muller-Karger FE. Influence of the spatial resolution of SeaWiFS, Landsat 7, SPOT and International Space Station data on landscape parameters of Pacific Ocean atolls. *Canadian Journal of Remote Sensing*. 2003; 29:210–218.	2003	Observing the Earth and Educational Activities

PAYLOAD	CITATION	YEAR	DISCIPLINE
CEO	Stumpf RP, Holderied K, Robinson JA, Feldman G, Kuring N. Mapping water depths in clear water from space. Proceedings of the 13th Biennial Coastal Zone Conference, Baltimore MD July 13-17, 2003.	2003	Observing the Earth and Educational Activities
CEO	Stefanov WL, Robinson JA. Vegetation Density Measurements From Digital Astronaut Photography. International Archives of the Photogrammetry, Remote Sensing, and Spatial Information Sciences. 2003; 34:185–189.	2003	Observing the Earth and Educational Activities
CEO	Wilkinson, MJ, NASA Tech Brief 2004: Large Fluvial Fans and Exploration for Hydrocarbons, 2004, http://www.techbriefs.com/content/view/170/34/	2004	Observing the Earth and Educational Activities
CEO	Andrefouet S, Gilbert A, Yan L, Remoissenet G, Payri C, Chancerelle Y. The remarkable population size of the endangered clam *Tridacna maxima* assessed in Fangatau Atoll using in situ remote sensing data. *ICES Journal of Marine Science*. 2005; 62(6):1037–1048.	2005	Observing the Earth and Educational Activities
CEO	Scambos T, Sergienko O, Sargent A, MacAyeal D, Fastook J. 2005, ICES at profiles of tabular iceberg margins and iceberg breakups at low latitudes. *Geophys. Research Letters*, Vol 32, L23S09 2005.	2005	Observing the Earth and Educational Activities
CEO	Wilkinson MJ, Marshall LG, Lundberg JG. River behavior on megafans and potential influences on diversification and distribution of aquatic organisms. *Journal of South American Earth Sciences* 2006; 21:151–172.	2006	Observing the Earth and Educational Activities
CEO	Robinson JA, Slack KJ, Olsen V, Trenchard M, Willis K, Baskin P, Ritsher JB. Patterns in Crew-Initiated Photography of Earth From ISS — Is Earth Observation a Salutogenic Experience? Presented at the 57th International Astronautical Congress. Valencia, Spain. Oct 2–6, 2006 IAC-06-A1.1.4.	2006	Observing the Earth and Educational Activities
CEO	Wilkinson MJ, Marshall LG, Lundberg JG. River behavior on megafans and potential influences on diversification and distribution of aquatic organisms. *Journal of South American Earth Sciences* 2006; 21:151–172.	2006	Observing the Earth and Educational Activities
CEO	M.J. Wilkinson, 2006, Method for Identifying Sedimentary bodies from images and its application to mineral exploration, USPTO Patent # 6,985,606 , http://patft.uspto.gov/netacgi/nph-Parser?Sect1=PTO2&Sect2=HITOFF&p=1&u=%2Fnetahtml%2FPTO%2Fsearch-bool.html&r=1&f=G&l=50&co1=AND&d=PTXT&s1=6985606.PN.&OS=PN/6985606&RS=PN/6985606	2006	Observing the Earth and Educational Activities
CEO	Elvidge CD, Cinzano P, Pettit DR, Arvesen J, Sutton P, Small C, Nemani R, Longcore T, Rich C, Safran J, Weeks J, Ebener S. 2007a. The Nightsat mission concept. *Int. J. of Remote Sensing*, V.28, 2645–2677.	2007	Observing the Earth and Educational Activities
CEO	Elvidge CD, Safran TB, Sutton P, Cinzano P, Pettit DR, Arvesen J, Small C, 2007b. Potential for Global Mapping of development via a nightsat mission. *GeoJournal* 69: 45–53.	2007	Observing the Earth and Educational Activities
CEO	Kohlmann B, Wilkinson MJ, 2007. The Tarcoles Line: biogeographic effects of the Talamanca Range in lower Central America, *Giornale Italiano di Entomologia* 12: 1–30.	2007	Observing the Earth and Educational Activities
CEO	Wilkinson MJ, Allen CC, Oheler DZ, Salvotore MR. A new Fluvial Analog for the Ridge-forming unit, Northern Sinus Meridiani, Southwest Arabia Terra, *Mars, Lunar and Planetary Science* XXXIX (2008).	2008	Observing the Earth and Educational Activities

PAYLOAD	CITATION	YEAR	DISCIPLINE
CEO	Lulla K. 2003 Nighttime Urban Imagery from International Space Station: Potential Applications for Urban Analyses and Modeling. Photogrammetric Engineering and Remote Sensing. 2003; 69:941–942.	2003	Observing the Earth and Educational Activities
CEO-IPY	Scambos T, Ross R, Bauer R, Yermolin Y, Skvarca P, Long D, Bohlander J, Haran T. Calving and ice-shelf break-up processes investigated by proxy: Antarctic tabular iceberg evolution during northward drift; Volume 54(187), 2008; 579–591.	2008	Observing the Earth and Educational Activities
CEO-IPY	Ingrid Sandahl, Christer Fuglesang, The Network for Optical Auroral Research in the Arctic Region, and The Upper Atmospheric Physics Group at the National Institute of Polar Research, Japan. Auroral Observations from Space Shuttle Discovery, 34th Annual European Optical Meeting, Aug 2007.	2007	Observing the Earth and Educational Activities
EarthKAM	Hurwicz M. Case Study: Attack Of The Space Data — Down To Earth Data Management At ISS EarthKAM. *New Architect*. Aug 1, 2002; 38.	2002	Observing the Earth and Educational Activities
Education-SEEDS	Levine HG, Norwood KLL, Tynes GK, Levine LH. Soybean and Corn Seed Germination in Space: The First Plant Study Conducted on Space Station Alpha. Proceedings of the 38th Space Congress, Cape Canaveral, Fla. May 2001; 181–187.	2001	Observing the Earth and Educational Activities
EPO	National Aeronautics and Space Administration Educational Product. International Toys in Space – Science on the Station. DVD. 2004; ED-2004-06-001-JSC.	2004	Observing the Earth and Educational Activities
Physical Sciences in Microgravity			
APCF-Camelids	Lorber B. The crystallization of biological macromolecules under microgravity: a way to more accurate three dimensional structures? *Biochimica et Biophysica Acta*. 2002; 1599(1-2):1–8.	2002	Physical Sciences in Microgravity
APCF-Camelids	Vergara A, Lorber B, Zagari A, Giege R. Physical aspects of protein crystal growth investigated with the Advanced Protein Crystallization Facility in reduced gravity environments. *Acta Crystallographica*, Section D, Biological Crystallography. 2003; 59:2–15.	2003	Physical Sciences in Microgravity
APCF-Camelids	Vergara A, Lober B, Sauter C, Giege R, Zagari A. Lessons from crystals grown in the Advanced Protein Crystallization Facility for conventional crystallization applied to structural biology. *Biophysical Chemistry*. 2005; 118:102–112.	2005	Physical Sciences in Microgravity
APCF-PPG10	Berisio R, Vitagliano L, Vergara A, Sorrentino G, Mazzarella L, Zagari A. Crystallization of the collagen-like polypeptide (PPG)10 aboard the International Space Station. 2. Comparison of crystal quality by X-ray diffraction. *Acta Crystallographica*, Section D, Biological Crystallography. 2002; 58:1695–1699.	2002	Physical Sciences in Microgravity
APCF-PPG10	Vergara A, Corvino E, Sorrentino G, Piccolo C, Tortora A, Caritenuto L, Mazzarella L, Zagari A. Crystallization of the collagen-like polypeptide (PPG)10 aboard the International Space Station. 1. Video observation. *Acta Crystallographica*, Section D, Biological Crystallography. 2002; 58:1690–1694.	2002	Physical Sciences in Microgravity
APCF-PPG10	Castagnolo D, Piccolo C, Carotenuto L, Vegara A, Zagari A. Crystalization of the collagen-like polypeptide (PPG)10 aboard the International Space Station. 3. Analysis of residual acceleration-induced motion. *Acta Crystallographica*, Section D, Biological Crystallography. 2003; 59(pt4):773–776.	2003	Physical Sciences in Microgravity

PAYLOAD	CITATION	YEAR	DISCIPLINE
BCAT-3-4-CP	Lu PJ, Weitz DA, Foale MC, Fincke M, Chiao LN, Meyer WV, Owens JC, Hoffmann MI, Sicker RJ, Rogers R, Havenhill MA, Anzalone SM, Yee H. Microgravity Phase Separation near the Critical Point in Attractive Colloids. 45th AIAA Aerospace Sciences Meeting and Exhibit. Reno, Nev. Jan 8 – 11, 2007; AIAA-2007-1152.	2007	Physical Sciences in Microgravity
CFE	Weislogel MM. Preliminary Results from the Capillary Flow Experiment Aboard ISS: The Moving Contact Line Boundary Condition. Proceedings of the 43rd AIAA Aerospace Sciences Meeting and Exhibit, Reno, Nev. Jan 10 – 13, 2005; AIAA 2005-1439.	2005	Physical Sciences in Microgravity
CFE	Weislogel MM, Jenson R, Klatte J, Dreyer M. Interim Results from the Capillary Flow Experiment Aboard ISS; the Moving Contact Line Boundary Condition. 45th AIAA Aerospace Sciences Meeting and Exhibit. Reno, Nev. Jan 8 – 11, 2007; AIAA-2007-747.	2007	Physical Sciences in Microgravity
CFE	Weislogel MM, Jenson RM, Klatte J, Dreyer ME. The Capillary Flow Experiments aboard ISS: Moving Contact Line Experiments and Numerical Analysis. 46th AIAA Aerospace Sciences Meeting and Exhibit, Reno, Nev. 2008, 7 – 10 Jan; AIAA 2008-816.	2008	Physical Sciences in Microgravity
CFE	Weislogel MM, Jenson RM, Chen Y, Collicott S, Bunnell CT, Klatte J, Dreyer ME. Postflight summary of the Capillary Flow Experiments aboard the International Space Station, No. IAC-08-A2.6.A8, 59th International Astronautical Congress-2008, Glasgow, Scotland, Sep 29–Oct 3, 2008.	2008	Physical Sciences in Microgravity
CFE	Chen Y, Jenson RM, Weislogel MM, Collicott SH. Capillary Wetting Analysis of the CFE-Vane Gap Geometry. 46th AIAA Aerospace Sciences Meeting and Exhibit, Reno, Nev. 2008, 7 – 10 Jan; AIAA 2008-817.	2008	Physical Sciences in Microgravity
CPCG-H	Vallazza M, Banumathi S, Perbandt M, Moore K, DeLucas L, Betzel C, Erdmann V. Crystallization and Structure Analysis of *Thermus flavus* 5S rRNA helix B. *Acta Crystallographica*, Section D, Biological Crystallography. 2002; 58:1700–1703.	2002	Physical Sciences in Microgravity
CPCG-H	Krauspenhaar R, Rypniewski W, Kalkura N, Moore K, DeLucas L, Stoeva S, Mikhailov A, Voelter W, Betzel C. Crystallisation under microgravity of mistletoe lectin I from *Viscum album* with adenine monophosphate and the crystal structure at 1.9 angstrom resolution. *Acta Crystallographica*, Section D, Biological Crystallography. 2002; 58:1704–1707.	2002	Physical Sciences in Microgravity
CPCG-H	Nardini M, Spano S, Cericola C, Pesce A, Damonte G, Luini A, Corda D, Bolognesi M. Crystallization and preliminary X-ray diffraction analysis of brefeldin A-ADP ribosylated substrate (BARS). *Acta Crystallographica*, Section D, Biological Crystallography. 2002; 58:1068–1070.	2002	Physical Sciences in Microgravity
CPCG-H	Miele AE, Federici L, Sciara G, Draghi F, Brunori M, Vallone B. Analysis of the effect of microgravity on protein crystal quality: the case of a myoglobin triple mutant. *Acta Crystallographica*, Section D, Biological Crystallography. 2004; D59: 928–988.	2004	Physical Sciences in Microgravity
CSLM-2	Kammer D, Genau A, Voorhees PW, Duval WM, Hawersaat RW, Hickman JM, Lorik T, Hall DG, Frey CA. Coarsening In Solid-Liquid Mixtures: A Reflight. 46th AIAA Aerospace Sciences Meeting and Exhibit, Reno, Nev. 2008, 7 – 10 Jan; AIAA 2008-813.	2008	Physical Sciences in Microgravity
EXPPCS	Weitz D, Bailey A, Manley A, Prasad V, Christianson R, Sankaran S, Doherty M, Jankovsky A, Lorik T, Shiley W, Bowen J, Kurta C, Eggers J, Gasser U, Serge P, Cipelletti L, Schofield, Pusey P. Results From the Physics of Colloids Experiment on ISS. NASA TM. 2002; 2002-212011:IAC-02-J.6.04.	2002	Physical Sciences in Microgravity
EXPPCS	Manley S, Cipelletti L, Trappe V, Bailey AE, Christianson RJ, Gasser U, Prasad V, Segre PN, Doherty MP, Sankaran S, Jankovsky AL, Shiley B, Bowen J, Eggers J, Kurta C, Lorik T, Weitz DA. Limits to Gelation in Colloidal Aggregation. *Physical Review Letters*. 2004; 93(10):108302.	2004	Physical Sciences in Microgravity

PAYLOAD	CITATION	YEAR	DISCIPLINE
EXPPCS	Manley S, Davidovitch B, Davies NR, Cipelletti L, Bailey AE, Christianson RJ, Gasser U, Prasad V, Segre PN, Dohert MP, Sankaran S, Jankovsky AL, Shiley B, Bowen J, Eggers J, Kurta C, Lorik T, Weitz DA. Time-Dependent Strength of Colloidal Gels. The American Physical Society – *Physical Review Letters*. 2005; 95(4);048302(4).	2005	Physical Sciences in Microgravity
EXPPCS	Bailey AE, Poon WC, Christianson RJ, Schofield AB, Gasser U, Prasad V, Manley S, Segre PN, Cipelletti L, Meyer WV, Doherty MP, Sankaran S, Jankovsky AL, Shiley WL, Bowen JP, Eggers JC, Kurta C, Lorik T Jr, Pusey PN, Weitz DA. Spinodal decomposition in a model colloid-polymer mixture in microgravity. *Physical Review Letters*. 2007;Nov 16; 99(20):205701.	2007	Physical Sciences in Microgravity
EXPPCS	Doherty MP, Bailey AE, Jankovsky AL, Lorik T. Physics of Colloids in Space: Flight Hardware operations on ISS. AIAA 2002-0762, 40th Aerospace Sciences meeting, Jan 14–17, 2002, Reno, Nev.	2002	Physical Sciences in Microgravity
FMVM	Ethridge E, Kaukler W, Antar B. Preliminary Results of the Fluid Merging Viscosity Measurement Space Station Experiment. Proceedings of the 44th AIAA Aerospace Sciences Meeting and Exhibit, Reno, Nev. Jan 9 – 12, 2006; AIAA 2006-1142.	2006	Physical Sciences in Microgravity
FMVM	Antar BN, Ethridge E, Lehman D. Fluid Merging Viscosity Measurement (FMVM) Experiment on the International Space Station. 45th AIAA Aerospace Sciences Meeting and Exhibit. Reno, Nev. Jan 8 – 11, 2007; AIAA-2007-1151.	2007	Physical Sciences in Microgravity
Foam	Veazy, C; Demetriou, MD; Schroers, J; Hanan, JC; Dunning, LA; Kaukler, WF; Johnson, WL. Foaming of Amorphous Metals Approached the Limit of Microgravity Foaming, *J. Adv. Materials*, v40, no 1 (2008) 7-11	2008	Physical Sciences in Microgravity
InSPACE	Vasquez PA, Furst EM, Agui J, Williams J, Pettit D, Lu E. Structural Transitions of Magnetoghreological Fluids in Microgravity. 46th AIAA Aerospace Sciences Meeting and Exhibit, Reno, Nev. 2008, 7–10 Jan; AIAA 2008-815.	2008	Physical Sciences in Microgravity
MFMG	Pojman JA. Miscible Fluids in Microgravity (MFMG): A Zero-Upmass Experiment on the International Space Station. Proceedings of the 43rd AIAA Aerospace Sciences Meeting and Exhibit, Reno, Nev. Jan 10–13, 2005; AIAA 2005-718.	2005	Physical Sciences in Microgravity
MFMG	Pojman JA, Bessonov N, Volpert V. "Miscible Fluids in Microgravity (MFMG): A Zero-Upmass Investigation on the International Space Station," *Microgravity Sci. Tech*. 2007, XIX, 33–41.	2007	Physical Sciences in Microgravity
PCG-EGN	Barnes CL, Snell EH, Kundrot CE. Thaumatin crystallization aboard the International Space Station using liquid-liquid diffusion in the Enhanced Gaseous Nitrogen Dewar (EGN). *Acta Crystallographica*, Section D, Biological Crystallography. 2002; 58(Pt 5): 751–760.	2002	Physical Sciences in Microgravity
PCG-EGN	Ciszak E, Hammons AS, Hong YS. Use of Capillaries for Macromolecular Crystallization in a Cryogenic Dewar. *Crystal Growth & Design*. 2002; 2(3):235–238.	2002	Physical Sciences in Microgravity
PCG-STES-MM	Vahedi-Faridi A, Porta J, Borgstahl G. Improved three-dimensional growth of manganese superoxide dismutase crystals on the International Space Station. *Acta Crystallographica*, Section D, Biological Crystallography. 2003;59:385–388.	2003	Physical Sciences in Microgravity
PFMI	Grugel RN, Anilkumar AV. Bubble Formation and Transport during Microgravity Materials Processing: Model Experiments on the Space Station. Proceedings of the 42nd AIAA Aerospace Sciences Meeting and Exhibit, Reno, Nev. Jan 5–8, 2004; AIAA 2004-0627.	2004	Physical Sciences in Microgravity

PAYLOAD	CITATION	YEAR	DISCIPLINE
PFMI	Grugel RN, Anilkumar AV, Lee CP. Direct Observation of Pore Formation and bubble mobility during controlled melting and re-solidification in microgravity, Solidification Processes and Microstructures. A Symposium in Honor of Wilfried Kurz. The Metallurgical Society, Warrendale, Penn. 2004; 111–116.	2004	Physical Sciences in Microgravity
PFMI	Strutzenberg LL, Grugel RN, Trivedi R. Observation of an Aligned Gas - Solid Eutectic during Controlled Directional Solidification aboard the International Space Station – Comparison with Ground-based Studies. Proceedings of the 42nd AIAA Aerospace Sciences Meeting and Exhibit, Reno, Nev. 2005; AIAA 2005-919.	2005	Physical Sciences in Microgravity
PFMI	Grugel RN, Anilkumar AV, Cox MC. Morphological Evolution of Directional Solidification Interface in Microgravity: An Analysis of Model Experiments Performed on the International Space Station. Proceedings of the 43rd AIAA Aerospace Sciences Meeting and Exhibit, Reno, Nev. Jan 10–13, 2005; AIAA 2005-917.	2005	Physical Sciences in Microgravity
PFMI	Grugel RN, Luz P, Smith G, Spivey R, Jeter L, Gillies D, Hua F, Anilkumar AV. Materials research conducted aboard the International Space Station: Facilities overview, operational procedures, and experimental outcomes. *Acta Astronautica.* 2008; 62;491–498 (Also presented at the 57th International Astronautical Congress IAC-06-A2.2.10).	2008	Physical Sciences in Microgravity
SUBSA	Spivey RA, Gilley S, Ostrogorsky A, Grugel R, Smith G, Luz P. SUBSA and PFMI Transparent Furnace Systems Currently in use in the International Space Station Microgravity Science Glovebox. 41st Aerospace Sciences Meeting and Exhibit. 2003; AIAA 2003-1362.	2003	Physical Sciences in Microgravity
SUBSA	Ostrogorsky A, Marin C, Churilov A, Volz M, Bonner WA, Spivey RA, Smith G. Solidification Using the Baffle in Sealed Ampoules. 41st Aerospace Sciences Meeting and Exhibit, Reno, Nev. 2003; AIAA 2003-1309.	2003	Physical Sciences in Microgravity
SUBSA	Churilov AV, Ostrogorsky AG. Model of Tellurium- and Zinc-Doped Indium Antimonide Solidification in Space. Journal of Thermophysics and Heat Transfer. 2005; 19(4);542–547 (Also published at the 42nd AIAA Meeting, 2004-1388, 2004).	2004	Physical Sciences in Microgravity
SUBSA	Churilov AV, Ostrogorsky AG. Solidification of Te and Zn doped InSb in space. 42nd AIAA Aerospace Sciences Meeting and Exhibit, Reno, Nev. 2004; AIAA 2004-1388 (Also published in the *Journal of Thermophysics and Heat Transfer*, 19(4);547-547, 2005).	2005	Physical Sciences in Microgravity
ZCG	Akata B, Yilmaz B., Jirapnogphan SS, Warzywoda J, Sacco Jr. A. Characterization of zeolite Beta grown in microgravity. Microporous and Mesoporous Materials. 2004; 71:1–9.	2004	Physical Sciences in Microgravity
Technology Development			
ARIS-ICE	Bushnell GS, Fialho IJ, Allen JL, Quraishi N. Microgravity Flight Characterization of the International Space Station Active Rack Isolation System. AIAA Microgravity Measurements Group Meeting, The World Space Congress, Houston. Oct 10 – 11, 2002.	2002	Technology Development
ARIS-ICE	Bushnell GS, Fialho IJ, McDavid T, Allen JL, Quraishi N. Ground And On-Orbit Command and Data Handling Architectures For The Active Rack Isolation System Microgravity Flight Experiment. AIAA 53rd International Astronautical Congress, The World Space Congress, Houston. Oct 10 – 19, 2002; IAC-02-J.5.07.	2002	Technology Development
ARIS-ICE	Fialho IJ, Bushnell GS, Allen JL, Quraishi N. Taking H-infinity To The International Space Station: Design, Implementation and On-orbit Evaluation of Robust Controllers For Active Microgravity Isolation. AIAA Guidance, Navigation and Control Conference, Austin, Texas. Aug 2003.	2003	Technology Development

PAYLOAD	CITATION	YEAR	DISCIPLINE
DAFT	Urban D, Griffin D, Ruff G, Cleary T, Yang J, Mulholland G, Yuan Z. Detection of Smoke from Microgravity Fires. Proceedings of the International Conference on Environmental Systems. 2005; 2005-01-2930.	2005	Technology Development
ISSI	Grugel R, Cotton LJ, Segre PN, Ogle JA, Funkhouser G, Parris F, Murphy L, Gillies D, Hua F, Anilkumar AV. The In-Space Soldering Investigation (ISSI): Melting and Solidification Experiments Aboard the International Space Station. Proceedings of the 44th AIAA Aerospace Sciences Meeting and Exhibit, Reno, Nev. Jan 9 – 12, 2006; AIAA 2006-521.	2006	Technology Development
MACE-II	Davis L. Economical and Reliable Adaptive Disturbance Cancellation. AFRL-VS-TR-2002-1118. Sep 2002 ;Vol. I – II: Pt. 1–3 (DoD clearance is needed to view this paper).	2002	Technology Development
MACE-II	Ninneman R, Founds D, Davis L, Greeley S, King J. Middeck Active Control Experiment Reflight (MACE II) Program: Adventures in Space. AIAA Space 2003 Conference and Exhibition, Long Beach, Calif. 2003; AIAA 2003-6243.	2003	Technology Development
MAMS	DeLombard R, Kelly EM, Foster, K, Hrovat K, McPherson, KM; Schafer, CP. Microgravity Acceleration Environment of the International Space Station, AIAA 2001-5113.	2001	Technology Development
MAMS and SAMS-II	DeLombard R, Hrovat K, Kelly EM, McPherson K, Jules K. An Overview of the Microgravity Environment of the International Space Station Under Construction. AIAA-2002-0608.	2002	Technology Development
MAMS and SAMS-II	Del Basso S, Laible M, O'Keefe E, Steelman A, Scheer S, Thampi S. Capitalization of Early ISS Data for Assembly Complete Microgravity Performance. Proceedings of the 40th AIAA Aerospace Sciences Meeting and Exhibit, Reno, Nev. Jan 14 – 17, 2002; AIAA 2002-606.	2002	Technology Development
MAMS and SAMS-II	Jules K, McPherson K, Hrovat K, Kelly E, Reckart T. A Status on the Characterization of the Microgravity Environment of the International Space Station. 54th International Astronautical Congress, 29 Sep to 3 Oct 2003, Bremen, Germany, IAC-03-J.6.01.	2003	Technology Development
MAMS and SAMS-II	DeLombard R, Kelly EM, Hrovat K, McPherson. Microgravity Environment of the International Space Station. AIAA-2004-125.	2004	Technology Development
MAMS and SAMS-II	Jules K, Hrovat K, Kelly EM. The Microgravity Environment of the International Space Station during the Buildup Period: Increments 2 to 8. 55th Int. Astronautical Congress 2004 Vancouver, Canada, IAC-04-J.6.01.	2004	Technology Development
MAMS and SAMS-II	DeLombard R, Hrovat K, Kelly E, Humphreys B. Interpreting the International Space Station Microgravity Environment. 43rd AIAA Aerospace Sciences Meeting and Exhibit, 10–13 Jan 2005, Reno, Nev., AIAA 2005-727.	2005	Technology Development
MAMS and SAMS-II	DeLombard R, Kelly EM, Hrovat K, Nelson ES, Pettit DR. Motion of Air Bubbles in Water Subjected to Microgravity Accelerations. Proceedings of the 43rd AIAA Aerospace Sciences Meeting and Exhibit, Reno, Nev. Jan 10 – 13, 2005; AIAA 2005-722.	2005	Technology Development
MISSE 5	Walters RJ, Garner JC, Lam SN, Vazquez JA, Braun WR, Ruth RE, Warne JH, Lorentzen JR, Messenger SR, Bruninga Cdr R, Jenkins PP, Flatico JM, Wilt DM, Piszczor MF, Greer LC, Krasowski MJ. Forward Technology Solar Cell Experiment First On-Orbit Data, 19th Space Photovoltaic Research and Technology Conference, (2007)NASA/CP-2007-214494, 79–94.	2007	Technology Development
MISSE 5	Krasowski M, Greer L, Flatico J, Jenkins P, Spina D. Big Science, Small-budget Space Experiment Package aka MISSE-5: A Hardware and Software perspective. 19th Space Photovoltaic Research and Technology Conference, (2007)NASA/CP-2007-214494, 95–117.	2007	Technology Development

PAYLOAD	CITATION	YEAR	DISCIPLINE
MISSE 5	Kinard WH. Materials Experiment Flown on MISSE 5. 2007 National Space & Missile Materials Symposium.	2007	Technology Development
MISSE 5	Finckenor M, Zweiner JM, Pippin G, Thermal Control Materials on MISSE-5 with Comparison to Earlier Flight Data, 25–29 Jun 2007; Keystone, Colo.; 2007 National Space & Missile Materials Symposium.	2007	Technology Development
MISSE 5	Simburger EJ, Matsumoto JH, Giants TW, Garcia III A, Liu S, Rawal SP, Perry AR, Marshall CH, John K. Lin JK, Scarborough SE, Curtis HB, Kerslake TW, Peterson TT. Development of a thin film solar cell interconnect for the PowerSphere concept, *Materials Science and Engineering*, Volume 116, Issue 3, 15 Feb 2005, pp. 321-325.	2005	Technology Development
MISSE 5	Walters RJ, Garner JC, Lam SN, Vasquez JA, Braun WR, Ruth RE, Warner JH, Lorentzen JR, Messenger SR, Bruninga CDR R, Jenkins PP, Flatico JM, Wilt DM, Piszczor MF, Greer LC, Krasowski MJ. Materials on the International Space Station Experiment-5, Forward Technology Solar Cell Experiment: First On-Orbit Data, Conference Record of the 2006 IEEE 4th World Conference on Photovoltaic Energy Conversion, May 2006; Volume 2, pp. 1951–1954.	2006	Technology Development
MISSE 5	Wilt DM. Clark EB, Ringel SA, Andre CL, Smith MA, Scheiman DA, Jenkins PP, Maurer WF, Fitzgerald EA, Walters RJ, et al. LEO Flight Testing of GaAs on Si Solar Cells Aboard MISSE 5. 19th European Photovoltaic Solar Energy Conference and Exhibition.	2006	Technology Development
MISSE 5	Pippin G, deGroh K, Finckenor M, Minton T. Post-Flight Analysis of Selected Fluorocarbon and Other Thin Film Polymer Specimens Flown on MISSE-5, 25–29 Jun 2007; Keystone, Colo.; 2007 National Space & Missile Materials Symposium.	2007	Technology Development
MISSE 5	deGroh K, Finckenor M, Minton T, Brunsvold, A, Pippin G. Post-Flight Analysis of Selected Fluorocarbon and Other Thin Film Polymer Specimens Flown on MISSE-5. 25–29 Jun 2007; Keystone, Colo.; 2007 National Space & Missile Materials Symposium.	2007	Technology Development
MISSE-1 and -2	Finckenor MM. The Materials on International Space Station Experiment (MISSE): First Results from MSFC Investigations. Proceedings of the 44th AIAA Aerospace Sciences Meeting and Exhibit, Reno, Nev. Jan 9–12, 2006; AIAA 2006-472.	2006	Technology Development
MISSE-1 and -2	Snyder A, Banks BA, Waters DL. Undercutting Studies of Protected Kapton H Exposed to In-Space and Ground-Based Atomic Oxygen. NASA TM-2006-214387 (2006).	2006	Technology Development
MISSE-1 and -2	de Groh K, Banks B. MISSE PEACE Polymers Atomic Oxygen Erosion Results. NASA/TM-2006-214482 (2006).	2006	Technology Development
MISSE-1 and -2	Harvey GA Kinard WH, MISSE 1 and 2 Tray Temperature measurements. Proceedings of MISSE Post Retrieval Conference and the 2006 National Space & Missile Materials Symposium, Orlando, Fla.	2006	Technology Development
MISSE-1 and -2	de Groh K, Banks B. Materials International Space Station Experiment (MISSE) Polymers Degradation. 9th International Conference on "Protection of Materials and Structures from Space Environment," May 20–23, 2008, Toronto, Canada.	2008	Technology Development
MISSE-1 and -2	de Groh K, Banks B, Stambler AH, Roberts LM, Inoshita KE, Barbagallo CE. Ground-Laboratory to In-Space Atomic Oxygen Correlation for the PEACE Polymers, 9th International Conference on "Protection of Materials and Structures from Space Environment," May 20–23, 2008, Toronto, Canada.	2008	Technology Development
MISSE-1 and -2	Tomczak SJ, Marchant D, Mabry JM, Vij V, Yandek GR, Minton TK, Brunsvold AL, Wright ME, Petteys BJ, Guenthner AJ. Studies of POSS-Polyimides Flown on MISSE-1. 25–29 Jun 2007; Keystone, Colo., 2007 National Space & Missile Materials Symposium.	2007	Technology Development

PAYLOAD	CITATION	YEAR	DISCIPLINE
MISSE-1 and -2	Juhl SB, Akinlemibola B, Kasten L, Vaia R. Durability of Poly(Caprolactam) (Nylon 6) and Poly(Caprolactam) Nanocomposites in Low Earth Orbit. 25–29 Jun 2007; Keystone, Colo., 2007 National Space & Missile Materials Symposium.	2007	Technology Development
MISSE-1 and -2	Rice N, Shepp A, Haghighat R, Connell J. Durable TOR Polymers on MISSE, 25–29 Jun 2007; Keystone, Colo., 2007 National Space & Missile Materials Symposium.	2007	Technology Development
MISSE-1 and -2	Watson KA, Ghose S, Lillehei PT, Smith Jr JG, Connell JW, Effect of LEO Exposure on Aromatic Polymers Containing Phenylphosphine Oxide Groups, SAMPE Proceedings Vol. 52, Jun 6 2007, Baltimore, Md.	2007	Technology Development
SAME	Urban DL, Ruff GA, Brooker JE, Cleary T, Yang J, Mulholland G, Yuan Z-G. Spacecraft Fire Detection: Smoke Properties and Transport in Low-Gravity. 46th AIAA Aerospace Sciences Meeting and Exhibit, Reno, Nev. 2008, 7 – 10 Jan; AIAA 2008-806.	2008	Technology Development
Results from ISS Operations			
Clinical Nutrition Assessment	Smith S, Zwart SR, Block G, Rice BL, Davis-Street JE. The nutritional status of astronauts is altered after long-term space flight aboard the International Space Station. *Journal of Nutrition.* 2005; 135(3):437–443.	2005	Results from ISS Operations
Clinical Nutrition Assessment	Smith SM, Zwart SR. Nutrition issues for space exploration. *Acta Astronautica.* 2008; 63: 609–613.	2008	Results from ISS Operations
Clinical Nutrition Assessment	Smith S, Zwart, SR, Block G, Rice BL, Davis-Street JE. The Nutritional Status of Astronauts is Altered After Long-Term Space Flight Aboard the International Space Station, in Cavanagh P, Rice AJ (eds), Bone Loss During Spaceflight: Etiology, Countermeasures, and Implications for Bone Health on Earth. Cleveland Clinic Press, Cleveland, Ohio (2007), pp. 133–147.	2007	Results from ISS Operations
Environmental Monitoring	Plumlee D, Mudgett PD, Schultz JR. Chemical Sampling and Analysis of ISS Potable Water: Expeditions 1–3. 32nd International Conference on Environmental Systems, San Antonio, Texas. 2002; SAE Technical Paper 2002-01-2537.	2002	Results from ISS Operations
Environmental Monitoring	Plumlee D, Mudgett P, Schultz J. ISS Potable Water Sampling and Chemical Analysis: Expeditions 4 & 5. 33rd International Conference on Environmental Systems, Vancouver, Canada. Jul 2003; SAE Technical Paper 2003-01-2401.	2003	Results from ISS Operations
Environmental Monitoring	James J, Limero T, Beck S, Martin B, Covington P, Boyd J, Peters R. Toxicological Assessment of the International Space Station Atmosphere with Emphasis on Metox Canister Regeneration. 33rd International Conference on Environmental Systems, Vancouver, Canada. Jul 2003; SAE Technical Paper 2003-01-2647.	2003	Results from ISS Operations
Environmental Monitoring	Perry J, Peterson B. Cabin air quality Dynamics on Board the International Space Station. 33rd International Conference on Environmental Systems, Vancouver, Canada. Jul 2003; SAE Technical Paper 2003-01-2650.	2003	Results from ISS Operations
Environmental Monitoring	Castro VA, Thrasher AN, Healy M, Ott CM, Pierson DL. Microbial Characterization during the Early Habitation of the International Space Station. *Microbial Ecology.* 2004; 47:119–126.	2004	Results from ISS Operations
Environmental Monitoring	La Duc MT, Sumner R, Pierson D, Venkat P, Venkateswaran K. Evidence of pathogenic microbes in the International Space Station drinking water: reason for concern? *Habitation.* 2004; 10:39–48.	2004	Results from ISS Operations

PAYLOAD	CITATION	YEAR	DISCIPLINE
Environmental Monitoring	Vesper SJ, Wong W, Kuo CM, Pierson DL. Mold species in dust from the International Space Station identified and quantified by mold-specific quantitative PCR. *Research in Microbiology*, (2008) doi 10.1016/j.resmic.2008.06.001.	2008	Results from ISS Operations
ISS Acoustics	Goodman JR. International Space Station Acoustics. Noise Conference, Cleveland Ohio. Jun 23 –25, 2003	2003	Results from ISS Operations
ISS Acoustics	Pilkinton GD. ISS Acoustics Mission Support. Noise Conference, Cleveland, Ohio. Jun 23 –25, 2003.	2003	Results from ISS Operations
Other ISS Research	Hall PS, Past and Current Practice in Space Nutrition, in Cavanagh P, Rice AJ (eds), Bone Loss During Spaceflight: Etiology, Countermeasures, and Implications for Bone Health on Earth. Cleveland Clinic Press, Cleveland, Ohio (2007), pp. 125–132.	2007	Results from ISS Operations
Other ISS Research	Schmidle W, Alred J, Mikatarian R, Soares C, Miles E, Howorth I, Mishina L, Murtazan R. Characterization of On-orbit US Lab Condensate Vacuum Venting, Int Astronautical Congress, 10–19 Oct 2002, Houston, IAF-02-T.P.06.	2002	Results from ISS Operations
Other ISS Research	Soares C, Mikatarian R, Scharf R, Miles E. International Space Station 1A/R - 6A External Contamination Observations and Surface Assessment. Int. Symposium on Optical Science and Technology, SPIE 47th Annual Meeting, Jul 7–11 2002, Seattle, Wash.	2002	Results from ISS Operations
Other ISS Research	D'Aunno DS, Dougherty AH, DeBlock HF, Meck JV. Effect of Short- and Long-Duration Spaceflight on QTc Intervals in Healthy Astronauts. *The American Journal of Cardiology*. 2003; 91:494–497.	2003	Results from ISS Operations
Other ISS Research	Alred J, Boeder P, Mikatarian R, Pnakop C, Schmidl W. Modeling of the thruster plume induced erosion. 9th International Symposium on Materials in Space, ESTEC Noordwijk, Netherlands, Jun 2003 (2003).	2003	Results from ISS Operations
Other ISS Research	Koontz S, Boeder PA, Pankop C, Reddell B. The Ionizing Radiation Environment on the International Space Station: Performance vs. expectations for avionics and materials,, in Radiation Effects Data Workshop, 2005, IEEE, 11-15 July 2005, p 110-116	2005	Results from ISS Operations
Other ISS Research	Schmidl W, Mikatarian R, Lam C-W, West B, Buchanan V, Dee L., Baker D., Koontz S. Fuel Oxidizer Reaction Products (FORP) Contamination of Service Module and Release of N-nitrosodimethylamine in a humid environment from crew EVA suits contaminated with FORP, in Space Technology Proceedings, V. 6, *Protection of Materials and Structurs from the Space Environment*, Springer Netherlands , 2006, p. 193–204.	2004?	Results from ISS Operations
Other ISS Research	Schmidl W, Visentine J, and Mikatarian R. Mitigation of Damage to the International Space Station (ISS) from water dumps, in Space Technology Proceedings, V. 6, Protection of Materials and Structurs from the Space Environment , Springer Netherlands , 2006, pp. 93-105.	2004?	Results from ISS Operations
Other ISS Research	Pankop C, Alred J, Boeder P. Mitigation of thruster plume-induced erosion of ISS sensitive hardware, in Protection of Materials and Structures from the Space Environment. Edited by Jacob I. Kleiman, Integrity Testing Laboratory Inc. Markham, Toronto, Canada. ISBN 978-1-4020-4319-2 (e-book). Series: Space Technology Proceedings, Springer, Dordrecht, 2006, p. 71.	2006	Results from ISS Operations
Other ISS Research	Boeder P, Mikatarian R, Lorenz MJ, Koontz S, Albyn K, Finckenor M. Effects of Space Environment Exposure on the Blocking Force of Silicone Adhesive., in Protection of Materials and Structures from the Space Environment. Edited by Jacob I. Kleiman, Integrity Testing Laboratory Inc. Markham, Toronto, Canada. ISBN 978-1-4020-4319-2 (e-book). Series: Space Technology Proceedings, Springer, Dordrecht, 2006, p. 295.	2006	Results from ISS Operations

PAYLOAD	CITATION	YEAR	DISCIPLINE
Other ISS Research	de Groh KK, Dever JA, Snyder A, Kaminski S, McCarthy CE, Rapoport AL, Rucker RN. "Solar Effects on Tensile and Optical Properties of Hubble Space Telescope Silver-Teflon Insulation," in "Materials in Extreme Environments," edited by Mailhiot C, Saganti PB, Ila D (Mater. Res. Soc. Symp. Proc. 929, Warrendale, Penn., 2006), 0929-II05-08.	2006	Results from ISS Operations
Other ISS Research	Boeder PA, Visentine JT, Shaw CG, Carniglia CK, Alred JW, Soares CE. Effect of silicone contaminant film on the transmittance properties of AR-coated fused silica., SPIE-04, 2004.	2004	Results from ISS Operations
Plasma Interaction Model	Reddell B, Alred J, Kramer L, Mikatarian R. Analysis of ISS Plasma Interaction. Proceedings of the 44th AIAA Aerospace Sciences Meeting and Exhibit, Reno, Nev. Jan 9–12, 2006; AIAA 2006-865.	2006	Results from ISS Operations
Saturday Morning Science	Grunsfeld JM. Shooting the Heavens from Space. *Sky and Telescope*. 2003; 128–132.	2003	Results from ISS Operations
Saturday Morning Science	Love SG, Pettit DR. Fast, Repeatable Clumping of Solid Particles in Microgravity. Lunar and Planetary Science XXXV. 2004; 1119.	2004	Results from ISS Operations
Saturday Morning Science	Tytell D. Building Planets in Plastic Bags. *Sky and Telescope*. Apr 13, 2004 [http://www.skyandtelescope.com/news/3308986.html?page=1&c=y].	2004	Results from ISS Operations
ZPM	Kang W, Bedrossian N. Pseudospectral Optimal Control Theory Makes Debut Flight, Saves NASA 1M dollars in Under Three Hours. *Society for Industrial and Applied Mathematics News*. 2007, Sep; 40(7).	2007	Results from ISS Operations
ZPM	Bedrossian N, Bhatt S, Lammers M, Nguyen L. Zero Propellant Maneuver Flight Results for 180 deg ISS Rotation. 20th International Symposium on Space Flight Dynamics. 2007, Sep; NASA/CP-2007-214158.	2007	Results from ISS Operations
ZPM	Bhatt S. Optimal Reorientation of Spacecraft Using Only Control Moment Gyroscopes. Master's Thesis, Department of Computational and Applied Mathematics, Rice University, Houston. 2007.	2007	Results from ISS Operations
ZPM	Bedrossian N, Bhatt S, Lammers M, Nguyen L, Zhang Y. First Ever Flight Demonstration of Zero Propellant Maneuver Attitude Control Concept. AIAA GN and C Conference. 2007, Aug. AIAA Paper 2007-6734.	2007	Results from ISS Operations

APPENDIX C: ACRONYMS AND ABBREVIATIONS

ACK	Acoustic Countermeasure Kit
ADF	Avian Development Facility
ADF-Otolith	Development and Function of the Avian Otolith System in Normal and Altered Gravity Environments
ADF-Skeletal	Skeletal Development in Embryonic Quail on the ISS
ADUM	Advanced Diagnostic Ultrasound in Microgravity
AdvAsc	Advanced Astroculture
AEA	ancillary equipment area
AEM	animal enclosure module
AFRL	Air Force Research Laboratory
ALARA	as low as reasonably achievable
ALS	advanced life support
ALTEA	Anomalous Long-term Effects in Astronauts' Central Nervous System
ANDE	Atmospheric Neutral Density Experiment
ANITA	Analyzing Interferometer for Ambient Air
AO	atomic oxygen
APCF	Advanced Protein Crystallization Facility
APS	Antibiotic Production in Space
ARIS	Active Rack Isolation System
ARISS	Amateur Radio on the International Space Station
ASD	air supply diffuser
ASI	Italian Space Agency
ATU	audio terminal unit
BARS	brefeldin A-ADP ribosylated substrate
BBND	Bonner Ball Neutron Detector
BBT	Beacon-Beacon Test
BCAT-3-4-CP	Binary Colloidal Alloy Test -3 and 4: Critical Point
bFGF	basic fibroblast growth factor
BMD	bone mineral density
BP/ECG	blood pressure/electrocardiograph
BPS	Biomass Production System
BSTC	biotechnology specimen temperature controller
CBOSS	Cellular Biotechnology Operations Support System
CBPD	continuous blood pressure device
CBTM	Commercial Biomedical Testing Module
CCM	Cell Culture Module
CCISS	Cardiovascular and Cerebrovascular Control on Return from ISS
cDNA	complementary deoxyribonucleic acid
CDRA	carbon dioxide removal assembly
CEA	carcinoembryonic antigen
CeMM	*C. elegans* maintenance medium
CEO	Crew Earth Observations
CEV	crew exploration vehicle
CEVIS	Cycle Ergometer with Vibration Isolation System
CFE	Capillary Flow Experiment
CGBA	Commercial Generic Bioprocessing Apparatus
CHab	*C. elegans* habitat
CHAMP	challenging minisatellite payload
CIR	combustion integrated rack
CL	Contact Line
cm^2	square centimeter
CMG	control moment gyroscope

CNRS	Centre National de la Recherche Scientifique
CNS	central nervous system
CO	carbon monoxide
CO_2	carbon dioxide
COTS	commercial off-the-shelf
CPCG-H	Commercial Generic Protein Crystal Growth-High-density
CPDS	charged-particle directional spectrometer
CSI-01	Commercial Generic Bioprocessing Apparatus Science Insert-01
CSI-02	Commercial Generic Bioprocessing Apparatus Science Insert-02
CSLM-2	Coarsening in Solid Liquid Mixtures-2
CT	computed tomography
CTP	citrate transporter protein
DAFT	Dust and Aerosol Measurement Feasibility Test
DCAM	Diffusion-Controlled Crystallization Apparatus for Microgravity
DcoH	4a-hydroxy-tetrahydropterin dehydratase
DCPCG	Dynamically Controlled Protein Crystal Growth
DCS	decompression sickness
DEXA	dual-energy X-ray absorptiometry
DGGE	denaturing gradient gel electrophoresis
DHEA	dihydroergocryptine
DHEAS	dehydroepiandrosterone
DIDO	digital input digital output
DLR	Deutsches Zentrum fur Luft und Raumfahrt
DNA	deoxyribonucleic acid
DOD	Department of Defense
DOSMAP	Dosimetric Mapping
DOSTEL	silicon dosimetry telescope
DU	detector unit
DUST	Dust and Aerosol Measurement Feasibility Test
E. coli	Escherichia coli
EAP	Educator Astronaut Program
EarthKAM	Earth Knowledge Acquired by Middle School Students
EBV	Epstein-Barr Virus
ECLSS	Environmental Control and Life Support System
EDA	education demonstration activity
EDTA	ethylenediamine-tetraacetic acid
EEG	electroencephalograph
EGN	enhanced gaseous nitrogen
EMA	epithelial membrane antigen
EMC	elastic memory composite
EMCH	elastic memory composite hinge
EMCS	European Modular Cultivation System
EMG	electromyography
EMU	extravehicular mobility unit
EPO	Education Payload Operations
ESA	European Space Agency
ESM	experiment support module
ETS	*Engelhard titanosilicate* structure
EU	endotoxin unit
eV	electron volt
EVA	extravehicular activity
EVARM	EVA radiation monitoring
EXPPCS	ExPRESS Physics of Colloids in Space
ExPRESS	Expedite the Processing of Experiments to Space Station

FCal	Fence Calibration
FDA	Food and Drug Administration
FDI	Fluid Dynamics Investigation
FFQ	Food Frequency Questionnaire
FIR	fluids integrated rack
FIT	Fungal Pathogenesis, Tumorigenesis, and Effects of Host Immunity in Space
FMVM	Fluid Merging Viscosity Measurement
FORP	fuel oxidizer reaction product
FPMU	floating potential measurement unit
FPP	floating potential probe
FTSCE	Forward Technology Solar Cell Experiment
GAP-FPA	Group Activation Pack-Fluid Processing Apparatus
GASMAP	gas analyzer system for metabolic analysis physiology
GCR	galactic cosmic ray
GDS	Gas Delivery System
GHab	garden habitat
GOES	geostationary operational environmental satellite
HDPCG	High-Density Protein Crystal Growth
HDTV	high-definition television
HEPA	high-efficiency particular accumulator
HiRAP	high-resolution accelerometer package
HIV	Human Immunodeficiency Virus
HPA	Hand Posture Analyzer
HRF	Human Research Facility
HRP	Human Research Program
IAA	International Academy of Astronautics
IBMP	Institute of Biomedical Problems
ICE	ISS Characterization Experiment
ICES	International Conference on Environmental Systems
ICF	Interior Corner Flow
ICM	isothermal containment module
ICU	internal cargo unit
IFRECOR	l'Initiative Française pour les Récifs Corallines
IGF-1	insulin-like growth factor 1
IGRF	International Geomagnetic Reference Field
InSb	indium antimonide
InSPACE	Investigating the Structure of Paramagnetic Aggregates from Colloidal Emulsion
IPY	International Polar Year
IRI	International Reference Ionosphere
ISR	Incoherent Scatter Radar
ISS	International Space Station
ISSI	In-space Soldering Experiment
IZECS	Improved Zeolite Electronic Control System
JAXA	Japan Aerospace Exploration Agency
JES	joint excursion sensor
JSC	Johnson Space Center
K-cit	potassium citrate
KCGE	Kidney Cell Gene Expression
KSS	Karolinska Sleepiness Score

LAN	local area network
LBNP	lower body negative pressure
LCD	liquid crystal display
LDL	low-density lipoprotein
LED	light-emitting diode
LEMS	lower extremity monitoring suit
LEO	low Earth orbit
LET	linear energy transfer
LFSAF	lightweight flexible solar array hinge
LOCAD-PTS	Lab-on-a-Chip Application Development-Portable Test System
MAA	Mock ANDE Active
MACE-II	Middeck Active Control Experiment-II
MAMS	Microgravity Acceleration Measurement System
MAUI	Maui analysis of upper atmospheric injections
Mb-YQR	triple mutant myoglobin
mBAND	Multicolor Banding Fluorescence In-Situ Hybridization
MBP	multi-body platform
MDU	mobile detector unit
MEMS	Microelectromechanical System
MEPS	Microencapsulation Electrostatic Processing System
MEPSI	Microelectromechanical System-based Picosat Inspector
MER	Mars exploration rover
MESA	miniature electro-static accelerometer
MeV	mega electron volt
mFISH	Multicolor Fluorescence In-Situ Hybridization
MFMG	Miscible Fluids in Microgravity
MISSE	Materials International Space Station Experiment
MIT	Massachusetts Institute of Technology
ML-I	mistletoe lectin-I
MLI	multilayer insulation
MMOD	micrometeoroid and orbital debris
MnSOD	manganese superoxide dismutase
MOBIAS	Multiple Orbital Bioreactor with Instrumentation and Automated Sampling
MOSFET	metal oxide semiconductor field effect transistor
MPV	Meerwein-Pohhdorf-Verley
MR	magnetorheological
mRNA	messenger RNA
MSG	microgravity sciences glovebox
MSRR	materials science research rack
MSSS	Maui Space Surveillance Site
NaCl	sodium chloride
NEEMO	NASA Extreme Environment Mission Operations
NOAA	National Oceanic and Atmospheric Administration
NSSS	Navy Space Surveillance System
NTDP	nuclear track detectors with and without converter
OH	hydroxide
OMS	Orbital Maneuvering System
OPE	on-board proficiency enhancer
OPG	osteoprotegerin
ORZS	Optimization of Root Zone Substrates
OUM	oxygen uptake measurement
PAM	Photoacoustic Analyzer Module

Pb	lead
PCAM	Protein Crystallization Apparatus for Microgravity
PCG-EGN	Protein Crystal Growth-Enhanced Gaseous Nitrogen
PCG-STES	Protein Crystal Growth-Single Locker Thermal Enclosure System
PCR	polymerase chain reaction
PCS	Physics of Colloids in Space
PCSat	Prototype Communications Satellite
PCU	plasma conductor unit
PEACE	Polymer Erosion and Contamination Experiment
PEC	passive experiment container
PESTO	Photosynthesis Experiment and System Testing and Operation
PFE	periodic fitness evaluation
PFM	Pulmonary Function Module
PFMI	Pore Formation in Microgravity
PFS	Pulmonary Function System
PGBA	Plant Generic Bioprocessing Apparatus
PGC	plant growth chamber
PIM	plasma integration model
PM2	pressurized module 2
PMC	polar mesopheric cloud
PMZ	Promethazine
POEMS	Passive Observatories for Experimental Microbial Systems
PRDX5	peroxiredoxin 5
PuFF	Pulmonary Function in Flight
PVA	polyvinyl alcohol
QCT	quantitative computed tomography
QTCMA	Quad Tissue Culture Module Assembly
QUS	quantiative ultrasound
RAFT	radar fence transponder
RAMBO	Ram Burn Observations
RANKL	receptor activator of NF-B ligand
RCS	Reaction Control System
RNA	ribonucleic acid
rRNA	ribosomal ribonucleic acid
RPA	Replication Protein A
RPI	Rensselaer Polytechnic Institute
RPM	revolutions per minute
RTS	remote triaxial sensor
RWV	rotating wall vessel
SAA	South Atlantic Anomaly
SAME	Smoke and Aerosol Measurement Experiment
SAMS-II	Space Acceleration Measurement Systems-II
SARJ	solar array rotary joint
SCN	succinonitrile
SDBI	Short-duration Biological Investigation
SDL	Space Dynamics Laboratory
SDS	sodium dodecyl sulfate
SDTO	Station Development Test Objective
SeaWiFS	sea-viewing wide field-of-view sensor
SEE	single event effect
SEEDS	Space Exposed Experiment Development for Students
SEM	scanning electron microscope
	Space Experiment Module

SGSM	slow growth sample module
SiO$_2$	silicon dioxide
SiOX	silicon oxide
SLAMMD	space linear acceleration mass measurement device
SLM	sound level meter
SLR	Satellite Laser Ranging
SM	Synaptogenesis in Microgravity
Sn	tin
SNFM	serial network flow monitor
SoRGE	Soldering in Reduced Gravity Experiment
SPE	solar particle event
SPEGIS	*Streptococcus pneumoniae* Expression of Genes in Space
SPENVIS	Space Environment Information System
SPHERES	Synchronized Position Hold, Engage, Reorient, Experimental Satellites
SSN	U.S. Space Surveillance Network
STEM	science, technology, engineering, and mathematics
STES	Single-locker Thermal Enclosure System
STP-H2	Space Test Program-H2
SUBSA	Solidification Using Baffle in Sealed Ampoules
Sv	Sievert
TCM	tissue culture module
TCS	Thermal Control System
TEPC	tissue-equivalent proportional counter
TeSS	temporary sleep station
TF-FGI	total force-foot ground interface
TFS	Teaching From Space (Office)
TID	total ionizing dose
TiO$_2$	titanium dioxide
TLD	thermo-luminescence dosimeter
TVIS	Treadmill Vibration Isolation System
TVT	Technology Validation Test
VEE	Venezuelan equine encephalitis
VG	Vane Gap
VOA	volatile organic analyzer
VTR	video tape recorder
VZV	Varicella zoster virus
WORF	Window Observation Research Facility
ZCG	Zeolite Crystal Growth
ZPM	zero-propellant maneuver

www.ingramcontent.com/pod-product-compliance
Lightning Source LLC
Chambersburg PA
CBHW081234180526
45171CB00005B/427